REAL NUMBERS AND SOME OF THEIR IMPORTANT PROPERTIES

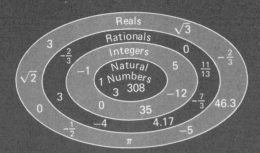

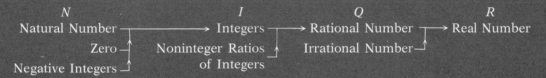

N	I	Q	R
Natural Number	Integers	Rational Number	Real Number
Zero	Noninteger Ratios of Integers	Irrational Number	
Negative Integers			

For a, b, c, and k any real numbers:

Commutative Property

$a + b = b + a$
$ab = ba$

Distributive Property

$a(b + c) = ab + ac$

Zero Property

$ab = 0$ if only if $a = 0$ or $b = 0$

Associative Property

$(a + b) + c = a + (b + c)$
$(ab)c = a(bc)$

Fundamental Principle of Fractions

$\dfrac{a}{b} = \dfrac{ka}{kb}$ $b, k \neq 0$

ELEMENTARY ALGEBRA
Structure and Use

RAYMOND A. BARNETT
Merritt College, California

ELEMENTARY ALGEBRA
Structure and Use

THIRD EDITION

McGRAW-HILL BOOK COMPANY

New York   St. Louis   San Francisco   Auckland   Bogotá   Hamburg
Johannesburg   London   Madrid   Mexico   Montreal   New Delhi   Panama   Paris
São Paulo   Singapore   Sydney   Tokyo   Toronto

ELEMENTARY ALGEBRA: Structure and Use

Copyright © 1980, 1975, 1968 by McGraw-Hill, Inc. All rights reserved. Printed in the United States of America. No part of this publication may be reproduced, stored in a retrieval system, or transmitted, in any form or by any means, electronic, mechanical, photocopying, recording, or otherwise, without the prior written permission of the publisher.

    5 6 7 8 9 0 DODO 8 9 8 7 6 5 4 3 2 1

This book was set in Aster by York Graphic Services, Inc. The editors were Carol Napier and James S. Amar; the designer was Anne Canevari Green; the production supervisor was Dominick Petrellese.
R. R. Donnelly & Sons Company was printer and binder.

Library of Congress Cataloging in Publication Data

Barnett, Raymond A
    Elementary algebra, structure and use.

    Includes index.
    1. Algebra.   I.   Title.
QA152.2B37 1980   512.9′042   79-17588
ISBN 0-07-003840-6

CONTENTS

In the final analysis a text succeeds or fails in the classroom. To succeed, a text must work for both instructors and students. It is gratifying to know that the earlier editions of this text have worked for large numbers of students and instructors. But there is always room for improvement—improvement resulting from continued classroom use and feedback. The improvements contained in this third edition evolved out of the generous response from the many users of the earlier editions.

This is still an introductory text in algebra, written for students with no background in algebra and for those who need a review before proceeding further.

Principle Changes from the Second Edition

Theoretical versus Practical

There is now a better balance between the *how* and the *why* of a given mathematical process. Retained are the intuitive discussions on *why* certain processes work based on carefully stated mathematical definitions and theorems. These are interlaced with substantially expanded discussions on *how* operations are performed or equations solved. Additional examples and annotation have been added throughout the text. In addition, exercise sets have been expanded to provide more practice in basics.

Applications

Significant applications are introduced earlier and are liberally distributed throughout the book. The approach has been modified to make more applications accessible to more students. More simple applications of the same type are included, and exercise sets have more matched pairs. A solution strategy is introduced in Section 2.8 and reinforced in subsequent chapters. By the end of Chapter 3 the student will have had much experience in translating verbal forms into symbolic forms and experience in solving significant applications (see Sections 1.4, 2.8, 3.5–3.9, 4.7, and 8.7).

Topic Organization

Some material (such as treatments of inequalities and the real numbers in the first three chapters of the second edition) has been consolidated and moved to later chapters. The chapters are now better balanced for a more evenly paced course.

PREFACE

New Material

Section 6.7 on complex fractions has been added. Section 1.1 on sets has been expanded slightly to include set-builder notation. The treatment of sets remains informal and set ideas and forms are used only where clarity results, but not otherwise. The treatment of factoring trinomials has been improved and expanded (see Sections 5.5–5.7).

Spiral Techniques

The text continues to use spiraling techniques for difficult topics; that is, a topic is introduced in a relatively simple framework, then returned to one or more times in successively more complex forms. Consider the following:

Factoring: Sections 1.6, 5.4–5.7, and 6.2–6.4

Word Problems: Sections 1.3, 1.4, 2.7, 2.8, 3.5–3.9, 4.5, and 8.7

Fractional Forms: Chapters 3 and 6

Examples and Matched Problems

As in the second edition, following each example is a matched problem to encourage active involvement in reading the text. The answers to these matched problems, however, have been moved to the end of each section just before the exercise set for the section.

Student and Instructor Aids

Student Aids

Common student errors are clearly identified at places where they naturally occur (see Sections 3.4, 6.2, and 6.4).

Dotted **"think boxes"** are used to enclose steps that are usually performed mentally (see Sections 2.7, 3.4, and 5.4).

Functional use of a **second color** (see Sections 1.3, 2.7, 3.4, and 5.4).

Annotated examples and developments have been substantially expanded to help students through critical stages (see Sections 1.3, 2.7, 3.4, and 5.4).

Chapter review exercises have been carefully reviewed and expanded where necessary. The review problems are purposely not arranged by sections so that students will get practice in recognizing problem types as well as solving them. However, following each answer in the answer section is a decimal number in italics that indicates the section in which that type of problem is discussed.

A **practice test for each chapter** now follows each chapter review exercise. It is designed to be worked in 50 minutes or less. Answers to all chapter review exercises and practice tests are in the book.

Concise summaries of formulas and symbols (keyed to the sections where they are introduced), the real number system and its basic properties, and the metric system are included in the end pages for easy reference. Many problems in the text utilize metric units.

An **arithmetic review,** covering fractions, decimals, and percent is included in the appendixes.

Instructor Aids

A uniquely designed **test battery** is included in the Instructor's Manual. The battery includes quizzes, chapter tests (which parallel chapter practice tests in the text), accumulative tests, and final examinations. There are two forms of each and all have easy-to-grade solution keys and sample student solution sheets. The format is $8\frac{1}{2}$ by 11 inches for ease of reproduction.

Answers to even numbered problems not included in the text are also in the Instructor's Manual.

The author wishes to thank the many users of the second edition for their kind remarks and helpful suggestions that were incorporated into the third edition. I particularly wish to thank Donna J. Bonorden, Southwest Texas State College; W. Homer Carlisle, Southwest Texas State College; Bob Finnell, Portland Community College; David A. Petrie, Cypress College; Roger H. Pitasky, Marietta College; and Lynn Tooley, Bellevue Community College for their very helpful detailed reviews.

A special thanks is due to Charles Burke, City College of San Francisco, for his very careful checking of answers to all exercises and examples; to Ray Westergard, Merritt College, for his preparation of an outstanding Instructor's Manual; and to Ikuko Workman, for her expert typing of the final manuscript.

Raymond A. Barnett

The following suggestions are made to help you get the most out of the course and the most out of your efforts.

Using the text is essentially a five-step process.

For each section:

1   Read a mathematical development.   ⎫
2   Read an illustrative example.   ⎬ Repeat 1-2-3 cycle until
3   Work the matched problem.   ⎭ section is finished.
4   Review the main ideas in the section.
5   Work the assigned exercises at end of the section.

All of the above should be done with plenty of inexpensive paper, pencils, and a waste basket. No mathematics text should be read without pencil and paper in hand. This is not a spectator's sport!

If you have difficulty with the course, then, in addition to the regular assignments:

1   Spend more time on the examples and matched problems.
2   Work more **A** exercises, even if they are not assigned.
3   If the **A** exercises continue to be difficult for you, see your instructor.

If you find the course too easy:

1   Work more problems from the **C** exercises, even if they are not assigned.
2   If the **C** exercises are consistently easy for you, then you should probably be in an intermediate algebra class. See your instructor.

Raymond A. Barnett

TO THE STUDENT

Use of the A, B, and C Exercises

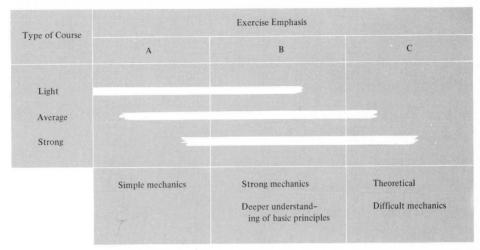

Type of Course	Exercise Emphasis		
	A	B	C
Light			
Average			
Strong			
	Simple mechanics	Strong mechanics / Deeper understanding of basic principles	Theoretical / Difficult mechanics

Testing Materials and Programs

A comprehensive test booklet is available to instructors through the publisher. There are two forms of each test with easy-to-use solution keys.

Quizzes (10–20 minutes)	Chapter Tests	Accumulative Tests	Final Examination (2 hours)
Each quiz covers	1 T	1 ⎫ T	1 ⎫
1 to 4 sections	2 T	2 ⎭	2
	3 T	3 ⎫ T	3
	4 T	4 ⎭	4 ⎬ Final
	5 T	5 ⎫ T	5
	6 T	6 ⎭	6
	7 T	7 ⎫ T	7
	8 T	8 ⎭	8 ⎭

Sample Testing Programs

   I   Quizzes + Chapter Tests + Final

  II Quizzes + Accumulative Tests + Final

III Your Own Combination

Raymond A. Barnett

TO THE INSTRUCTOR

ELEMENTARY ALGEBRA
Structure and Use

1

NATURAL NUMBERS

2

1: NATURAL NUMBERS

1.1 SETS AND SUBSETS

To begin the study of algebra, we will start with a very simple but important mathematical idea, the concept of **set.** Our use of this word will not differ appreciably from the way it is used in everyday language. Words such as "set," "collection," "bunch," and "flock" all convey the same idea. Thus, we think of a set as any collection of objects with the important property that given any object, it is either a member of the set or it is not. If an object a is in set A, we say that a **is an element of** or **is a member of** set A and write

$a \in A$

If an object is **not an element of** set A, we write

$a \notin A$

Sets are often specified by **listing** their elements between braces $\{ \ \}$. For example,

$\{2, 3, 5, 7\}$

represents the set with elements 2, 3, 5, and 7.

Two sets A and B are said to be **equal,** and we write

$A = B$

if the two sets have exactly the same elements. We write

$A \neq B$

if sets A and B are **not equal.**

The order of listing the elements in a set does not matter; thus,

$\{3, 4, 5\} = \{4, 3, 5\} = \{5, 4, 3\}$

Also, elements in a set are not listed more than once. For example, the set of letters in the word "letter" is

$\{e, l, t, r\}$

EXAMPLE 1   If $A = \{2, 4, 6\}$, $B = \{3, 5, 7\}$, and $C = \{4, 6, 2\}$, then

$4 \in A \qquad 3 \notin A \qquad 7 \in B$

$2 \notin B \qquad A = C \qquad A \neq B$

PROBLEM 1†   If $P = \{1, 3, 5\}, Q = \{2, 3, 4\}$, and $R = \{3, 4, 2\}$, replace each question mark with $\in$, $\notin$, $=$, or $\neq$, as appropriate:

(A) $1 ? P$ (B) $3 ? Q$ (C) $5 ? R$
(D) $4 ? P$ (E) $P ? Q$ (F) $Q ? R$

†Answers to matched problems following examples are located at the end of each section before the exercise set for the section. For this section, see page 4.

From time to time we will be interested in sets, called subsets, that are parts of sets. We say that a set A is a **subset** of set B if every element in set A is in set B. For example, the set of all women in a mathematics class would form a subset of all students in the class.

A set with no elements is called the **empty** or **null** set. It is symbolized by

$$\emptyset$$

For example, the set of all months of the year beginning with B is an empty or null set, and would be designated by $\emptyset$.

The method of specifying sets by **listing** the elements, such as found in Example 1, is clear and convenient for small sets. However, if we are interested in specifying a set with a large number of elements, say the set of all whole numbers from 10 to 10,000, then listing these elements would be tedious and wasteful of space. The **rule method** for specifying sets takes care of situations of this type, as well as others. Using the rule method we would write

$$\{x \mid x \text{ is a whole number from 10 to 10,000}\}$$

which is read, "the set of all elements x such that x is a whole number from 10 to 10,000." The vertical bar represents "such that."

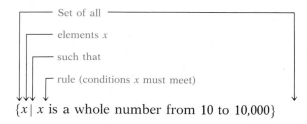

$$\{x \mid x \text{ is a whole number from 10 to 10,000}\}$$

EXAMPLE 2   If $M = \{3, 4, 5, 6\}$ and $N = \{4, 5, 6, 7, 8\}$, then the set of all elements in M that are also in N can be written using either the rule method or the listing method as follows:

Rule method: $\{x \mid x \in M \text{ and } x \in N\}$

Listing method: $\{4, 5, 6\}$

PROBLEM 2   Using the sets M and N in Example 2, write the set

$$\{x \mid x \in M \text{ or } x \in N\}$$

using the listing method.

The letter x introduced above is a variable. A **variable** is a symbol used as a placeholder for elements out of a set with two or more elements. A **constant** is a symbol that names exactly one object. The symbol "4" is a constant, since it always names the number four. We will have more to say about variables and constants in Section 1.3.

Additional ideas about sets will be introduced as needed; however, our approach will remain informal. We will use set concepts and notation where clarity results, but not otherwise.

ANSWERS TO
MATCHED
PROBLEMS

1. (A) $1 \in P$;   (B) $3 \in Q$;   (C) $5 \notin R$;   (D) $4 \notin P$;   (E) $P \neq Q$;
(F) $Q = R$
2. $\{3, 4, 5, 6, 7, 8\}$

EXERCISE 1.1

A   *In Problems 1 to 8, indicate which statements are true (T) or false(F).*

1. $4 \in \{2, 3, 4\}$ 2. $7 \notin \{2, 3, 4\}$
3. $6 \notin \{2, 3, 4\}$ 4. $7 \in \{2, 3, 4\}$
5. $\{3, 4, 5\} = \{5, 3, 4\}$ 6. $\{1, 2, 3, 4\} = \{4, 2, 3, 1\}$
7. $\{3, 5, 7\} \neq \{4, 7, 5, 3\}$ 8. $\{4, 6, 3\} \neq \{6, 3, 4\}$

Given sets

$P = \{1, 3, 5, 7\}$

$Q = \{2, 4, 6, 8\}$

$R = \{5, 1, 7, 3\}$

replace each question mark with $\in$, $\notin$, $=$, or $\neq$, as appropriate.

9. $5 ? P$ 10. $6 ? Q$
11. $6 ? R$ 12. $4 ? P$
13. $P ? R$ 14. $Q ? R$
15. $P ? Q$ 16. $R ? P$

B   *Indicate the following sets using the listing method. If the set is empty, write $\emptyset$.*

17. $\{x \mid x$ is a whole number between 5 and 10$\}$
18. $\{x \mid x$ is a whole number between 10 and 15$\}$
19. $\{x \mid x$ is a whole number between 7 and 8$\}$
20. $\{x \mid x$ is a whole number between 10 and 11$\}$
21. $\{x \mid x$ is a day of the week$\}$
22. $\{x \mid x$ is a month in the year$\}$
23. $\{x \mid x$ is a letter in *alababa*$\}$
24. $\{x \mid x$ is a letter in *millimeter*$\}$
25. $\{u \mid u$ is a Native American president of the United States$\}$
26. $\{u \mid u$ is a day of the week starting with the letter "k"$\}$

C *If* $A = \{1, 2, 3, 4\}$

$\qquad B = \{2, 4, 6, 8\}$

$\qquad C = \{1, 3, 5, 7\}$

indicate each set using the listing method.

27. $\{x \mid x \in A \text{ and } x \in B\}$

28. $\{x \mid x \in A \text{ and } x \in C\}$

29. $\{x \mid x \in B \text{ and } x \in C\}$

30. $\{x \mid x \in A \text{ or } x \in B\}$

31. $\{x \mid x \in B \text{ or } x \in C\}$

32. $\{x \mid x \in A \text{ or } x \in C\}$

1.2 THE SET OF NATURAL NUMBERS

Since algebra has to do with manipulating symbols that represent numbers, it is essential that we go back and take a careful look at some of the properties of numbers that you might have overlooked or have taken for granted.

The set of **counting numbers**

$$N = \{1, 2, 3, \ldots\}$$

will be our starting place. (The three dots tell us that the numbers go on without end following the pattern indicated by the first three numbers. This is a convenient way to represent some **infinite sets**— sets whose elements cannot be counted.) The set of counting numbers is also referred to as the set of **natural numbers,** and these two names will be used interchangeably:

EXAMPLE 3   Examples of natural numbers: 1, 5, 17, 83, 674

Examples of other numbers: $\frac{2}{3}$, $5\frac{7}{8}$, 7.63, $\sqrt{2}$, π, e

PROBLEM 3   Select the natural numbers out of the following list: 4, $\frac{3}{4}$, 19, 305, $4\frac{2}{3}$, 7.32, $\sqrt{3}$.

ASSUMPTION

We assume that you know what natural numbers are, how to add and multiply them, and how to subtract and divide them when the result is a natural number.

Important subsets of the natural numbers

The natural numbers can be divided into two subsets called even numbers and odd numbers. A natural number is an **even** number if it is exactly divisible by 2; if it is not exactly divisible by 2, then it is an **odd** number.

EXAMPLE 4   The set of even numbers: $\{2, 4, 6, \ldots\}$

The set of odd numbers: $\{1, 3, 5, \ldots\}$

PROBLEM 4   Separate the following list into even and odd numbers: 8, 13, 7, 32, 57, 625, 532.

When we add or subtract two or more numbers, the numbers are called **terms;** when we multiply two or more numbers, the numbers are called **factors.** In the sum

$3 + 5 + 8$

3, 5, and 8 are terms; in the product

$3 \times 5 \times 8$

3, 5, and 8 are factors.

In mathematics, at the level of algebra and higher, parentheses ( ) or the dot "·" are usually used in place of the times sign "×," since the latter is easily confused with the letter "x," a letter that finds frequent use in algebra. Thus

$(3)(5)(8)$

$3 \cdot 5 \cdot 8$

also represent the product of 3, 5, and 8.

The natural numbers, excluding 1, can also be divided into two other important subsets called composite numbers and prime numbers. A natural number is called a **composite number** if it can be represented as a product of two or more natural numbers other than itself and 1; it is called a **prime number** if its only natural number factors are itself and 1.

REMARKS

1. The natural number 1 is neither composite nor prime.
2. The natural number 2 is the only even prime number.

EXAMPLE 5   **(A)** The first eight composite numbers are: 4, 6, 8, 9, 10, 12, 14, 15 since

$4 = 2 \cdot 2$ $10 = 2 \cdot 5$

$6 = 2 \cdot 3$ $12 = 3 \cdot 4 = 2 \cdot 6$

$8 = 2 \cdot 4$ $14 = 2 \cdot 7$

$9 = 3 \cdot 3$ $15 = 3 \cdot 5$

(B) The first eight prime numbers are: 2, 3, 5, 7, 11, 13, 17, 19.

PROBLEM 5 Separate the following list into composite and prime numbers: 6, 9, 11, 21, 23, 25, 27, 29.

A fundamental theorem of arithmetic states that every composite number has, except for order, a unique (one and only one) set of prime factors. A natural number represented as a product of prime factors is said to be **completely factored.**

EXAMPLE 6 Write each number in completely factored form, that is, as a product of prime factors: 8, 36, 60.

Solution $8 = 2 \cdot 4 = 2 \cdot 2 \cdot 2$

$36 = 6 \cdot 6 = 2 \cdot 3 \cdot 2 \cdot 3 = 2 \cdot 2 \cdot 3 \cdot 3$

$60 = 10 \cdot 6 = 2 \cdot 5 \cdot 2 \cdot 3 = 2 \cdot 2 \cdot 3 \cdot 5$

PROBLEM 6 Repeat Example 6 using 12, 26, and 72.

Least common multiple

We can use prime factorization to aid us in finding the least common multiple (LCM) of two or more natural numbers, a process we will need to know later when dealing with fractions and certain types of equations.

The **least common multiple** of two or more natural numbers is defined to be the smallest natural number exactly divisible by each of the numbers. Often one can find the LCM by inspection. For example, the LCM of 3 and 4 is obviously 12, since 12 is the smallest natural number exactly divisible by 3 and 4. But what is the LCM of 15 and 18? Since the LCM is not obvious by inspection, we proceed as follows:

Finding the LCM

1. Factor each number in the set completely.

2. Identify the different prime factors.

3. The LCM contains each different prime factor as many times as the most number of times it appears in any one factorization.

To find the LCM of 15 and 18, we start by writing 15 and 18 in completely factored forms:

$15 = 3 \cdot 5$

$18 = 2 \cdot 9 = 2 \cdot 3 \cdot 3$

The different prime factors are 2, 3, and 5. The most that 2 appears in any one factorization is once; the most that 3 appears in any one factorization is twice; and the most that 5 appears is once. Thus, the LCM must contain one 2, two 3s, and one 5.

LCM of 15 and $18 = 2 \cdot 3 \cdot 3 \cdot 5 = 90$

and 90 is the smallest natural number exactly divisible by 15 and 18.

EXAMPLE 7   Find the LCM for 8, 6, and 9.

Solution   First, write each number as a product of prime factors:

$8 = 2 \cdot 2 \cdot 2$

$6 = 2 \cdot 3$

$9 = 3 \cdot 3$

The different prime factors are 2 and 3. The most that 2 appears in any one factorization is three times, and the most that 3 appears in any one factorization is twice; thus, the LCM must contain three 2s and two 3s.

LCM $= 2 \cdot 2 \cdot 2 \cdot 3 \cdot 3 = 72$

and 72 is the smallest natural number exactly divisible by 8, 6, and 9.

PROBLEM 7   Find the LCM for 10, 12, and 15.

ANSWERS TO MATCHED PROBLEMS

3.   4, 19, 305
4.   Even: 8, 32, 532;   Odd: 13, 7, 57, 625
5.   Composite: 6, 9, 21, 25, 27;   Prime: 11, 23, 29
6. $12 = 2 \cdot 2 \cdot 3$, $26 = 2 \cdot 13$, $72 = 2 \cdot 2 \cdot 2 \cdot 3 \cdot 3$
7.   LCM $= 2 \cdot 2 \cdot 3 \cdot 5 = 60$

EXERCISE 1.2

A   *Select the natural numbers out of each list.*

1.   6, 13, 3.5, $\frac{2}{3}$
2.   4, $\frac{1}{8}$, 22, 6.5
3. $3\frac{1}{2}$, 67, 402, 22.35
4.   203.17, 63, $\frac{33}{5}$, 999

Separate each list into even and odd numbers.

5.   9, 14, 28, 33
6.   8, 24, 1, 41
7.   23, 105, 77, 426
8.   68, 530, 421, 72

Separate each list into composite and prime numbers.

9. 2, 6, 9, 11
10. 3, 4, 7, 15
11. 12, 17, 23, 27
12. 16, 19, 25, 39

B *Let M be the set of natural numbers from 20 to 30 and N the set of natural numbers from 40 to 50. List the following:*

13. Even numbers in M
14. Even numbers in N
15. Odd numbers in M
16. Odd numbers in N
17. Composite numbers in M
18. Composite numbers in N
19. Prime numbers in M
20. Prime numbers in N

Write each of the following composite numbers as a product of prime factors.

21. 10	22. 21	23. 30
24. 90	25. 84	26. 72
27. 60	28. 120	29. 108
30. 112	31. 210	32. 252

Find the LCM for each group of numbers.

33. 9, 12	34. 9, 15	35. 6, 16
36. 12, 16	37. 3, 8, 12	38. 4, 6, 18
39. 4, 10, 15	40. 10, 12, 9	41. 10, 15, 18
42. 6, 15, 18	43. 8, 15, 20, 24	44. 6, 8, 15, 20

C

45. Is every even number a prime number? Is every odd number a prime number? Is every prime number an odd number? Is every prime number except 2 an odd number?

46. Is every even number a composite number? Is every odd number a composite number? Is every even number except 2 a composite number?

Intuitively, a set is said to have a finite number of elements if the number of elements is a natural number; otherwise, the set is said to be infinite. Tell which of the following sets are finite or infinite.

47. The set of natural numbers between 1 and 1 million
48. The set of even numbers between 1 and 1 million
49. The set of all natural numbers
50. The set of all even numbers
51. The set of all the grains of sand on all the beaches in the world

1.3 ALGEBRAIC EXPRESSIONS—THEIR FORMULATION AND EVALUATION

Consider the statement: "The perimeter of a rectangle is twice its length plus twice its width." If we let

P = perimeter

a = length

b = width

then the formula

$$P = 2a + 2b$$

has the same meaning as the original statement, but with increased clarity and a substantial decrease in the number of symbols used: 7 in the formula as compared to 57 in the written statement.

By the use of symbols that name numbers or are placeholders for numerals, we can form general statements relating many particular facts. The perimeter formula holds for *all* rectangles, not just one particular rectangle. This formula is an example of an algebraic form.

Variables and constants

In the perimeter formula above, the three letters P, a, and b can be replaced with many different numerals, depending on the size of the rectangle; hence, these letters are called variables. The symbol "2" names only one number and is consequently called a constant. In general, a **constant** is defined to be any symbol that names one particular thing; a **variable** is a symbol that holds a place for two or more constants.

EXAMPLE 8 In the formula

$$F = \tfrac{9}{5}C + 32$$

(for the conversion of Celsius degrees to Fahrenheit degrees), $\frac{9}{5}$ and 32 are constants and C and F are variables.

PROBLEM 8 List the constants and variables in each formula:

(A) $P = 4s$ perimeter of a square

(B) $A = s^2$ area of a square (Note: $s^2 = s \cdot s$)

NOTE: A number of useful formulas are listed inside the back cover of the text for convenient reference.

The introduction of variables into mathematics occurred about A.D. 1600. A French mathematician, François Vieta (1540–1603), is singled out as the one mainly responsible for this new idea. Many mark this point as the beginning of modern mathematics.

Algebraic expressions

An **algebraic expression** is a symbolic form involving constants; variables; mathematical operations such as addition, subtraction, multiplication, and division (other operations will be added later); and grouping symbols such as parentheses ( ), brackets [ ], and braces { }. For example,

$$8 + 7 \qquad 3 \cdot 5 - 6 \qquad 12 - 2(8 - 5)$$

$$3x - 5y \qquad 8(x - 3y) \qquad 3\{x - 2[x + 4(x + 3)]\}$$

are all algebraic expressions.

Two or more algebraic expressions joined using plus (+) or minus (−) signs are called **terms;** two or more algebraic expressions joined by multiplication are called **factors.**

$$5 - 2 \cdot 3 \qquad \text{2 terms; second term has 2 factors}$$

$$10 + 2(6 - 3) \qquad \text{2 terms; second term has 2 factors}$$

$$2x + 3y - 6z \qquad \text{3 terms; each term has 2 factors}$$

$$5[x - 3(x + 5)] \qquad \text{2 factors; second factor has 2 terms}$$

When evaluating numerical expressions involving various operations and symbols of grouping, we follow the convention:

Order of Operations

---

1. Simplify inside the innermost symbols of grouping first, then the next innermost, and so on.

2. Multiplication and division are performed before addition and subtraction. (In both cases, we proceed from left to right.)

EXAMPLE 9   Evaluate each expression:

(A) $8 - 2 \cdot 3 = 18$ (B) $9 - 2(5 - 3) = 14$

(C) $(9 - 2)(5 - 3) = 14$ (D) $2[12 - 3(8 - 5)]$

Solution   (A) $8 - 2 \cdot 3 = 8 - 6$ Multiplication precedes addition and subtraction.

$$= 2$$

(B) $9 - 2(5 - 3) = 9 - 2 \cdot 2$ Perform operation inside parentheses first,
then multiply,

$$= 9 - 4 \qquad \text{then subtract.}$$

$$= 5$$

(C) $(9 - 2)(5 - 3) = 7 \cdot 2$ Parentheses first, then multiply.

$$= 14 \qquad \text{Note how (B) and (C) differ.}$$

(D) $2[12 - 3(8 - 5)] = 2[12 - 3 \cdot 3]$ Parentheses first.

$$= 2(12 - 9)$$

$$= 2 \cdot 3$$

$$= 6$$

Parentheses ( ) can always replace brackets [ ] and braces { } when the latter are used as symbols of grouping.

PROBLEM 9 Evaluate each expression:

(A) $2 \cdot 10 - 3 \cdot 5$

(B) $11 - 3(7 - 5)$

(C) $(11 - 3)(7 - 5)$

(D) $6[13 - 2(14 - 8)]$

EXAMPLE 10 Evaluate each algebraic expression for $x = 10$ and $y = 3$.

(A) $2x - 3y$

(B) $x - 3(2y - 4)$

(C) $(x - 3)(2y - 4)$

(D) $5[32 - x(x - 7)]$

Solution Substitute $x = 10$ and $y = 3$ into each expression, then evaluate, following order of operations described above.

(A) $2x - 3y$

$$2(10) - 3(3) = 20 - 9 = 11$$

(B) $x - 3(2y - 4)$

$$10 - 3(2 \cdot 3 - 4) = 10 - 3(6 - 4) = 10 - 3 \cdot 2 = 10 - 6 = 4$$

(C) $(x - 3)(2y - 4)$

$$(10 - 3)(2 \cdot 3 - 4) = 7(6 - 4) = 7 \cdot 2 = 14$$

Note how parts **B** and **C** differ.

(D) $5[32 - x(x - 7)]$

$$5[32 - 10(10 - 7)] = 5(32 - 10 \cdot 3) = 5(32 - 30) = 5 \cdot 2 = 10$$

PROBLEM 10 Evaluate each algebraic expression for $x = 12$ and $y = 3$:

(A) $x - 3y$

(B) $x - 4(y - 1)$

(C) $(x - 4)(y - 1)$

(D) $3[x - 2(x - 9)]$

From English to algebra EXAMPLE 11 If x represents a natural number, write an algebraic expression that represents each of the expressed numbers:

(A) A number 3 times as large as x

(B) A number 3 more than x

(C) A number 7 less than the product of 4 and x

(D) A number 3 times the quantity 2 less than x

Solution **(A)** $3x$ "times" corresponds to "multiply"

           (B) $x + 3$ "more than" corresponds to "added to"

           (C) $4x - 7$ (not $7 - 4x$)   "less than" corresponds to "subtracted from"

           (D) $3(x - 2)$ [not $3(2 - x)$]

PROBLEM 11 If y represents a natural number, write an algebraic expression that represents each of the expressed numbers:

(A) A number 7 times as large as y

(B) A number 7 less than y

(C) A number 9 more than the product of 4 and y

(D) A number 5 times the quantity 4 less than y

ANSWERS TO MATCHED PROBLEMS

8. (A) Constants: 4; Variables: P, s

     (B) Constants: 2; Variables: A, s

9. (A) 5;   (B) 5;   (C) 16;   (D) 6

10. (A) 3;   (B) 4;   (C) 16;   (D) 18

11. (A) $7y$;   (B) $y - 7$;   (C) $4y + 9$;   (D) $5(y - 4)$

EXERCISE 1.3

A *Evaluate each expression.*

1. $7 + 3 \cdot 2$ 2. $5 + 6 \cdot 3$ 3. $8 - 2 \cdot 3$

4. $20 - 5 \cdot 3$ 5. $7 \cdot 6 - 5 \cdot 5$ 6. $8 \cdot 9 - 6 \cdot 11$

7. $(2 + 9) - (3 + 6)$ 8. $(8 - 3) + (7 - 2)$ 9. $8 + 2(7 + 1)$

10. $3 + 8(2 + 5)$ 11. $(8 + 2)(7 + 1)$ 12. $(3 + 8)(2 + 5)$

13. $10 - 3(7 - 4)$ 14. $20 - 5(12 - 9)$ 15. $(10 - 3)(7 - 4)$

16. $(20 - 5)(12 - 9)$ 17. $12 - 2(7 - 5)$ 18. $15 - 3(9 - 5)$

Evaluate each algebraic expression for $x = 8$ and $y = 3$.

19. $x + 2$ 20. $y + 5$ 21. $x - y$

22. $22 - x$ 23. $x - 2y$ 24. $6y - x$

25. $3x - 2y$ 26. $9y - xy$ 27. $y + 3(x - 5)$

28. $5 + y(x - y)$ 29. $x - 2(y - 1)$ 30. $x - y(x - 7)$

If x and y represent natural numbers, write an algebraic expression that represents each of the following numbers.

31. A number 5 times as large as x

32. A number 7 times as large as y

33. A number 5 more than x

34. A number 12 more than y

35. A number 5 less than x

36. A number 8 less than y

37. A number x less than 5

38. A number y less than 8

B *Identify the constants and variables in each algebraic expression.*

39. $A = \frac{1}{2}bh$ (area of a triangle)

40. $A = ab$ (area of a rectangle)

41. $d = rt$ (distance-rate-time formula)

42. $C = \frac{5}{9}(F - 32)$ (Fahrenheit-Celsius formula)

43. $I = prt$ (simple interest)

44. $A = P(1 + rt)$ (simple interest)

45. $y = 2x + 3$

46. $3x + 2y = 5$

47. $3(u + v) + 2u$

48. $2(x + 1) + 3(w + 5z)$

In Problems 49–52 find the area and perimeter for each rectangle ($A = ab$ and $P = 2a + 2b$). Example: If $a = 5$ meters and $b = 3$ meters, then $A = 5 \cdot 3 = 15$ square meters and $P = 2 \cdot 5 + 2 \cdot 3 = 10 + 6 = 16$ meters.

49. $a = 6$ cm, $b = 3$ cm

50. $a = 12$ ft, $b = 4$ ft

51. $a = 10$ km, $b = 8$ km

52. $a = 9$ meters, $b = 6$ meters

Evaluate each.

53. $4[15 - 10(9 - 8)]$

54. $6[22 - 3(13 - 7)]$

55. $7 \cdot 9 - 6(8 - 3)$

56. $5(8 - 3) - 3 \cdot 6$

57. $2[(7 + 2) - (5 - 3)]$

58. $6[(8 - 3) + (4 - 2)]$

Evaluate each for $w = 2$, $x = 5$, $y = 1$, and $z = 3$.

59. $w(y + z)$

60. $wy + wx$

61. $wy + z$

62. $y + wz$

63. $(z - y) + (z - w)$

64. $4(y + w) - 2z$

65. $2[x + 3(z - y)]$

66. $6[(x + z) - 3(z - w)]$

67. How far can you travel in 12 hr at 57 km/hr? ($d = rt$)

68. How far can you travel in 9 hr at 43 km/hr? ($d = rt$)

69. How many words can a typist type in 10 min if he or she can type 60 words per min? ($Q = rt$)

70. How many gallons can a pipe fill in 20 min if it fills at the rate of 10 gal per min? ($Q = rt$)

If x represents a given natural number, write an algebraic expression that represents each of the following numbers.

71. A number 3 more than twice the given number

72. A number 3 more than the product of 12 and the given number

73. A number 3 less than the product of 12 and the given number

74. A number 3 less than twice the given number

75. A number 3 times the quantity 8 less than the given number

76. A number 6 times the quantity 4 less than the given number

C *Evaluate each expression.*

77. $3[(6 - 4) + 4 \cdot 3 + 3(1 + 6)]$

78. $2[(3 + 2) + 2(7 - 4) + 6 \cdot 2]$

79. $2\{26 - 3[12 - 2(8 - 5)]\}$

80. $5\{32 - 5[(10 - 2) - 2 \cdot 3]\}$

Evaluate for $u = 2$, $v = 3$, $w = 4$, and $x = 5$.

81. $2\{w + 2[7 - (u + v)]\}$

82. $3\{(u + v) + 3[x - 2(w - u)] + uv\}$

If t represents an even number, write an algebraic expression that represents each of the following:

83. A number 3 times the first even number larger than t

84. A number 5 times the first even number smaller than t

85. The sum of three consecutive even numbers starting with t

86. The sum of three consecutive odd numbers following t

87. The distance s in feet that an object falls in t sec (Figure 1) is 16 times the square of the time.   (A) Write a formula that indicates the distance s that the object falls in t sec.   (B) Identify the constants and variables. (C) How far will the object have fallen at the end of 8 sec?

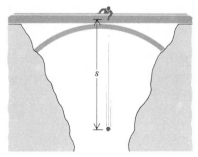

Figure 1

1.4 EQUALITY

In the preceding sections the equal sign "=" was used in a number of places. You are probably most familiar with its use in formulas such as

$$d = rt$$

$$A = ab$$

$$I = prt$$

The equal sign is very important in mathematics, and you will be using it frequently. Its mathematical meaning, however, is not as obvious as it first might seem. For this reason we are devoting one section of this chapter solely to this sign so that you will use it correctly from the beginning.

An **equal sign** will be used to join two expressions if the two expressions are names or descriptions of exactly the same thing. Since

$$a = b$$

means a and b are names for the same object, it is natural that we define

$$a \neq b$$

to mean a and b do not name the same thing; that is, a is not equal to b.

EXAMPLE 12   **(A)** $VI = 6$

(B) $4 - 3 \neq 2$

(C) $3 + 4 \cdot 2 = 11$

PROBLEM 12   True or false?

(A) $III = 3$ **(B)** $V \neq 6$

(C) $8 - 6 = 2$ **(D)** $8 - 3 \cdot 2 = 10$

Algebraic equations

If two algebraic expressions involving at least one variable are joined with an equal sign, the resulting form is called an algebraic equation. The following are **algebraic equations** in one or more variables:

$$x + 3 = 8$$

$$x + 2 = 2 + x$$

$$2x + 3y = 12$$

Since a variable is a placeholder for constants, an equation is neither true nor false as it stands; it does not become so until the variable has been replaced by a constant. Formulating algebraic equations is an important first step in solving certain types of real world problems using algebraic methods. We address ourselves to this problem now.

EXAMPLE 13   Translate each statement into an algebraic equation using only one variable:

(A)   15 is 9 more than a certain number.

Solution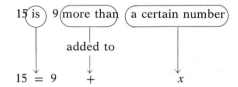

Thus,

$$15 = 9 + x \quad \text{or} \quad 15 = x + 9$$

(B)   3 times a certain number is 7 less than twice the number.

Solution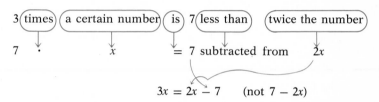

$$3x = 2x - 7 \quad (\text{not } 7 - 2x)$$

Compare the symbolic and verbal forms on the right side carefully and note how 7 and $2x$ reverse.

PROBLEM 13   Repeat Example 13 for:

(A)   8 is 5 more than a certain number.

(B)   6 is 4 less than a certain number.

(C)   4 times a certain number is 3 less than twice the number.

(D)   If 12 is added to a certain number, the sum is twice a number that is 3 less than the certain number.

EXAMPLE 14   If x represents an even number, write an algebraic equation that is equivalent to "the sum of three consecutive even numbers is 84."

Solution   The set of even numbers is

$$\{2, 4, 6, \ldots, x, \ldots\}$$

where x is an unspecified even number. We note that even numbers increase 2 at a time; hence,

$$\left. \begin{array}{l} x \\ x + 2 \\ x + 4 \end{array} \right\}$$ represent 3 consecutive (one following the other) even numbers, starting with x

Thus,

$$\left(\begin{array}{c}\text{The sum of three}\\\text{consecutive even numbers}\end{array}\right) \text{ is } 84$$

$$x + (x + 2) + (x + 4) = 84$$

PROBLEM 14 If x represents an odd number, write an algebraic equation that is equivalent to "the sum of four consecutive odd numbers is 160."

From the logical meaning of the equal sign, a number of rules or properties can easily be established for its use. We state these properties below for completion, and will return to them later when we discuss equations. They control a great deal of the activity related to the solving of equations.

Properties of equality

If a, b, and c are names of objects, then:

1. $a = a$ reflexive property

2. If $a = b$, then $b = a$.   symmetric property

3. If $a = b$ and $b = c$, then $a = c$.   transitive property

4. If $a = b$, then either may replace the other in any expression without changing the truth or falsity of the statement.   substitution principle

Property **2** allows, for example,

$5 = x$ to become $x = 5$

and

$A = P + Prt$ to become $P + Prt = A$

Property **3** allows us to conclude, for example, that

$8 - 3 \cdot 2 = 2$

since

$$\overbrace{8 - 3 \cdot 2}^{a} = \overbrace{8 - 6}^{b} \quad \text{and} \quad \overbrace{8 - 6}^{b} = \overbrace{2}^{c}$$

therefore,

$$\overbrace{8 - 3 \cdot 2}^{a} = \overbrace{2}^{c}$$

Property **4** allows us to conclude, for example, that

$8 - 3 \cdot 2 = 8 - 6$ since $3 \cdot 2 = 6$

that is, $3 \cdot 2$ in the first expression is replaced, using the substitution principle, by 6 to obtain the second expression.

The importance of the four properties of equality will not be fully appreciated until we start solving equations and simplifying algebraic expressions.

ANSWERS TO MATCHED PROBLEMS

12. (A) T; (B) T; (C) T; (D) F
13. (A) $8 = 5 + x$ or $8 = x + 5$; (B) $6 = x - 4$ (not $6 = 4 - x$); (C) $4x = 2x - 3$; (D) $x + 12 = 2(x - 3)$
14. $x + (x + 2) + (x + 4) + (x + 6) = 160$

EXERCISE 1.4

A *Indicate which of the following are true (T) or false (F).*

1. $10 - 6 = 4$
2. $12 + 7 = 19$
3. $6 \cdot 9 = 56$
4. $8 \cdot 7 = 54$
5. $12 - 9 \neq 2$
6. $15 - 8 \neq 6$
7. $\text{II} = 2$
8. $\text{IX} \neq 11$
9. $9 - 2 \cdot 3 \neq 21$
10. $4 + 2 \cdot 5 \neq 30$

Translate each statement into an algebraic equation using x as the only variable.

11. 5 is 3 more than a certain number.
12. 10 is 7 more than a certain number.
13. 8 is 3 less than a certain number.
14. 14 is 6 less than a certain number.
15. 18 is 3 times a certain number.
16. 25 is 5 times a certain number.

B **17.** 49 is 7 more than twice a certain number.
18. 27 is 3 more than 6 times a certain number.
19. 52 is 8 less than 5 times a certain number.
20. 103 is 7 less than 10 times a certain number.
21. 4 times a given number is 3 more than 3 times that number.
22. 8 times a number is 20 more than 4 times the number.
23. 5 more than a certain number is 3 times a number that is 4 less than the certain number.
24. 6 less than a certain number is 5 times a number that is 7 more than the certain number.
25. The sum of three consecutive natural numbers is 90.
26. The sum of four consecutive natural numbers is 54.
27. The sum of two consecutive even numbers is 54.
28. The sum of three consecutive odd numbers is 105.

C **29.** What is wrong with the following argument? John is a human and Mary is a human. Hence, we can write John = human and Mary = human. By the symmetric property for equality we can write human = Mary and conclude, using the transitive property for equality (since John = human and human = Mary), that John = Mary.

30. What is wrong with the following argument? The number 5 is a prime number and 7 is a prime number. Hence, we can write 5 = prime number and 7 = prime number. By the symmetric property for equality we can write prime number = 7 and conclude, using the transitive property for equality (since 5 = prime number and prime number = 7), that 5 = 7.

APPLICATIONS

31. Pythagoras found that the octave chord could be produced (Figure 2) by placing the movable bridge so that a taut string is divided into two parts with the longer piece twice the length of the shorter piece. If the total string is 54 cm long and we let x represent the length of the shorter piece, write an equation relating the lengths of the two pieces and the total length of the string.

**Figure 2
Monochord**

32. A steel rod 7 meters long is cut into two pieces so that the longer piece is 1 meter less than twice the length of the shorter piece. Write an equation relating the lengths of the two pieces with the total length.

33. In a rectangle of area 50 cm² (square centimeters) the length is 10 cm more than the width. Write an equation relating the area with the length and the width.

34. In a rectangle of area 75 yd² the length is 5 more than 3 times the width. Write an equation relating the area with the length and the width.

1.5 PROPERTIES OF ADDITION AND MULTIPLICATION; EXPONENTS

Algebra in many ways can be thought of as a game, a game that requires the manipulation of symbols to change algebraic expressions from one form to another. As is the case in any game, to play, one must know the rules.

In this section we will discuss some of the basic rules in the "game of algebra." These, along with others that we will add later, govern the manipulation of symbols that represent numbers in algebra.

Properties of addition and multiplication

We now discuss a couple of basic properties of natural numbers that you have been using in arithmetic for a long time. Assume we are given the set of natural numbers

$$N = \{1, 2, 3, \ldots\}$$

and the operations

$$+ \quad - \quad \cdot \quad \div$$

Is the result of applying any of these operations on any two natural numbers the same no matter what order the numbers are written? For example, which of the following are true?

$$8 + 4 = 4 + 8$$

$$8 - 4 = 4 - 8$$

$$8 \cdot 4 = 4 \cdot 8$$

$$8 \div 4 = 4 \div 8$$

We see that the first and third equations are true and the second and fourth are false. In fact, we will not be able to find two natural numbers in which the order in which we add or multiply them will make a difference. But since we are not in a position to prove this for all natural numbers, we state the result as an **axiom** (a property accepted for a mathematical system without proof) and give it a name. We say that the natural numbers are **commutative** relative to addition and multiplication. By this we simply mean that the order of operation in addition or in multiplication doesn't matter. More formally, we state:

Commutative Property

For all natural numbers a and b

$$a + b = b + a$$

$$ab = ba$$

The order in addition doesn't matter.
The order in multiplication doesn't matter.

On the other hand, subtraction and division are not commutative. The order in subtraction and division does matter.

EXAMPLE 15   If x and y are natural numbers, then

(A) $x + 7 = 7 + x$

(B) $y5 = 5y$

(C) $yx = xy$

(D) $3 + 5x = 5x + 3$

(E) $7 + x2 = 7 + 2x$

Each example illustrates the use of the commutative property. Notice that the order is reversed in each case.

PROBLEM 15　If a and b are natural numbers, use the commutative properties for addition and multiplication to write each of the following in an equivalent form:

(A) $a + 3$ (B) $b5$

(C) ba (D) $b + a$

Now we turn to another property of the natural numbers. Suppose you are given the following four problems:

$8 + 4 + 2$

$8 - 4 - 2$

$8 \cdot 4 \cdot 2$

$8 \div 4 \div 2$

We notice that in the addition problem if we add 4 to 8 first, then add 2 we get the same result as adding 2 to 4 first, then adding 8. That is,

$(8 + 4) + 2 = 8 + (4 + 2)$

It appears that grouping the terms relative to addition doesn't seem to make a difference. Does grouping make a difference for any of the other three operations? That is, which of the following is true?

$(8 - 4) - 2 = 8 - (4 - 2)$

$(8 \cdot 4) \cdot 2 = 8 \cdot (4 \cdot 2)$

$(8 \div 4) \div 2 = 8 \div (4 \div 2)$

The only equation that is true is the one involving multiplication. Once again we will not be able to find even one case where changing grouping relative to addition or relative to multiplication will make a difference. Since we are not able to prove this for all natural numbers, we state the result as an axiom and give it a name. We say that the natural numbers are **associative** relative to addition and multiplication. By this we simply mean that we may insert or remove parentheses at will relative to addition and insert or remove parentheses at will relative to multiplication. More formally, we state:

Associative Property

For all natural numbers a, b, and c

$$(a + b) + c = a + (b + c)$$

$$(ab)c = a(bc)$$

Grouping doesn't matter in addition.
Grouping doesn't matter in multiplication.

On the other hand, subtraction and division are not associative. Grouping relative to both of these operations does matter.

We now have another tool for transforming algebraic expressions into other equivalent forms.

EXAMPLE 16 If x, y, and z are natural numbers, then

(A) $(x + 3) + 5 = x + (3 + 5) † = x + (3 + 5) = x + 8$

(B) $2(3x) = (2 \cdot 3)x = (2 \cdot 3)x = 6x$

(C) $(x + y) + z = x + (y + z) = x + (y + z)$

(D) $(xy)z = x(y z) = x(yz)$

Notice how the parentheses are moved in each case. No order is changed. Each example illustrates the use of the associative property.

PROBLEM 16 If a, b, and c are natural numbers, replace each question mark with an appropriate algebraic expression.

(A) $(a + 5) + 7 = a + (?)$

(B) $5(9b) = (?)b$

(C) $(a + b) + c = a + (?)$

(D) $(ab)c = a(?)$

We now use both properties to simplify

$$(x + 3) + (y + 5)$$

We will first go through the process carefully, justifying each step, then we will indicate how most of the steps can be done mentally.

†Dotted boxes are used throughout the text to indicate steps that are usually done mentally.

$$(x + 3) + (y + 5) = [(x + 3) + y] + 5 \qquad \text{Associative property for } +$$
$$= [y + (x + 3)] + 5 \qquad \text{Commutative property for } +$$
$$= [(y + x) + 3] + 5 \qquad \text{Associative property for } +$$
$$= [(x + y) + 3] + 5 \qquad \text{Commutative property for } +$$
$$= (x + y) + (3 + 5) \qquad \text{Associative property for } +$$
$$= (x + y) + 8 \qquad \text{Substitution property for } =$$
$$= x + y + 8 \qquad \text{Order of operation}$$

Normally, we do most of these steps mentally and simply write

$$(x + 3) + (y + 5) = x + 3 + y + 5$$
$$= x + y + 8$$

However, even though we did not write each step as above, you should not lose sight of the fact that the associative and commutative properties are behind the mental steps taken in the simpler version.

Use of Commutative and Associative Properties—Conclusion

Relative to addition, commutativity and associativity permit us to change the order of addition at will and insert or remove parentheses as we please. The same thing is true for multiplication, but not for subtraction and division.

EXAMPLE 17 **(A)** $(a + 5) + (b + 2) + (c + 4) = a + 5 + b + 2 + c + 4$
$$= a + b + c + 5 + 2 + 4$$
$$= a + b + c + 11$$

(B) $(2x)(3y) = 2x3y$
$$= (2 \cdot 3)(xy)$$
$$= 6xy$$

PROBLEM 17 Simplify as in Example 17:

(A) $(u + 4) + (v + 5) + (w + 3)$ **(B)** $(4m)(8n)$

Exponents

Exponent forms play an important role in algebra and we will be using them almost daily. Here we will see the use of the commutative and associative properties in several important ways.

In a preceding section we encountered the form

$$r^2 = rr$$

There is obviously no reason to stop here: you no doubt can guess how r^3 and r^4 should be defined. If you guessed

$$r^3 = rrr$$

$$r^4 = rrrr$$

then you have anticipated the following general definition of b^n where n is any natural number and b is any number:

$$b^n = \underbrace{bbb \ldots b}_{n \text{ factors of } b}$$

b is called the **base** and n the **power** or **exponent.** In addition, we define

$$b^1 = b$$

and usually use b in place of b^1.

EXAMPLE 18 **(A)** $x^2 = xx$ $\qquad\qquad$ $t^1 = t$

$3^4 = 3 \cdot 3 \cdot 3 \cdot 3$ $\quad$ $5x^3y^5 = 5xxxyyyyy$

(B) $\qquad\qquad xxx = x^3$ $\quad$ $2xxy = 2x^2y$

$2 \cdot 2 \cdot 2 \cdot 2 = 2^4$ $\quad$ $3xxxyy = 3x^3y^2$

PROBLEM 18 **(A)** Write in nonpower form: y^3, 2^4, $3x^3y^4$

(B) Write in exponent form: uu, $5 \cdot 5 \cdot 5 \cdot 5$, $7xxxxyyy$

Something interesting happens if we multiply two exponent forms with the same base:

$$x^3x^5 = (xxx)(xxxxx)$$

$$= xxxxxxxx$$

$$= x^8$$

which we could get by simply adding the exponents in x^3x^5.

In general, for any natural numbers m and n and any number b

$$b^m b^n = b^{m+n}$$

This is the **first law of exponents,** one of five very important exponent laws you will get to know well before the end of this book.

EXAMPLE 19 **(A)** $x^3 x^4 = x^{3+4} = x^7$

(B) $5^{10} \cdot 5^{23} = 5^{10+23} = 5^{33}$ not 25^{33}

(C) $(2y^2)(3y^5) = (2 \cdot 3)(y^2 y^5) = 6y^7$

(D) $(3x^2 y)(4x^4 y^5) = (3 \cdot 4)(x^2 x^4)(yy^5) = 12x^6 y^6$

Note how commutative and associative properties are used in parts **C** and **D** where we rearranged the factors and regrouped them.

PROBLEM 19 Simplify as in Example 19:

(A) $y^5 y^3$ **(B)** $3^{17} \cdot 3^{20}$
(C) $(3a^6)(5a^3)$ **(D)** $(2x^2 y^4)(3xy^2)$

ANSWERS TO MATCHED PROBLEMS
15. (A) $3 + a$;   (B) $5b$;   (C) ab;   (D) $a + b$
16. (A) $5 + 7$;   (B) $5 \cdot 9$;   (C) $b + c$;   (D) bc
17. (A) $u + v + w + 12$;   (B) $32mn$
18. (A) yyy, $2 \cdot 2 \cdot 2 \cdot 2$, $3xxxyyyy$;   (B) u^2, 5^4, $7x^4 y^3$
19. (A) y^8;   (B) 3^{37} (not 9^{37});   (C) $15a^9$;   (D) $6x^3 y^6$

EXERCISE 1.5   **A**   *Remove parentheses and simplify.*

1. $(7 + x) + 3$ 2. $(5 + z) + 12$
3. $(7a)(4b)$ 4. $(3x)(4y)$
5. $(7 + a) + (9 + b)$ 6. $(x + 7) + (y + 8)$

Write in nonexponent form.

7. x^3 8. y^4 9. $2x^3 y^2$
10. $5a^2 b^3$ 11. $3w^2 xy^3$ 12. $7ab^3 c^2$

Write in exponent form.

13. xxx 14. $yyyy$ 15. $2xxxyy$
16. $7uuvvvvv$ 17. $3xyyzzz$ 18. $9aabccc$

Multiply, using the first law of exponents.

19. $u^{10}u^4$ **20.** m^8m^7 **21.** aa^5

22. b^7b **23.** $w^{12}w^7$ **24.** $n^{23}n^{10}$

25. $y^{12}y^4$ **26.** u^4u^{44} **27.** $3^{10} \cdot 3^{20}$

28. $7^8 \cdot 7^5$ **29.** $9^5 \cdot 9^6$ **30.** $2^5 \cdot 2^{12}$

B *Remove parentheses and simplify.*

31. $(3a)(5b)(2c)$ **32.** $(2x)(8y)(3z)$

33. $(4u)(5v)(3w)$ **34.** $(2x)(3y)(4z)$

35. $(x + 2) + (y + 4) + (z + 8)$ **36.** $(a + 3) + (b + 5) + (c + 2)$

37. $(u + 5) + (v + 10) + (w + 4)$ **38.** $(r + 6) + (s + 8) + (t + 10)$

Multiply, using the first law of exponents.

39. x^2xx^4 **40.** mm^3m^4

41. yyy^6y^2 **42.** uu^2uu^4

43. $(2x^3)(3x)(4x^5)$ **44.** $(3u^4)(2u^5)(u^7)$

45. $(a^2b)(ab^2)$ **46.** $(cd^2)(c^2d^3)$

47. $(4x)(3xy^2)$ **48.** $(5b)(2a^2b^3)$

49. $(2xy)(3x^3y)$ **50.** $(3xy^2z^3)(5xyz^2)$

C **51.** If a statement is not true for all natural numbers a and b, find replacements for a and b that show that the statement is false.

(A) $a + b = b + a$ (B) $ab = ba$

(C) $a - b = b - a$ (D) $a \div b = b \div a$

52. Repeat the preceding problem for

(A) $(a + b) + c = a + (b + c)$ (B) $(ab)c = a(bc)$

(C) $(a - b) - c = a - (b - c)$ (D) $(a \div b) \div c = a \div (b \div c)$

Each statement illustrates either the commutative property or the associative property. State which.

53. $5 + z = z + 5$ **54.** $bc = cb$

55. $(5x)y = 5(xy)$ **56.** $(a + 5) + 7 = a + (5 + 7)$

57. $3x + x5 = 3x + 5x$ **58.** $5(x8) = 5(8x)$

59. $3 + (x + 2) = 3 + (2 + x)$ **60.** $(5x)y = y(5x)$

61. $5 + (x + 3) = (x + 3) + 5$

62. $(x + 2) + (y + 3) = (x + 2) + (3 + y)$

63. $(x + 3) + (y + 2) = (y + 2) + (x + 3)$

64. $(x + 3) + (y + 2) = x + [3 + (y + 2)]$

1.6 A PROPERTY INVOLVING BOTH MULTIPLICATION AND ADDITION

We now introduce another important property of the natural numbers, a property that involves both multiplication and addition, called the distributive property. To discover this property let us compute $3(5 + 2)$ and $3 \cdot 5 + 3 \cdot 2$:

$$3(5 + 2) = 3 \cdot 7 = 21$$

$$3 \cdot 5 + 3 \cdot 2 = 15 + 6 = 21$$

Thus

$$3(5 + 2) = 3 \cdot 5 + 3 \cdot 2$$

Note that the right side of the last equality is obtained from the left by multiplying each term within the parentheses by 3. Is the fact that these are equal just a coincidence? Let us try another set of numbers:

$$7(2 + 6) = 7 \cdot 8 = 56$$

$$7 \cdot 2 + 7 \cdot 6 = 14 + 42 = 56$$

Thus

$$7(2 + 6) = 7 \cdot 2 + 7 \cdot 6$$

Again we see that if we multiply each term within the parentheses by 7 first and add, we get the same result as adding the terms first, and then multiplying by 7. If we continue testing this apparent relationship for various other sets of natural numbers, we will not be able to find any for which it does not hold. But once again we are not in a position to prove it; therefore, we state the result as an axiom.

Distributive Property

For all natural numbers a, b, and c

$$a(b + c) = ab + ac$$

Multiplication distributes over addition.

EXAMPLE 20

(A) $3(x + y) = 3x + 3y$

(B) $4(w + 2) = 4w + 4 \cdot 2 = 4w + 8$

(C) $x(x + 1) = x \cdot x + x \cdot 1 = x^2 + x$

(D) $2x^2(3x + 2y) = 2x^2 \cdot 3x + 2x^2 \cdot 2y$

$$= (2 \cdot 3)(x^2 x) + (2 \cdot 2)(x^2 y)$$

$$= 6x^3 + 4x^2 y$$

PROBLEM 20 Multiply using the distributive property:

(A) $2(a + b)$ (B) $5(x + 3)$

(C) $u(u^2 + 1)$ (D) $3n^2(2m^2 + 3n)$

It is important to realize that the distributive property holds in both directions. That is, if $a(b + c) = ab + ac$, then using the symmetric law of equality (Section 1.4) we can write

$$ab + ac = a(b + c)$$

The expression on the left is converted to a factored form by taking the common factor a out:

a is a common factor   common factor taken out

$$ab + ac \quad = \quad a(b + c)$$

2 terms on left   2 factors on right

EXAMPLE 21 Taking out common factors.

(A) $2x + 2y = 2(x + y)$

(B) $3w + 6 = 3w + 3 \cdot 2 = 3(w + 2)$

(C) $x^2 + x = x \cdot x + 1 \cdot x = x(x + 1)$ 1 is a factor of every number

(D) $6y + 4y^2 = 2y \cdot 3 + 2y \cdot 2y = 2y(3 + 2y)$

NOTE: Multiplying the expression on the right should take you back where you started on the left.

PROBLEM 21 Take out factors common to all terms:

(A) $5x + 5y$ (B) $2w + 8$

(C) $u^2 + u$ (D) $6y^2 + 4y^3$

Since $a(b + c) = (b + c)a$ (why?), a factor common to two terms may be either taken out on the left or on the right. That is, $ab + ac = (b + c)a$ as well as $a(b + c)$. Consider the following example.

EXAMPLE 22 (A) $3x + 5x = (3 + 5)x = 8x$

(B) $7y + 2y = (7 + 2)y = 9y$

(C) $3x^2y^2 + 4x^2y^2 = (3 + 4)x^2y^2 = 7x^2y^2$

Notice the marked simplification obtained in each case. The result is significant and will be expanded upon in the next section.

PROBLEM 22   Simplify as in Example 22:

(A) $2x + 3x$ **(B)** $8u + 3u$ **(C)** $5uv^2 + 7uv^2$

By repeated use of the distributive property we can show that multiplication distributes over any finite sum. Thus,

---

Extended Distributive Property

$$a(b + c + d) = ab + ac + ad$$

$$a(b + c + d + e) = ab + ac + ad + ae$$

and so on.

EXAMPLE 23
(A) $3(x + y + z) = 3x + 3y + 3z$ multiplication
(B) $ma + mb + mc = m(a + b + c)$ factoring
(C) $2x(x + 3y + 2) = 2x^2 + 6xy + 4x$ multiplication
(D) $4x^2 + 2xy + xz = x(4x + 2y + z)$ factoring
(E) $3x^2y(2xy^3 + 3xy + x^2y^2) = 6x^3y^4 + 9x^3y^2 + 3x^4y^3$ multiplication

PROBLEM 23
(A)   Multiply: $5(a + b + c)$
(B)   Take out factors common to all terms: $ax + ay + az$
(C)   Multiply: $3x(2x + 3y + 5)$
(D)   Take out factors common to all terms: $4x^3 + 2xy + 6x^2$
(E)   Multiply: $2u^2v^3(4u^2v + 2uv + uv^2)$

The properties we have considered in this and the last sections regulate a considerable amount of activity in algebra. These are, so to speak, some of the rules of the game of algebra. However, as is the case in many games (such as chess), one must practice using the rules to become good at algebra.

ANSWERS TO MATCHED PROBLEMS

20.   (A) $2a + 2b$;   (B) $5x + 15$;   (C) $u^3 + u$;   (D) $6m^2n^2 + 9n^3$
21.   (A) $5(x + y)$;   (B) $2(w + 4)$;   (C) $u(u + 1)$;   (D) $2y^2(3 + 2y)$
22.   (A) $5x$;   (B) $11u$;   (C) $12uv^2$
23.   (A) $5a + 5b + 5c$;   (B) $a(x + y + z)$;   (C) $6x^2 + 9xy + 15x$;
   (D) $2x(2x^2 + y + 3x)$;   (E) $8u^4v^4 + 4u^3v^4 + 2u^3v^5$

EXERCISE 1.6   **A** *Compute:*

1. $2(1 + 5)$ and $2 \cdot 1 + 2 \cdot 5$ 2. $3(4 + 2)$ and $3 \cdot 4 + 3 \cdot 2$
3. $5(2 + 7)$ and $5 \cdot 2 + 5 \cdot 7$ 4. $7(3 + 2)$ and $7 \cdot 3 + 7 \cdot 2$

Multiply, using the distributive axiom.

5. $4(x + y)$ 6. $5(a + b)$
7. $7(m + n)$ 8. $9(u + v)$
9. $6(x + 2)$ 10. $3(y + 7)$
11. $5(2 + m)$ 12. $8(3 + n)$

Take out factors common to all terms.

13. $3x + 3y$ 14. $2a + 2b$
15. $5m + 5n$ 16. $7u + 7v$
17. $ax + ay$ 18. $mu + mv$
19. $2x + 4$ 20. $3y + 9$

Multiply, using the distributive property.

21. $2(x + y + z)$ 22. $5(a + b + c)$
23. $3(x + y + z)$ 24. $4(a + b + 3)$

Take out factors common to all terms.

25. $7x + 7y + 7z$ 26. $9a + 9b + 9c$
27. $2m + 2n + 6$ 28. $3x + 3y + 12$

B *Multiply, using the distributive property.*

29. $x(1 + x)$ 30. $y(y + 7)$
31. $y(1 + y^2)$ 32. $x(x^2 + 3)$
33. $3x(2x + 5)$ 34. $5y(2y + 7)$
35. $2m^2(m^2 + 3m)$ 36. $3a^2(a^3 + 2a^2)$
37. $3x(2x^2 + 3x + 1)$ 38. $2y(y^2 + 2y + 3)$
39. $5(2x^3 + 3x^2 + x + 2)$ 40. $4(y^4 + 2y^3 + y^2 + 3y + 1)$
41. $3x^2(2x^3 + 3x^2 + x + 2)$ 42. $7m^3(m^3 + 2m^2 + m + 4)$

Write as a single term (see Example 22).

43. $3x + 7x$ 44. $4y + 5y$
45. $2u + 9u$ 46. $8m + 5m$
47. $2xy + 3xy$ 48. $5mn + 7mn$
49. $2x^2y + 8x^2y$ 50. $3uv^2 + 5uv^2$
51. $7x + 2x + 5x$ 52. $8y + 3y + 4y$

Take out factors common to all terms.

53. $x^2 + 2x$ **54.** $y^2 + 3y$

55. $u^2 + u$ **56.** $m^2 + m$

57. $(2x^3 + 4x)$ **58.** $3u^5 + 6u^3$

59. $x^2 + xy + xz$ **60.** $y^3 + y^2 + y$

61. $3m^3 + 6m^2 + 9m$ **62.** $12x^3 + 9x^2 + 3x$

63. $u^2v + uv^2$ **64.** $2x^3y^2 + 4x^2y^3$

C   *Multiply, using the distributive property.*

65. $4m^2n^3(2m^3n + mn^2)$ **66.** $5uv^2(2u^3v + 3uv^2)$

67. $3x^2y(2xy^3 + 4x + y^2)$ **68.** $2cd^3(c^2d + 2cd + 4c^3d^2)$

69. $4x^2yz^3(3x^2z + yz)$ **70.** $3u^2v^3w(5uw^4 + vw^3)$

71. $(u + v)(c + d) = (u + v)c + (u + v)d = ?$ (finish multiplication)

72. $(m + n)(x + y) = (m + n)x + (m + n)y = ?$ (finish multiplication)

73. $(x + 3)(x + 2)$ **74.** $(m + 5)(m + 3)$

Take out factors common to all terms.

75. $a^2bc + ab^2c + abc^2$ **76.** $m^3n + mn^2 + m^2n^2$

77. $16x^3yz^2 + 4x^2y^2z + 12xy^2z^3$

78. $27u^5v^2w^2 + 9u^2v^3w^4 + 12u^3v^2w^5$

1.7 COMBINING LIKE TERMS

The constant factor in a term is called the **numerical coefficient** (or simply the **coefficient**) of the term. If no constant factor appears in the term, then the coefficient is understood to be 1. (For example, x^2y has a coefficient of 1, since $x^2y = 1 \cdot x^2y$.)

EXAMPLE 24   In the expression

$$2x^3 + x^2y + 3xy^2 + y^3$$

the coefficient of the first term is 2, the second 1, the third 3, and the fourth 1.

PROBLEM 24   Given the algebraic expression

$$5x^4 + 2x^3y + x^2y^2 + 4xy^3 + y^4$$

what is the coefficient of each term?

     If two or more terms are exactly alike, except for possibly their numerical coefficients or the order in which the factors are multiplied, then they are called **like terms.** Thus, like terms must have the same variables to the same powers, but their coefficients do not have to be the same.

EXAMPLE 25 **(A)** In $4x + 2y + 3x$, $4x$ and $3x$ are like terms.

(B) In $9x^2y + 3xy + 2x^2y + x^2y$, the first, third, and fourth terms are like terms.

PROBLEM 25 List the like terms in

(A) $5m + 6n + 2n$ **(B)** $2xy + 3xy^3 + xy + 2xy^3$

If an algebraic expression contains two or more like terms, these terms can always be combined into a single term. The distributive axiom (as was seen in the last section) is the principal tool behind the process.

EXAMPLE 26 **(A)** $3x + 5x = (3 + 5)x = 8x$

(B) $5t + 4s + 7t + s = 5t + 7t + 4s + s = (5 + 7)t + (4 + 1)s$

$$= 12t + 5s$$

PROBLEM 26 Combine like terms proceeding as in Example 26:

(A) $6y + 5y$ **(B)** $4x + 7y + x + 2y$

We can mechanize the above process of combining like terms as follows:

Mechanical Rule for Combining Like Terms

To combine like terms add their numerical coefficients.

EXAMPLE 27 Combine like terms mentally:

(A) $7x + 2y + 3x + y = 7x + 3x + 2y + 1y$

$$= 10x + 3y$$

(B) $2u^2 + 3u + 4u^2 = 2u^2 + 4u^2 + 3u$

$$= 6u^2 + 3u$$

(C) $(3x^2 + x + 2) + (4x^2 + 2x + 1) = 3x^2 + 1x + 2 + 4x^2 + 2x + 1$

$$= 3x^2 + 4x^2 + 1x + 2x + 2 + 1$$

$$= 7x^2 + 3x + 3$$

PROBLEM 27   Combine like terms mentally:

(A) $4x + 7y + 9x$ (B) $3x^2 + y^2 + 2x^2 + 3y^2$

(C) $(2m^2 + 3m + 5) + (m^2 + 4m + 2)$

In the next example we will use most of what we have been discussing in this chapter up until now.

EXAMPLE 28   Multiply as indicated and combine like terms.

(A) $3x(x + 5) + 4x(2x + 3) = 3x^2 + 15x + 8x^2 + 12x$

$$= 11x^2 + 27x$$

(B) $2x(x^2 + 2x + 1) + x(3x^2 + x + 2) = 2x^3 + 4x^2 + 2x + 3x^3 + x^2 + 2x$

$$= 5x^3 + 5x^2 + 4x$$

(C) $3x(2x + 4y) + 2y(3x + y) + 2x^2 + 3y^2$

$$= 6x^2 + 12xy + 6xy + 2y^2 + 2x^2 + 3y^2 \qquad \text{Note: } 6yx = 6xy$$

$$= 8x^2 + 18xy + 5y^2$$

PROBLEM 28   Multiply as indicated and combine like terms.

(A) $4m(m + 3) + m(6m + 1)$

(B) $3x(2x^3 + x + 1) + 2x(x^3 + 3x^2 + 2)$

(C) $4x^2 + 3y(2x + y) + 2x(x + 3y) + y^2$

ANSWERS TO MATCHED PROBLEMS

24. 5, 2, 1, 4, and 1

25. (A) $6n, 2n$;   (B) $2xy, xy$; $3xy^3, 2xy^3$

26. (A) $11y$;   (B) $5x + 9y$

27. (A) $13x + 7y$;   (B) $5x^2 + 4y^2$;   (C) $3m^2 + 7m + 7$

28. (A) $10m^2 + 13m$;   (B) $8x^4 + 6x^3 + 3x^2 + 7x$;
(C) $6x^2 + 12xy + 4y^2$

EXERCISE 1.7

A   *Indicate the numerical coefficient of each term.*

   1. $4x$ **2.** $7ab$ **3.** $8x^2y$

   4. $9uv^2$ **5.** x^3 **6.** y^5

   7. u^2v^3 **8.** m^3n^5

Given the algebraic expression $2x^3 + 3x^2 + x + 5$, indicate:

   9.   The coefficient of the second term.

  10.   The coefficient of the first term.

  11.   The exponent of the variable in the second term.

  12.   The exponent of the variable in the first term.

13. The coefficient of the third term.
14. The exponent of the variable in the third term.

Select like terms in each group of terms.

15. $3x,\ 2y,\ 4x,\ 5y$ 16. $3m,\ 2n,\ 5m,\ 7n$

17. $6x^2,\ x^3,\ 3x^2,\ x^2,\ 4x^3$ 18. $2y^2,\ 3y^4,\ 5y^4,\ y^2,\ y^4$

19. $2u^2v,\ 3uv^2,\ u^2v,\ 5uv^2$ 20. $5mn^2,\ m^2n,\ 2m^2n,\ 3mn^2$

Combine like terms.

21. $5x + 4x$ 22. $2m + 3m$

23. $3u + u$ 24. $x + 7x$

25. $7x^2 + 2x^2$ 26. $4y^3 + 6y^3$

27. $2x + 3x + 5x$ 28. $4u + 5u + u$

29. $2x + 3y + 5x + y$ 30. $m + 2n + 3m + 4n$

31. $2x + 3y + 5 + x + 2y + 1$ 32. $3a + b + 1 + a + 4b + 2$

B   *Select like terms in each group.*

33. $m^2n,\ 4mn^2,\ 2mn,\ 3mn,\ 5m^2n,\ mn^2$

34. $3u^2v,\ 2uv,\ u^2v,\ 2uv^2,\ 4uv,\ uv^2$

Combine like terms.

35. $2t^2 + t^2 + 3t^2$

36. $6x^3 + 3x^3 + x^3$

37. $3x + 5y + x + 4z + 2y + 3z$

38. $2r + 7t + r + 4s + r + 3t + s$

39. $9x^3 + 4x^2 + 3x + 2x^3 + x$

40. $y^3 + 2y + 3y^2 + 4y^3 + 2y^2 + y + 5$

41. $x^2 + xy + y^2 + 3x^2 + 2xy + y^2$

42. $3x^2 + 2x + 1 + x^2 + 3x + 4$

43. $(2x + 1) + (2x + 3) + (2x + 5)$

44. $(4x + 1) + (3x + 2) + (2x + 5)$

45. $(t^2 + 5t + 3) + (3t^2 + t) + (2t + 7)$

46. $(4x^4 + 2x^2 + 3) + (x^4 + 3x^2 + 1)$

47. $(x^3 + 3x^2y + xy^2 + y^3) + (2x^3 + 3xy^2 + y^3)$

48. $(2u^3 + uv^2 + v^3) + (u^3 + v^3) + (u^3 + 3u^2v)$

Multiply, using the distributive property, and combine like terms.

49. $2(x + 5) + 3(2x + 7)$

50. $5(m + 7) + 2(3m + 6)$

51. $x(x + 1) + x(2x + 3)$

52. $2t(3t + 5) + 3t(4t + 1)$

53. $5(t^2 + 2t + 1) + 3(2t^2 + t + 4)$

54. $4(u^2 + 3u + 2) + 2(2u^2 + u + 1)$

55. $y(y^2 + 2y + 3) + (y^3 + y) + y^2(y + 1)$

56. $2y(y^2 + 2y + 5) + 7y(3y + 2) + y(y^2 + 1)$

57. $2x(3x + y) + 3y(x + 2y)$

58. $3m(2m + n) + 2n(3m + 2n)$

59. $2x^2(2x^2 + y^2) + y^2(x^2 + 3y^2)$

60. $3u^2(u^2 + 2v^2) + v^2(2u^2 + v^2)$

61. $3m^4(m^2 + 2m + 1) + m^3(m^3 + 3m^2 + m)$

62. $4x^3(x^2 + 3x) + 2x^4(3x + 1)$

63. If x represents a natural number, write an algebraic expression for the sum of four consecutive natural numbers starting with x. Simplify the expression by combining like terms.

64. If t represents an even number, write an algebraic expression for the sum of three consecutive even numbers starting with t. Simplify.

C *Multiply, using the distributive property, and combine like terms.*

65. $2xy^2(3x + x^2y) + 3x^2y(y + xy^2)$

66. $3s^2t^3(2s^3t + s^2t^2) + 2s^3t^2(3s^2t^2 + st^3)$

67. $3u^2v(2uv^2 + u^2v) + 2uv^2(u^2v + 2u^3)$

68. $4m^3n^2(3mn^2 + n) + 2mn^2(2m^3n^2 + m^2n)$

69. $(2x + 3)(3x + 2) = (2x + 3)3x + (2x + 3)2 = ?$

70. $(x + 2)(2x + 3) = (x + 2)2x + (x + 2)3 = ?$

71. $(x + 2y)(2x + y)$

72. $(3x + y)(x + 3y)$

73. $(x + 3)(x^2 + 2x + 5)$

74. $(r + s + t)(r + s + t)$

75. If y represents an odd number, write an algebraic expression for the product of y and the next odd number. Write as the sum of two terms.

76. If y represents the first of four consecutive even numbers, write an algebraic expression that would represent the product of the first two added to the product of the last two. Simplify.

77. An even number plus the product of it and the next even number is 180. Introduce a variable, and write as an algebraic equation. Simplify the left and right sides of the equation where possible.

78. There exist at least two consecutive odd numbers such that 5 times the first plus twice the second is equal to twice the first plus 3 times the second. Introduce a variable and write as an algebraic equation. Simplify the left and right sides of the equation where possible.

1.8 CHAPTER REVIEW: IMPORTANT TERMS AND SYMBOLS, REVIEW EXERCISE, PRACTICE TEST

Important terms and symbols

set *(1.1)*   element of *(1.1)*   member of *(1.1)* $\in$, $\notin$ *(1.1)*   subset *(1.1)*   empty set *(1.1)*   null set *(1.1)* $\emptyset$ *(1.1)*   variable *(1.1, 1.3)*   constant *(1.1, 1.3)*   counting numbers *(1.2)*   natural numbers *(1.2)*   even numbers *(1.2)*   odd numbers *(1.2)*   terms *(1.2, 1.3)*   factors *(1.2, 1.3)*   composite number *(1.2)*   prime number *(1.2)*   completely factored form *(1.2)*   least common multiple, LCM *(1.2)*   equality *(1.4)*   =, $\neq$, *(1.4)*   algebraic equation *(1.4)*   properties of equality *(1.4)*   axiom *(1.5)*   commutative properties *(1.5)*   associative properties *(1.5)*   exponent *(1.5)*   power *(1.5)* b^n *(1.5)*   distributive property *(1.6)*   coefficient *(1.7)*   like terms *(1.7)*

Exercise 1.8 Review exercise

Work through all the problems in this chapter review and check answers in the back of the book. (Answers to all review problems are there, and following each answer is a number in italics indicating the section in which that type of problem is discussed.) Where weaknesses show up, review appropriate sections in the text. When you are satisfied that you know the material, take the practice test following this review. All variables represent natural numbers.

A   **1.**   Given $G = \{10, 11, 12, 13, 14, 15\}$
(A) Write the set of odd numbers in G.
(B) Write the set of prime numbers in G.

Evaluate:

2. $12 - 5 \cdot 2$

3. $5 + 3(7 - 5)$

4. $x - 4(x - 7)$ for $x = 9$

5. $(x + 4)(x - 4)$ for $x = 6$

Multiply:

6. $x^{12}x^{13}$

7. $(2x^3)(3x^5)$

8. $2^5 \cdot 2^{20}$

9. $x(x + 1)$

10. $5(2x + 3y + z)$

11. $3u(2u^2 + u)$

Combine like terms:

12. $3y + 6y$

13. $2m + 5n + 3m$

14. $3x^2 + 2x + 4x^2 + x$

15. $3x^2y + 2xy^2 + 5x^2y$

Write in a factored form by taking out factors common to all terms.

16. $3m + 3n$

17. $8u + 8v + 8w$

18. $xy + xw$

19. $4x + 8w$

If x represents a natural number, write an algebraic expression that represents each of the following.

20. A number 12 times as large as x

21. A number 3 more than 3 times x

22. A number 5 less than twice x

B **23.** Let A be the set of natural numbers starting at 21 and ending at 31.
(A) List the elements in the set $\{x \mid x \text{ is a prime number in } A\}$.
(B) Which is true: $20 \notin A$, $25 \in A$?

24. Given $5x^3 + 3x^2 + x + 7$
(A) What is the coefficient of the second term?
(B) What is the coefficient of the third term?
(C) What is the exponent of the variable in the third term?

25. Write 120 as a product of prime factors.

Find the LCM for:

26. 3, 4, 9

27. 6, 5, 9

28. 15, 18

29. 12, 18, 10

Evaluate:

30. $(8 + 10) - 3(7 - 3)$

31. $2 \cdot 9 - 6(8 - 2 \cdot 3)$

32. $2[12 - 2(6 - 3)]$

33. $2[(8 + 4) - (7 - 5)]$

34. $2[x + 3(x - 4)]$ for $x = 6$

35. $6[(x + y) - 3(x - y)]$ for $x = 7$ and $y = 5$

Multiply as indicated and combine like terms where possible.

36. $(2x^3)(3x)(3x^4)$

37. $(3xy^2z)(4x^2y^3z^3)$

38. $3y^3(2y^2 + y + 5)$

39. $2(5u^2 + 2u + 1) + 3(3u^2 + u + 5)$

40. $3x(x + 5) + 2x(2x + 3) + x(x + 1)$ **41.** $(x + 2y)3x + (x + 2y)y$

Write in a factored form by taking out factors common to all terms.

42. $u^3 + u^2 + u$

43. $6x^2y + 3xy^2$

44. $3m^5 + 6m^4 + 15m^2$

Translate each statement into an algebraic equation using only the variable x.

45. 24 is 6 less than twice a certain number.

46. 3 times a given number is 12 more than that number.

47. The sum of four consecutive natural numbers is 138. (Let x be the first of the four consecutive natural numbers.)

48. The sum of three consecutive even numbers is 78. (Let x be the first of the three consecutive even numbers.)

C *Given the set* $M = \{27, 51, 61\}$, *write as a set:*

49. The odd numbers in M **50.** The prime numbers in M

Evaluate:

51. $4\{20 - 4[(11 - 3) - 3 \cdot 2]\}$

52. $3\{x + 2[8 - 2(x - y)]\}$ for $x = 5$ and $y = 3$

Multiply and combine the terms where possible.

53. $5u^3v^2(2u^2v^2 + uv + 2)$

54. $2x^3(2x^2 + 1) + 3x^2(x^3 + 3x + 2)$

55. $(4x + 3)(2x + 1)$

Write in factored form by taking out factors common to all terms.

56. $12x^3yz^2 + 9x^2yz$ **57.** $20x^3y^2 + 5x^2y^3 + 15x^2y^2$

Each statement is justified by either the commutative or associative property. State which.

58. $x3 = 3x$ **59.** $(x + 3) + 2 = x + (3 + 2)$

60. $(3 + x) + 5 = (x + 3) + 5$

61. $(x + 3) + (x + 5) = x + [3 + (x + 5)]$

62. If x represents the first of three consecutive odd numbers, write an algebraic equation that represents the fact that 4 times the first is equal to the sum of the second and third.

Practice test —Chapter 1

Pretend this is an in-class test, and allow yourself 50 minutes for its completion. Work the problems straight through without looking back in the chapter. All answers are in the back of the book in the answer section. This provides good practice for tests taken in class. All variables represent natural numbers.

1. If $E = \{2, 4, 6, \ldots\}$, indicate true (T) or false (F) for each of the following:

(A) $7 \notin E$;   (B) $E = \{x \mid x$ is an odd number$\}$

2. (A) Write 180 as a product of prime factors.
   (B) Find the LCM for 15 and 18.

3. Evaluate:

(A) $3[10 - 2(5 - 2)]$

(B) $3[x + 2(x - y)]$ for $x = 8$ and $y = 5$

4. Given $6x^4 + 3x^2 + x + 6$:

(A) What is the coefficient of the third term?

(B) What is the exponent of the variable in the first term?

5. Combine like terms:

(A) $(3x^2 + x + 1) + (2x^2 + 3x + 5)$

(B) $6x^3y^2 + 2x^2y^3 + x^3y^2$

6. Multiply:

(A) $(5x^2y)(3x^2y^3)$

(B) $2x^2(3x^2 + x + 4)$

7. Write in factored form by taking out factors common to all terms:

(A) $6a + 6b + 6c$

(B) $8x^4 + 12x^3 + 4x^2$

8. Multiply and combine like terms where possible:

(A) $2u(u + 5) + u(3u + 1)$

(B) $(2x + 3y)3x + (2x + 3y)y$

9. If x represents a natural number, write an algebraic expression for a number:

(A) 5 less than twice x

(B) 6 more than 3 times the quantity x minus 2

10. Translate each statement into an algebraic equation using x as the only variable:

(A) 7 more than a certain number is 5 less than twice the number.

(B) The sum of three consecutive even numbers (starting with x) is 72.

2

INTEGERS

By limiting ourselves in the first chapter to the simplest number system within your experience, the natural numbers, we were able to develop many basic algebraic processes without the distracting influence of more complicated numbers such as decimals, fractions, and radicals. We will find that most of these processes carry on without change to the more involved number systems to be presented in this and the next chapter.

2.1 THE SET OF INTEGERS

The set of natural numbers has obvious restrictions. For example, neither

$$3 \div 5 \quad \text{nor} \quad 3 - 5$$

have answers that are natural numbers. As a step toward remedying these deficiencies and at the same time increasing our manipulative power, we now extend the natural numbers to the integers.

We start by giving the natural numbers another name. From now on they will also be called **positive integers.** To help us emphasize the difference between the positive integers (natural numbers) and the negative integers that are to be introduced shortly, we will often place a plus sign in front of a numeral used to name a natural number. Thus, we may use either

$$+3 \text{ or } 3 \qquad +25 \text{ or } 25 \qquad +372 \text{ or } 372$$

and so on.

If we form a **number line** (a line with numbers associated with points on the line) using the positive integers, and divide the line to the left into line segments equal to those used on the right, how should the endpoints of segments on the left be labeled?

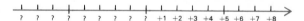

As you no doubt will guess, we label the first point to the left of $+1$ with *zero*

0

and the other points in succession with

$$-1, -2, -3, \dots$$

These last numbers are called **negative integers.**

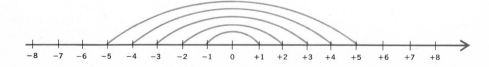

In general, to each positive integer there corresponds a unique (one and only one) number called a negative integer: -1 to $+1$, -2 to $+2$, -3 to $+3$, and so on. The minus sign is part of the number symbol. The elements in the integer pairs -1, $+1$; -2, $+2$, and so on; are often referred to as **opposites** of each other.

By collecting the positive integers, 0, and the negative integers into one set we obtain the set of **integers,** *I,* the subject matter of this chapter.

The Set of Integers I

$$\{\ldots, -4, -3, -2, -1, 0, +1, +2, +3, +4, \ldots\}$$

We do not attempt to give a precise definition of each integer. We do, however, postulate the existence of this set of numbers, and we will learn to manipulate the symbols that name them according to certain rules.

EXAMPLE 1   The points *a, b, c,* and *d* on the number line below are labeled -8, -6, 0, and $+7$, respectively.

PROBLEM 1   What numbers are associated with the points *a, b, c,* and *d* on the number line

Zero and the negative integers are relatively recent historically speaking. Both concepts were introduced as numbers in their own right between A.D. 600 and A.D. 700. Hindu mathematicians in India are given credit for their invention. The growing importance of commercial activities seemed to be the stimulus. Since business transactions involve decreases as well as increases, it was found that both transactions could be treated at once if the positive integers represented amounts received and the negative integers represented amounts paid out. Of course, since then negative numbers have been put to many other uses such as recording temperatures below 0, indicating altitudes below sea level, and representing deficits in financial statements, to name a few that the reader is probably already aware of. In addition, without negative numbers it is not possible to perform the operation

$7 - 12$

or to solve the simple equation

$8 + x = 2$

Before this course is over many more uses of negative numbers will be considered. In the next several sections we will learn how to add, subtract, multiply, and divide integers—an essential step to their many uses.

ANSWERS TO MATCHED PROBLEMS

1. $-13, -6, -1, +9$

EXERCISE 2.1

A

1. What numbers are associated with points a, b, c, and d?

2. What numbers are associated with points a, b, c, d, and e?

Locate each set of numbers on a number line.

3. $\{-4, -2, 0, +2, +4\}$
4. $\{-7, -4, 0, +4, +8\}$
5. $\{-25, -20, -15, +5, +15\}$
6. $\{-30, -20, -5, +10, +15\}$

Using the figure for Problem 2, write down the number associated with the point.

7. 3 units to the left of d
8. 4 units to the right of e
9. 4 units to the right of a
10. 2 units to the left of b
11. 10 units to the left of d
12. 20 units to the right of a

B

Let P = the set of positive integers
M = the set of negative integers
I = the set of integers

Indicate which are true (T) or false (F).

13. $+5 \in P$
14. $-3 \in M$
15. $-4 \in P$
16. $+3 \in M$
17. $-7 \in I$
18. $+4 \in I$
19. $0 \in P$
20. $0 \in M$

21. $0 \in I$

22. $-14 \in I$

23. P is a subset of I.

24. I is a subset of M.

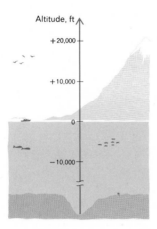

Figure 1

Referring to Figure 1, express each of the following quantities by means of an appropriate integer:

25. A mountain height of 20,270 ft (Mount McKinley, highest mountain in the United States)

26. A mountain peak 29,141 ft high (Mount Everest, highest point on earth)

27. A valley depth of 280 ft below sea level (Death Valley, the lowest point below sea level in the Western Hemisphere)

28. An ocean depth of 35,800 ft (Marianas Trench in the Western Pacific, greatest known depth in the world)

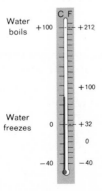

Figure 2

Referring to Figure 2, express each of the following quantities by means of an appropriate integer:

29. 5° below freezing on Celsius scale

30. 35° below freezing on Celsius scale

31. 5° below freezing on Fahrenheit scale

32. 100° below boiling on Fahrenheit scale

33. 35° below freezing on Fahrenheit scale

34. 220° below boiling on Fahrenheit scale

Express each of the following quantities by means of an appropriate integer:

35. A bank deposit of $25

36. A bank balance of $237

37. A bank withdrawal of $10

38. An overdrawn checking account of $17

39. A 9-yd loss in football

40. A 23-yd gain in football

C *In each problem start at 0 on a number line and give the number associated with the final position.*

41. Move 7 units in the positive direction, 4 units in the negative direction, 5 more units in the negative direction, and finally, 3 units in the positive direction.

42. Move 4 units in the negative direction, 7 units in the positive direction, and 13 units in the negative direction.

Express the net gain or loss by means of an appropriate integer.

43. In banking: a $23 deposit, a $20 withdrawal, a $14 deposit

44. In banking: a $32 deposit, a $15 withdrawal, an $18 withdrawal

45. In football: 5-yd gain, a 3-yd loss, a 4-yd loss, an 8-yd gain, a 9-yd loss

46. In an elevator: up 2 floors, down 7 floors, up 3 floors, down 5 floors, down 2 floors

2.2 THE OPPOSITE OF, AND ABSOLUTE VALUE OF A NUMBER

One of the important activities in algebra and in the uses of algebra in the real world is the evaluation of algebraic expressions for various replacements of variables by constants. An algebraic expression is like a recipe in that it contains symbolic instructions on how to proceed in its evaluation. For example, if we were to evaluate

$5(x + 2y)$

for $x = 10$ and $y = 3$, we would write

$5(10 + 2 \cdot 3)$

then we would multiply 2 and 3, add the product to 10, and then multiply the sum by 5.

In this section we are going to define two more operations on numbers called "the opposite of" and "the absolute value of," and you will get additional practice in following symbolic instructions. These two new operations are widely used in mathematics and its applications and will be used in the following sections to formulate definitions for addition, subtraction, multiplication, and division for integers. We start by defining "the opposite of a number."

The opposite of a number

By the **opposite of** a number x, we mean an operation on x, symbolized by

$-x$ opposite of x

that produces another number; namely, it changes the sign of x if x is not zero, and if x is zero it leaves it alone.

EXAMPLE 2

(A) $-(+3) = -3$

(B) $-(-5) = +5$

(C) $-(0) = 0$

(D) $-[-(+3)] = -(-3) = +3$

PROBLEM 2   Find:

(A) $-(+7)$ **(B)** $-(-6)$

(C) $-(0)$ **(D)** $-[-(-4)]$

Graphically, the opposite of a number is its "mirror image" relative to 0 (see Figure 3).

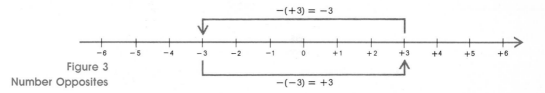

Figure 3
Number Opposites

As a consequence of the definition of "the opposite of a number," we note the following important properties:

1. The opposite of a positive number is a negative number.
2. The opposite of a negative number is a positive number.
3. The opposite of 0 is 0.

Thus, we see that $-x$ is not necessarily a negative number: $-x$ represents a positive number if x is negative and a negative number if x is positive.

You should be aware by now that the minus sign "−" is used in the following three distinct ways:

1. As the operation "subtract": $7 \overset{\downarrow}{-} 5 = 2$
2. As the operation "the opposite of": $\overset{\downarrow}{-}(-6) = +6$
3. As part of a number symbol: $\overset{\downarrow}{-}8$

NOTE: The opposite of a number x, $-x$, is sometimes referred to as "the negative of x." Since the latter designation often causes confusion at this stage of the development, we will avoid its use until later.

The absolute value of a number

The **absolute value of a number** x is an operation on x, denoted symbolically by

$|x|$ absolute value of x

(not square brackets) that produces another number. What number? If x is positive or 0 it leaves it alone; if x is negative it makes it positive. Symbolically, and more formally,

$$|x| = \begin{cases} x & \text{if } x \text{ is a positive number or 0} \\ -x & \text{if } x \text{ is a negative number} \end{cases}$$

Do not be afraid of this symbolic form of the definition. It represents a first exposure to more precise mathematical representations and it will take on more meaning with repeated exposure.

EXAMPLE 3   **(A)** $|+7| = +7$

(B) $|-7| = +7$

(C) $|0| = 0$

PROBLEM 3   Evaluate:

(A) $|+5|$ **(B)** $|-5|$ **(C)** $|0|$

Thus, we see that

1. The absolute value of a positive number is a positive number.
2. The absolute value of a negative number is a positive number.
3. The absolute value of 0 is 0.

and conclude that

The absolute value of a number is never negative.

"The absolute value" and "the opposite of" operations are often used in combination and it is important to perform the operations in the right order—generally, from the inside out.

EXAMPLE 4   (A) $|-(-3)| = |+3| = +3$

(B) $-|-3| = -(+3) = -3$

(C) $-(|-5| - |-2|) = -[(+5) - (+2)] = -(+3) = -3$

PROBLEM 4   Evaluate:

(A) $|-(+5)|$ (B) $-|+5|$ (C) $-(|-7| + |-3|)$

ANSWERS TO   2. (A) -7;   (B) $+6$;   (C) 0;   (D) -4
MATCHED   3. (A) $+5$;   (B) $+5$;   (C) 0
PROBLEMS   4. (A) $+5$;   (B) -5;   (C) -10

EXERCISE 2.2

A *Evaluate:*

1. $-(+9)$ 2. $-(+14)$ 3. $-(-2)$

4. $-(-3)$ 5. $|+4|$ 6. $|+10|$

7. $|-6|$ 8. $|-7|$ 9. $-(0)$

10. $|0|$

11. The opposite of a number is (*always, sometimes, never*) a negative number.

12. The opposite of a number is (*always, sometimes, never*) a positive number.

13. The absolute value of a number is (*always, sometimes, never*) a negative number.

14. The absolute value of a number is (*always, sometimes, never*) a positive number.

Replace each question mark with an appropriate integer.

15. $-(+11) = ?$ 16. $-(-15) = ?$

17. $-(?) = +5$ 18. $-(?) = -8$

19. $|-13| = ?$ 20. $|+17| = ?$

21. $|?| = +2$ 22. $|?| = +8$

23. $|?| = -4$ 24. $|?| = 0$

B *Evaluate:*

25. $-[-(+6)]$ 26. $-[-(-11)]$

27. $|-(-5)|$ 28. $|-(+7)|$

29. $-|-5|$
30. $-|+7|$
31. $(|-3| + |-2|)$
32. $(|-7| - |+3|)$
33. $-(|-12| - |-4|)$
34. $-(|-6| + |-2|)$

Evaluate for $x = +7$ and $y = -5$.

35. $-x$
36. $|x|$
37. $|y|$
38. $-y$
39. $-|x|$
40. $-|y|$
41. $-(-y)$
42. $-(-x)$
43. $|-y|$
44. $|-x|$
45. $|x| - |y|$
46. $-(|x| + |y|)$

Guess at the solution set of each equation from the set of integers.

47. $|+5| = x$
48. $|-7| = x$
49. $-x = -3$
50. $-x = +8$
51. $|x| = +6$
52. $|x| = +9$
53. $|x| = -4$
54. $-|x| = +4$

C *Describe the elements in each set.*

55. $\{x \in I \mid |x| = 0\}$
56. $\{x \in I \mid -x = 0\}$
57. $\{x \in I \mid -x = |x|\}$
58. $\{x \in I \mid x = |x|\}$
59. $\{x \in I \mid -(-x) = x\}$
60. $\{x \in I \mid |-x| = |x|\}$
61. $\{x \in I \mid |-x| = x\}$
62. $\{x \in I \mid -x = x\}$
63. $\{x \in I \mid -|x| = |x|\}$
64. $\{x \in I \mid |-x| = -|x|\}$

2.3 ADDITION OF INTEGERS

How should addition in the integers be defined so that we can assign numbers to each of the following sums?

$(+2) + (+5) = ?$

$(+2) + (-5) = ?$

$(-2) + (+5) = ?$

$(-2) + (-5) = ?$

$(+7) + \quad 0 = ?$

$\quad 0 + (-3) = ?$

To give us an idea, let us think of addition of integers in terms of deposits and withdrawals in a checking account, starting with a 0 balance. If we do this, then a deposit of $2 followed by another deposit of $5 would provide us with a balance of $7; thus, as we would expect from addition in the natural numbers,

$$(+2) + (+5) = +7$$

Similarly, a deposit of $2 followed by a withdrawal of $5 would yield an overdrawn account of $3, and we would write

$$(+2) + (-5) = -3$$

Continuing in the same way, we can assign to each sum the value that indicates the final status of our account after the two transactions have been completed. Hence,

$$(-2) + (+5) = +3$$
$$(-2) + (-5) = -7$$
$$(+7) + \quad 0 = +7$$
$$0 + (-3) = -3$$

We would like any formal definition of addition for integers to yield the same results as above and, also, to yield the same commutative and associative properties we had with the natural numbers. With these considerations in mind, it turns out that we have very little choice but to define addition as follows.

Definition of addition of integers

Numbers with like sign. If a and b are positive integers, add as in the set of natural numbers. If a and b are both negative, the sum is the opposite of the sum of their absolute values.

Numbers with unlike signs. The sum of two integers with unlike signs is a number of the same sign as the integer with the larger absolute value. The absolute value of the sum is the difference of the absolute values of the two numbers found by subtracting the smaller absolute value from the larger. If the numbers have the same absolute values, their sum is 0.

Zero. The sum of any integer and 0 is that integer; the sum of 0 and any integer is that integer.

This definition when applied to the deposit-withdrawal illustrations above will produce the same results. But you will no doubt object to the difficulty in its use. Fortunately, we will be able to mechanize the process so that you will be able to handle addition problems without difficulty. You should not forget, however, that these mechanical rules that are to be discussed shortly are justified on the basis of the above definition and not vice versa.

The following important properties of addition are an immediate consequence of the definition of addition above:

THEOREM 1 For all integers a, b, and c

(A) $a + b = b + a$ commutative property

(B) $(a + b) + c = a + (b + c)$ associative property

As a consequence of this theorem we will have essentially the same kind of freedom that we had with the natural numbers in rearranging terms and inserting or removing parentheses.

Now let us turn to the mechanics of adding signed numbers. It may relieve you to know that no one (not even the professional mathematician) in everyday routine calculations involving the addition of signed numbers goes through the steps precisely as they are described in the formal definition of addition; mechanical shortcuts soon take over. The following process, or something close to it, is very likely used. We will restrict our attention to nonzero quantities since addition involving zero seems to offer few difficulties.

MECHANICS OF ADDING SIGNED NUMBERS

Are the signs of the two numbers alike or unlike?

alike

(A)	Mentally block out the signs.
(B)	Add the two numbers as if they were natural numbers.
(C)	Prefix the common sign of the original numbers to the sum.

examples

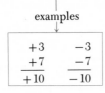

unlike

(A)	Mentally block out the signs.
(B)	Subtract the smaller unsigned number from the larger.
(C)	Prefix the sign associated with the larger of the two unsigned numbers.

examples

EXAMPLE 5 **(A)** Add:

$$
\begin{array}{ccccc}
-8 & +8 & -8 & +4 & 0 \\
-3 & -3 & +3 & 0 & -6 \\
\hline
-11 & +5 & -5 & +4 & -6
\end{array}
$$

(B) Add:

$(-4) + (-6) = -10$

$(-4) + (+6) = +2$

$$(+4) + (-6) = -2$$
$$0 + (-1) = -1$$

PROBLEM 5   **(A)**   Add:

$$
\begin{array}{cccc}
-4 & +4 & -4 & -9 \\
\underline{-5} & \underline{-5} & \underline{+5} & \underline{0}
\end{array}
$$

(B)   Add:

$(-2) + (+7), \quad (+2) + (-7),$
$(-2) + (-7), \quad 0 + (-5)$

To add three or more integers, add all of the positive integers together, add all of the negative integers together (the commutative and associative properties of integers justify this procedure), and then add the two resulting sums as above.

EXAMPLE 6   Add: $(+3) + (-6) + (+8) + (-4) + (-5)$

Solution $(+3) + (-6) + (+8) + (-4) + (-5)$
$$= [(+3) + (+8)] + [(-6) + (-4) + (-5)]$$
$$= (+11) + (-15)$$
$$= -4$$

or vertically,

$$
\begin{array}{c}
+3 \\
-6 \\
+8 \\
-4 \\
\underline{-5} \\
-4
\end{array}
$$

Done mentally or on scratchpaper

$$
\begin{array}{ccc}
+3 & -6 & +11 \\
\underline{+8} & -4 & \underline{-15} \\
+11 & \underline{-5} & -4 \\
 & -15 &
\end{array}
$$

PROBLEM 6   Add: $(-8) + (-4) + (+6) + (-3) + (+10) + (+1)$

We conclude this section by stating without proof another, but less obvious, property of addition that follows from the definition of addition. We will refer to this property in future developments.

THEOREM 2   **(A)**   For each integer a the sum of it and its opposite is 0; that is,

$$a + (-a) = 0$$

(B)   If the sum of two numbers is 0, then each must be the opposite of the other; symbolically, if

$$a + b = 0$$

then $a = -b$ and $b = -a$.

5. (A) −9, −1, +1, −9;   (B) +5, −5, −9, −5

6. +2

EXERCISE 2.3

A *Add:*

1. $\begin{array}{r} +5 \\ +6 \\ \hline \end{array}$	**2.** $\begin{array}{r} +7 \\ -4 \\ \hline \end{array}$	**3.** $\begin{array}{r} -9 \\ +6 \\ \hline \end{array}$	**4.** $\begin{array}{r} -6 \\ +8 \\ \hline \end{array}$
5. $\begin{array}{r} +6 \\ -8 \\ \hline \end{array}$	**6.** $\begin{array}{r} -3 \\ -4 \\ \hline \end{array}$	**7.** $\begin{array}{r} -7 \\ -1 \\ \hline \end{array}$	**8.** $\begin{array}{r} +8 \\ +2 \\ \hline \end{array}$
9. $\begin{array}{r} 0 \\ +3 \\ \hline \end{array}$		**10.** $\begin{array}{r} -4 \\ 0 \\ \hline \end{array}$	

11. $(+5) + (+4)$ **12.** $(-7) + (-3)$

13. $(-8) + (+2)$ **14.** $(+3) + (-7)$

15. $(-6) + (-3)$ **16.** $(+2) + (+3)$

17. $0 + (-9)$ **18.** $(+2) + 0$

19. $\begin{array}{r} +4 \\ -3 \\ -5 \\ -7 \\ +9 \\ \hline \end{array}$	**20.** $\begin{array}{r} -6 \\ -4 \\ +8 \\ +3 \\ -5 \\ \hline \end{array}$	**21.** $\begin{array}{r} -7 \\ +2 \\ -3 \\ -1 \\ +5 \\ \hline \end{array}$	**22.** $\begin{array}{r} +6 \\ -4 \\ -8 \\ -2 \\ +9 \\ \hline \end{array}$

23. $(+5) + (-8) + (-9) + (+7)$ **24.** $(-8) + (-7) + (+3) + (+9)$

25. $(-6) + 0 + (+5) + (-2) + (-1)$ **26.** $(+9) + (-3) + 0 + (-8)$

B *Add:*

27. $\begin{array}{r} +11 \\ -23 \\ \hline \end{array}$	**28.** $\begin{array}{r} -12 \\ -21 \\ \hline \end{array}$	**29.** $\begin{array}{r} -403 \\ -219 \\ \hline \end{array}$	**30.** $\begin{array}{r} -307 \\ +231 \\ \hline \end{array}$

31. $(-63) + (+25)$ **32.** $(-45) + (-73)$

33. $(-237) + (-431)$ **34.** $(-197) + (+364)$

35. $\begin{array}{r} +12 \\ -18 \\ -23 \\ +\ 4 \\ -11 \\ \hline \end{array}$	**36.** $\begin{array}{r} -63 \\ +45 \\ -\ 3 \\ +17 \\ +12 \\ \hline \end{array}$

37. $(+12) + (+7) + (-37) + (+14)$ **38.** $(-23) + (-35) + (+43) + (-33)$

Replace each question mark with an appropriate integer.

39. $(-3) + ? = -7$ **40.** $? + (-9) = -13$

41. $(+8) + ? = +3$ **42.** $(-12) + ? = +4$

43. $? + (-12) = -7$

44. $(+54) + ? = -33$

45. $(+33) + ? = -44$

46. $? + (-14) = +20$

Evaluate:

47. $|-8| + |+6|$

48. $|(-8) + (+6)|$

49. $(-|-3|) + (-|+3|)$

50. $|-5| + [-(-8)]$

Evaluate for $x = -5$, $y = +3$, and $z = -2$.

51. $x + y$ **52.** $y + z$ **53.** $|(-x) + z|$ **54.** $-(|x| + |z|)$

55. You own a stock that is traded on the New York Stock Exchange. On Monday it closed at $23 per share, it fell $3 on Tuesday and another $6 on Wednesday, it rose $2 on Thursday, and finished strongly on Friday by rising $7. Use addition of signed numbers to determine the closing price of the stock on Friday.

56. Your football team is on the opponent's 10-yd line and in 4 downs gains 8 yd, loses 4 yd, loses another 8 yd, and gains 13 yd. Use addition of signed numbers to determine if a touchdown was made.

57. A spelunker (cave explorer) had gone down 2,340 (vertical) ft into the 3,300-ft Gouffre Berger, the world's deepest pothole cave, located in the Isere province of France. On his ascent he climbed 732 ft, slipped back 25 ft and then another 60 ft, climbed 232 ft, and finally slipped back 32 ft. Use addition of signed numbers, starting with $-2,340$, to find his final position.

58. In a card game (such as rummy, where cards held in your hand after someone goes out are counted against you) the following scores were recorded after four hands of play. Who was ahead at this time and what was his or her score?

RUSS	JAN	PAUL	MEG
$+35$	$+80$	-5	$+15$
$+45$	$+5$	$+40$	-10
-15	-35	$+25$	$+105$
-5	$+15$	$+35$	-5

C   *Replace each question mark with an appropriate symbol (variables represent integers).*

59. $a + (-a) = ?$

60. $(-x) + x = ?$

61. $m + ? = 0$

62. $(-x) + ? = 0$

63. Give a reason for each step:

$$[a + b] + (-a) = (-a) + [a + b]$$
$$= [(-a) + a] + b$$
$$= 0 + b$$
$$= b$$

2.4 SUBTRACTION OF INTEGERS

From subtraction in the natural numbers we know that

$$(+8) - (+5) = +3$$

but what can we write for the differences

$$(+8) - (-5) = ?$$
$$(+5) - (+8) = ?$$
$$(-8) - (+5) = ?$$
$$(-8) - (-5) = ?$$
$$(-5) - (-8) = ?$$
$$0 - (-5) = ?$$

We are going to define subtraction in such a way that all of these problems will have answers. Where do we start? We start with some of the notions you learned about subtraction in elementary school and then generalize on these ideas.

You will recall that to check the subtraction problem

$$\frac{\begin{array}{r} 8 \\ -5 \end{array}}{3}$$

we add 3 to 5 to obtain 8. We can use this checking requirement to transform subtraction into addition: Instead of saying

Subtract 5 from 8.

we can ask

What must be added to 5 to produce 8?

Notice that the answer to each is the same, 3. The latter way of looking at subtraction is the most useful way of the two for its generalization to more involved number systems. And it motivates the following general definition that will not only apply to the integers, but also to any other number system we encounter.

Definition of Subtraction

We write

$$M - S = D \quad \text{if and only if} \quad M = S + D$$

The difference D is the number that must be added to S to produce M.

Let us use the definition to find answers to the problems stated earlier. For

$$(+8) - (-5) = ? \qquad \text{or} \qquad \begin{array}{r} (+8) \\ -(-5) \\ \hline ? \end{array}$$

we ask, "What must be added to (-5) to produce $(+8)$?" From our definition of addition we know the answer to be $(+13)$. Thus we write

$$(+8) - (-5) = +13 \qquad \text{or} \qquad \begin{array}{r} (+8) \\ -(-5) \\ \hline +13 \end{array}$$

since $(+13)$ added to (-5) produces $(+8)$.

Similarly,

$$\begin{array}{ccccc} (+5) & (-8) & (-8) & (-5) & 0 \\ -(+8) & -(+5) & -(-5) & -(-8) & -(-5) \\ \hline -3 & -13 & -3 & +3 & +5 \end{array}$$

since each difference added to the second number (subtrahend) produces the top number (minuend).

Fortunately, we will be able to mechanize the process above so that subtraction will be no more difficult than addition. The following rule is a direct consequence of the definition of subtraction given above:

Mechanical Rule for Subtraction

To subtract S from M, add the opposite of S to M. That is,

$M - S = M + (-S)$

Recall: The opposite of S is obtained by changing the sign of S.

Thus, any subtraction problem can be changed to an equivalent addition problem. This rule is actually a theorem and can be established by showing that if $M + (-S)$ is added to S, the result is M. We now use the rule in the following examples.

EXAMPLE 7   Subtract:

(A) $(+7)$

$$\begin{array}{r} \overset{+}{-}(\overset{+}{-}8) \\ \hline +15 \end{array} \qquad \text{Change the sign of } (-8) \text{ and add.}$$

(B) (-4)

$\underline{\overset{+}{-}(\overset{-}{+}5)}$ Change the sign of $(+5)$ and add.
-9

(C) $(-9) - (-4) = (-9) + (+4) = -5$ Change the sign of (-4) and add.

(D) $0 - (+8) = \quad 0 + (-8) = -8$ Change the sign of $(+8)$ and add.

PROBLEM 7 Subtract:

(A) $(+6)$
$\underline{-(-9)}$

(B) (-3)
$\underline{-(-5)}$

(C) $(-7) - (-2)$

(D) $(-3) - (+8)$

EXAMPLE 8 Evaluate:

(A) $[(-3) - (+2)] - [(+2) - (+5)]$

Solution $[(-3) - (+2)] - [(+2) - (+5)]$ Evaluate inside brackets first.

$= [(-3) + (-2)] - [(+2) + (-5)]$

$= (-5) - (-3)$ Now subtract.

$= (-5) + (+3)$

$= -2$

(B) $(-x) - y$ for $x = -3$ and $y = +5$

Solution $(-x) - y$ Substitute $x = -3$ and $y = +5$.

$[-(-3)] - (+5)$ Evaluate inside brackets first.

$= (+3) - (+5)$ Subtract.

$= (+3) + (-5)$

$= -2$

PROBLEM 8 Evaluate:

(A) $[(+3) + (-8)] - [(-2) - (+3)]$

(B) $x - (-y)$ for $x = (-3)$ and $y = (+8)$

ANSWERS TO MATCHED PROBLEMS

7. (A) $+15$;   (B) $+2$;   (C) -5;   (D) -11

8. (A) 0;   (B) $+5$

EXERCISE 2.4 **A** *Subtract as indicated.*

1. $(+9)$ 2. $(+10)$ 3. $(+9)$ 4. $(+10)$
$-(+4)$ $-(+7)$ $-(-4)$ $-(-7)$

5. $(+4)$ 6. $(+7)$ 7. (-9) 8. (-10)
$-(+9)$ $-(+10)$ $-(-4)$ $-(-7)$

9. (-4) 10. (-7) 11. 0 12. 0
$-(-9)$ $-(-10)$ $-(+6)$ $-(-4)$

13. (-2) 14. $(+7)$ 15. (-3) 16. $(+2)$
$-(+6)$ $-(+9)$ $-\ \ 0$ $-\ \ 0$

17. $(+6) - (-8)$ 18. $(-4) - (+7)$
19. $(+6) - (+10)$ 20. $(+6) - (-10)$
21. $(-9) - (-3)$ 22. $0 - (-7)$
23. $0 - (+5)$ 24. $(-1) - (+6)$

B 25. $(-12) - (-27)$ 26. $(+57) - (+92)$
27. $0 - (-87)$ 28. $0 - (+101)$
29. $(-271) - (+44)$ 30. $(+327) - (-73)$
31. $(-245) - 0$ 32. $(+732) - 0$

Perform the indicated operations.

33. $[(-2) - (+4)] - (-7)$ 34. $(-2) - [(+4) - (-7)]$
35. $(-23) - [(-7) + (-13)]$ 36. $[(-23) - (-7)] + (-13)$
37. $[(+6) - (-8)] + [(-8) - (+6)]$ 38. $[(+3) - (+5)] - [(-5) - (-8)]$

Evaluate for $x = +2$, $y = -5$, and $z = -3$.

39. $x - y$ 40. $y - z$
41. $(x + z) - y$ 42. $x - (y - z)$
43. $(-y) - (-z)$ 44. $(-z) - y$
45. $-(|x| - |y|)$ 46. $|(x - y) - (y - z)|$

Use Figure 4 (page 60) and subtraction of integers, subtracting the one lower on the scale from the higher one, to find each of the following:

47. The difference in altitude between the highest point on earth, Mount Everest, and the deepest point in the ocean, Marianas Trench

48. The difference in altitude between the highest point in the United States, Mount McKinley, and the lowest point in the United States, Death Valley

49. The difference in altitude between the Salton Sea and Death Valley (both in California)

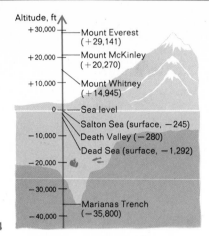

Altitude, ft
+30,000 — Mount Everest (+29,141)
+20,000 — Mount McKinley (+20,270)
+10,000 — Mount Whitney (+14,945)
0 — Sea level
— Salton Sea (surface, −245)
−10,000 — Death Valley (−280)
— Dead Sea (surface, −1,292)
−20,000
−30,000
−40,000 — Marianas Trench (−35,800)

Figure 4

50. The difference in altitude between the Dead Sea, the deepest fault in the earth's crust, and Death Valley

Which of the following statements hold for all integers a, b, and c? Illustrate each false statement with an example that shows that it is false.

51. $a + b = b + a$

52. $a + (-a) = 0$

53. $a - b = b - a$

54. $a - b = a + (-b)$

55. $(a + b) + c = a + (b + c)$

56. $(a - b) - c = a - (b - c)$

57. $|a + b| = |a| + |b|$

58. $|a - b| = |a| - |b|$

2.5 MULTIPLICATION AND DIVISION OF INTEGERS

Having discussed addition and subtraction of integers, we now turn to multiplication and division.

Multiplication

We already know that

$$(+3)(+2) = +6$$

but how shall we define

$$(+3)(-2) = ?$$

$$(-3)(+2) = ?$$

$$(-3)(-2) = ?$$

$$(0)(-4) = ?$$

We would like to define multiplication of integers in such a way that the commutative, associative, and distributive properties continue to hold for the integers. In order for this to happen we are led to the following formal definition of multiplication.

Definition of multiplication

Numbers with like signs. The product of two integers with like signs is a positive number and is found by multiplying the absolute values of the two numbers.

Numbers with unlike signs. The product of two integers with unlike signs is a negative integer and is found by taking the opposite of the product of the absolute values of the two integers.

Zero. The product of any integer and 0 is 0; the product of 0 and any integer is 0.

As in the case with addition we will be able to mechanize the process of multiplication. However, again it is important that you realize it is the definition above that is in back of any mechanical process as well as the many useful multiplication properties we will state. In particular, an immediate consequence of the definition is

THEOREM 3   For all integers a, b, and c

(A) $ab = ba$ commutative property

(B) $(ab)c = a(bc)$ associative property

(C) $a(b + c) = ab + ac$ distributive property

Because of Theorem 3 we will continue to have the same kind of freedom in rearranging factors and inserting and removing parentheses as we have with multiplication in the natural numbers.

Let us now mechanize the multiplication process and work some examples.

Mechanics of Multiplying Integers

To multiply two integers, mentally block out their signs (that is, take their absolute values), multiply as natural numbers, then prefix to the product a

+ sign (or no sign) if the original integers have like signs

− sign if the original integers have unlike signs

If one of the factors is zero, the product is zero.

EXAMPLE 9　**(A)** $(+5)(+3) = +15$ $\Big\}$ Product of two numbers with like signs is positive.

　　　　　(B) $(-8)(-6) = +48$

　　　　　(C) $(-7)(+3) = -21$ $\Big\}$ Product of two numbers with unlike signs is negative.

　　　　　(D) $(+9)(-4) = -36$

PROBLEM 9　Evaluate, using the mechanical rule:

　　　　　(A) $(+6)(+5)$ 　　　　　　　　　　**(B)** $(+7)(-6)$

　　　　　(C) $(-4)(-10)$ 　　　　　　　　　**(D)** $(-9)(+8)$

Additional properties of multiplication

Several important properties of multiplication follow from the definition of multiplication.

THEOREM 4　For all integers a and b

　　　　　(A) $(+1)a = a$

　　　　　(B) $(-1)a = -a$

　　　　　(C) $(-a)b = a(-b) = -(ab)$

　　　　　(D) $(-a)(-b) = ab$

In words, part **(B)** states that multiplying a number by -1 is the same as taking the opposite of the number. Part **(D)** states that the product of opposites of two numbers is the same as the product of the original numbers. Similar interpretations are given to parts **(A)** and **(C)**. It is also worth observing that parts **(A)** and **(B)** provide the justification for saying that a has a coefficient of $+1$ and $-a$ has a coefficient of -1.

EXAMPLE 10　Evaluate for $a = -5$ and $b = +4$:

　　　　　(A) $(+1)a$ and a

　　　　　(B) $(-1)a$ and $-a$

　　　　　(C) $(-a)b$, $a(-b)$, and $-(ab)$

　　　　　(D) $(-a)(-b)$ and ab

Solution　**(A)** $(+1)a = (+1)(-5) = -5$
　　　　　　　　$a = -5$

　　　　　(B) $(-1)a = (-1)(-5) = +5$
　　　　　　　　$-a = -(-5) = +5$

　　　　　(C) $(-a)b = [-(-5)](+4) = (+5)(+4) = +20$
　　　　　　　　$a(-b) = (-5)[-(+4)] = (-5)(-4) = +20$
　　　　　　　　$-(ab) = -[(-5)(+4)] = -(-20) = +20$

　　　　　(D) $(-a)(-b) = [-(-5)][-(+4)] = (+5)(-4) = -20$
　　　　　　　　$ab = (-5)(+4) = -20$

PROBLEM 10   Repeat Example 10 for $a = +3$ and $b = -2$.

Expressions of the form

$-ab$

occur frequently and at first glance are confusing to students. If you were asked to evaluate $-ab$ for $a = -3$ and $b = +2$, how would you proceed? Would you take the opposite of a and then multiply it by b, or multiply a and b first and then take the opposite of the product? Actually it does not matter! Because of Theorem 4 we get the same result either way since $(-a)b = -(ab)$. If, in addition, we consider other material in this section, we find that

$$-ab = \begin{cases} (-a)b \\ a(-b) \\ -(ab) \\ (-1)ab \end{cases}$$

and we are at liberty to replace any one of these five forms with another from the same group.

Division

We now consider division of integers when the quotient is an integer. Because of our knowledge of natural numbers, we know that

$(+8) \div (+4) = +2$

but how shall we define

$(+8) \div (-4) = ?$

$(-8) \div (+4) = ?$

$(-8) \div (-4) = ?$

$(0) \div (-4) = ?$

$(-8) \div \quad (0) = ?$

Division will be defined in such a way that all of these problems will have answers in the integers, except one. (Can you guess the exception?) Our approach here will parallel that used in defining subtraction in that we will investigate some elementary school notions about division to find motivation for a general definition of division for the integers.

You will recall that to check division in the problem

$$\begin{array}{r} 5 \\ 8\overline{)40} \end{array}$$

we multiply 8 by 5 to obtain 40. We will use this checking requirement to transform division into multiplication. Instead of saying,

Divide 8 into 40.

we can ask

What must 8 be multiplied by to produce 40?

Notice that the answer to each is the same, 5. The latter way of looking at division is the most useful of the two for its generalization to more involved number systems, and it provides the motivation for the following general definition that will not only apply to the integers, but also to any other number system we will encounter.

Defintion of Division

We write

$$a \div b = Q \qquad \text{if and only if} \qquad a = bQ \text{ and } Q \text{ is unique}$$

The quotient Q is the number that must be multiplied times b to produce a.

Let us use this definition to find answers to the problem stated earlier. For

$$(+8) \div (-4) = ? \qquad \text{or} \qquad -4\overline{)+8}^{\,?}$$

we ask, "What must (-4) be multiplied by to produce $(+8)$?" From the definition of multiplication we know the answer is (-2). Thus we write

$$(+8) \div (-4) = -2 \qquad \text{or} \qquad -4\overline{)+8}^{\,-2}$$

since -2 times -4 produces $+8$.

What about

$$0 \div (-3) \qquad (+12) \div 0 \qquad \text{and} \qquad 0 \div 0$$

The first is assigned a value of 0, since $0(-3) = 0$. The second is not defined, since no number times 0 is $+12$. In the third case, any number could be the quotient, since 0 times any number is 0, and so the quotient is not unique. We thus conclude that

Zero cannot be used as a divisor—ever!

The two division symbols "$\div$" and "$\overline{)}$" from arithmetic are not used a great deal in algebra and higher mathematics. The horizontal bar "$-$" and slash mark "$/$" are the symbols most frequently used.

Thus

$$a/b \qquad \frac{a}{b} \qquad a \div b \qquad \text{and} \qquad b\overline{)a} \qquad \text{In all cases } b \text{ is the divisor.}$$

all name the same number (assuming the quotient is defined), and we can write

$$a/b = \frac{a}{b} = a \div b = b\overline{)a}$$

Just as with the other operations, division can be mechanized.

Mechanical Rule for Division

If neither number is zero, mentally block out their signs (that is, take their absolute values) and divide as in the natural numbers. To determine the sign of the quotient use the same rule of signs as in multiplication (that is, quotients of numbers with like signs are positive and quotients of numbers with unlike signs are negative).

Zero divided by a nonzero number is always zero.

Zero can never be used as a divisor.

EXAMPLE 11

(A) $\dfrac{+27}{+9} = +3$

(B) $\dfrac{-27}{-9} = +3$
    } Quotients of numbers with like signs are positive.

(C) $\dfrac{-27}{+9} = -3$

(D) $(+27)/(-9) = -3$
    } Quotients of numbers with unlike signs are negative.

(E) $0/(-4) = 0$ } Zero divided by a nonzero number is zero.

(F) $(+7)/0$

(G) $0/0$
    } Not defined, since 0 is a divisor.

PROBLEM 11 Divide:

(A) $\dfrac{+18}{+6}$ **(B)** $\dfrac{-18}{-6}$ **(C)** $\dfrac{+18}{-6}$

(D) $(-18)/(+6)$ **(E)** $0/(-6)$ **(F)** $(-18)/0$

(G) $\dfrac{0}{0}$

Let us finish this section by considering examples involving all four arithmetic operations $(+, -, \cdot, \div)$. Recall that multiplication and division precede addition and subtraction unless grouping symbols indicate otherwise.

EXAMPLE 12 **(A)** $\dfrac{-18}{-3} + (-2)(+3) = (+6) + (-6) = 0$

(B) $(+6)(-3) - \dfrac{-20}{+4} = (-18) - (-5) = -13$

(C) $(-1)(-2)(-3) - \left[(-4) - \dfrac{-6}{+2}\right] = (-6) - [(-4) - (-3)]$
$$= (-6) - (-1)$$
$$= -5$$

PROBLEM 12 Evaluate:

(A) $(-4)(-3) + \dfrac{-16}{+2}$

(B) $\dfrac{(+2)(-9)}{-3} + \dfrac{(+4) - (-6)}{-5}$

(C) $(-3)[(-4) - (-2)] - \dfrac{-24}{-3}$

ANSWERS TO MATCHED PROBLEMS

9. (A) $+30$; (B) -42; (C) $+40$; (D) -72

10. (A) $+3$; (B) -3; (C) $+6$; (D) -6

11. (A) $+3$; (B) $+3$; (C) -3; (D) -3; (E) 0; (F) Not defined; (G) Not defined

12. (A) $+4$; (B) $+4$; (C) -2

EXERCISE 2.5

A *All variables represent integers.*
Multiply or divide as indicated.

1. $(-8)(-4)$ **2.** $(+8)(+4)$

3. $(+8)(-4)$ **4.** $(-8)(+4)$

5. $(0)(-7)$ **6.** $(-5)(0)$

7. $\dfrac{-4}{-2}$ **8.** $\dfrac{+14}{+7}$

9. $\dfrac{-6}{+2}$ **10.** $\dfrac{+8}{-4}$

11. $\dfrac{-6}{0}$ **12.** $\dfrac{0}{0}$

13. $(+2)(-7)$ **14.** $(-2)(-7)$

15. $(-2)(+7)$ **16.** $(+2)(+7)$

17. $(+1)(0)$ **18.** $(0)(+6)$

19. $(-9)/(-3)$ 20. $(+9)/(+3)$

21. $(-9)/(+3)$ 22. $(+9)/(-3)$

23. $0/(+3)$ 24. $(-9)/0$

B   *Evaluate:*

25. $(-2) + (-1)(+3)$ 26. $(-3)(-2) + (+4)$

27. $\dfrac{-9}{+3} + (-4)$ 28. $(-8) + \dfrac{-12}{-2}$

29. $(+2)[(+3) + (-2)]$ 30. $(+5)[(-4) + (+6)]$

31. $\dfrac{(-10) + (-6)}{-4}$ 32. $\dfrac{(-12) + (+4)}{+2}$

33. $(+4) - (-2)(-4)$ 34. $(-7) - (+4)(-3)$

35. $(-6)[(+3) - (+8)]$ 36. $(-3)[(-2) - (-4)]$

37. $(-7)(+2) - \dfrac{-20}{+5}$ 38. $\dfrac{+36}{-12} - (-4)(-8)$

39. $(-6)(+2) + (-2)^2$ 40. $(-3)^2 + (-2)(+1)$

41. $(-6)(+7) - (-3)^2$ 42. $(-6) - (-2)^2$

43. $(-4)(-3) - \left[(-8) - \dfrac{-6}{+2}\right]$ 44. $(+7)(-2) - \left[(+10) - \dfrac{+9}{-3}\right]$

Evaluate for $w = +2$, $x = -3$, $y = 0$, and $z = -24$.

45. wx 46. wz 47. z/x

48. z/w 49. xyz 50. wyz

51. y/x 52. y/z 53. w/y

54. z/y 55. $z/(wx)$ 56. xy/z

57. $wx - \dfrac{z}{w}$ 58. $\dfrac{z}{x} - wz$ 59. $\dfrac{z}{w} - \dfrac{z}{x}$

60. $\dfrac{+15}{x} - \dfrac{-12}{w}$ 61. $wxy - \dfrac{y}{z}$ 62. $\dfrac{xy}{w} - xyz$

C   63. Evaluate for $x = -4$ and $y = +3$:   (A) $(-x)y$;   (B) $x(-y)$; (C) $-(xy)$

64. Repeat Problem 63 for $x = -2$ and $y = -5$.

65. Evaluate for $x = -3$: (A) $(-1)x$;   (B) $-x$

66. Repeat Problem 65 for $x = +8$.

67. Evaluate for $x = +5$ and $y = -7$: (A) $(-x)(-y)$;   (B) xy

68. Repeat Problem 67 for $x = -3$ and $y = -4$.

What integer replacements for x will make each equation true?

69. $(-3)x = -24$ 70. $(+5)x = -20$

71. $\dfrac{(-24)}{x} = (-3)$ 72. $\dfrac{x}{(-4)} = (+12)$

73. $(-3)x = 0$

74. $\dfrac{x}{(+32)} = 0$

75. $\dfrac{x}{0} = (+4)$

76. $\dfrac{0}{x} = 0$

2.6 SIMPLIFYING ALGEBRAIC EXPRESSIONS

We are close to where we can use algebra to solve practical problems. First, however, we need to streamline our methods of representing algebraic expressions. For ease of reading and faster manipulation, it will be desirable to reduce the number of grouping symbols and plus signs to a minimum.

Dropping unnecessary plus signs

To start, we drop the plus sign from numerals that name positive integers unless a particular emphasis is desired. Thus, we will write

1, 2, 3, . . . instead of $+1, +2, +3, \ldots$

Addition and subtraction without grouping symbols

When three or more terms are combined by addition or subtraction and symbols of grouping are omitted, we convert (mentally) any subtraction to addition (Section 2.4) and add. Thus,

$$8 - 5 + 3 = 8 + (-5) + 3 = 6$$
$$\text{\textit{think}}$$

EXAMPLE 13 **(A)** $2 - 3 - 7 + 4 = 2 + (-3) + (-7) + 4 = -4$
$$\text{\textit{think}}$$

(B) $-4 - 8 + 2 + 9 = (-4) + (-8) + 2 + 9 = -1$
$$\text{\textit{think}}$$

PROBLEM 13 Evaluate:

(A) $5 - 8 + 2 - 6$ **(B)** $-6 + 12 - 2 - 1$

Building on these ideas, we formulate the following general theorem:

THEOREM 5 When two or more terms in an algebraic expression are combined by addition or subtraction:

(A) Any subtraction sign may be replaced with an addition sign if the term following it is replaced by its opposite [thus, $a - b - c = a + (-b) + (-c)$].

(B) The terms may be reordered without restriction as long as the sign preceding an involved term accompanies it in the process (thus, $a - b + c = a + c - b$).

Using Theorem 5, we extend the idea of a numerical coefficient and devise a simple mechanical rule for combining like terms in more involved algebraic expressions. The **numerical coefficient** of a given term in an algebraic expression is to include the sign that precedes it.

EXAMPLE 14   Since $3x^3 - 2x^2 - x + 3 = 3x^3 + (-2x^2) + (-1x) + 3$ the coefficient of

think

(A)   the first term is 3.

(B)   the second term is -2.

(C)   the third term is -1.

PROBLEM 14   In $2y^3 - y^2 - 2y + 4$, what is the coefficient of

(A)   the first term   (B)   the second term   (C)   the third term

Using Theorem 5 again, and distributive and commutative properties, we can write an expression such as

$3x - 2y - 5x + 7y$

in the form

$3x - 5x - 2y + 7y$

or

$3x + (-5x) + (-2y) + 7y$

or

$[3 + (-5)]x + [(-2) + 7]y$

or

$-2x + 5y$

which leads to the same **mechanical rule** we had before for combining like terms. That is,

Like terms are combined by adding their numerical coefficients.

EXAMPLE 15   Combine like terms:

(A) $2x - 3x = -x$

(B) $-4x - 7x = -11x$

(C) $3x - 5y + 6x + 2y = 9x - 3y$

(D) $2x^2 - 3x - 5 + 5x^2 - 2x + 3 = 7x^2 - 5x - 2$

PROBLEM 15   Combine like terms:

(A) $4x - 5x$ (B) $-8x - 2x$

(C) $7x + 8y - 5x - 10y$ (D) $4x^2 + 5x - 8 - 3x^2 - 7x - 2$

Removing symbols of grouping

How can we simplify expressions such as

$$2(3x - 5y) - 2(x + 3y)$$

We would like to multiply and combine like terms as we did when only $+$ signs were involved. Since any subtraction can be converted to addition (Theorem 5) we can proceed as follows:

$$2(3x - 5y) - 2(x + 3y) = 2[3x + (-5)y] + (-2)(x + 3y)$$
$$= 6x + (-10)y + (-2)x + (-6)y$$
$$= 6x - 10y - 2x - 6y$$
$$= 4x - 16y$$

Mechanically, we usually leave out the justifying steps in the dotted box and use the following process:

Mechanics of Removing Grouping Symbols

Parentheses (and other symbols of grouping) can be cleared by multiplying each term within the parentheses by the coefficient of the parentheses.

EXAMPLE 16   (A) $2(x - 3y) + 3(2x - y) = 2x - 6y + 6x - 3y$
$$= 8x - 9y$$

(B) $4x(2x + y) - 3x(x - 3y) = 8x^2 + 4xy - 3x^2 + 9xy$
$$= 5x^2 + 13xy$$

(C) $(x + 5y) + (x - 3y) = 1(x + 5y) + 1(x - 3y)$
think

Think of the coefficient of $(x + 5y)$ and $(x - 3y)$ as 1; then use the above rule.

$$= x + 5y + x - 3y$$
$$= 2x + 2y$$

(D) $(x + 5y) - (2x - 3y) = 1(x + 5y) - 1(2x - 3y)$
think

Think of the coefficient of $-(2x - 3y)$ as -1; then use the above rule.

$$= x + 5y - 2x + 3y$$
$$= -x + 8y$$

PROBLEM 16 Remove parentheses and simplify:

(A) $3(2x - 3y) + 2(x - 2y)$ ˙   (B) $5x(x - 2y) - 3x(2x + y)$

(C) $(3x + 2y) + (2x - 4y)$ (D) $(3x + 2y) - (2x - 4y)$

EXAMPLE 17 Replace each question mark with an appropriate algebraic expression.

(A) $5 + 2x - 3y = 5 + (\ ? \ )$

(B) $5 + 2x - 3y = 5 - (\ ? \ )$

(C) $3x + 4a - 6b = 3x + 2(\ ? \ )$

(D) $6a - 3x + 9y = 6a - 3(\ ? \ )$

Solution In these problems we are inserting symbols of grouping. The expression that replaces the question mark must be chosen so that when the parentheses on the right are removed we obtain the expression on the left.

(A) $5 + 2x - 3y = 5 + (2x - 3y)$

(B) $5 + 2x - 3y = 5 - (-2x + 3y)$

(C) $3x + 4a - 6b = 3x + 2(2a - 3b)$

(D) $6a - 3x + 9y = 6a - 3(x - 3y)$

We note again that if a parenthesis is to be preceded by a + sign, the terms placed within the parentheses remain unchanged. If a parenthesis is to be preceded with a − sign, then each term placed inside the parentheses undergoes a sign change.

PROBLEM 17 Repeat Example 17 for

(A) $2y + 3z - 4w = 2y + (\ ? \ )$

(B) $2y + 3z - 4w = 2y - (\ ? \ )$

(C) $a + 4b - 12c = a + 4(\ ? \ )$

(D) $u - 6v - 9w = u - 3(\ ? \ )$

EXAMPLE 18 Remove grouping symbols and combine like terms.

(A) $4x - 3[x - 2(x + 3)]$

(B) $u - \{u - [u - 2(u - 1)]\}$

Solution Generally, we remove ( ) first, then [ ], and finally, { }.

(A) $4x - 3[x - 2(x + 3)]$ Remove ( ) first.

$= 4x - 3[x - 2x - 6]$ Combine like terms.

$= 4x - 3[-x - 6]$ Remove [ ].

$= 4x + 3x + 18$ Combine like terms.

$= 7x + 18$

(B) $u - \{u - [u - 2(u - 1)]\}$

$\qquad = u - \{u - [u - 2u + 2]\}$ We can collect like

$\qquad\qquad\qquad\qquad\qquad\qquad$ terms as we go or

$\qquad = u - \{u - u + 2u - 2\}$ wait until the end.

$\qquad = u - u + u - 2u + 2$

$\qquad = -u + 2$

PROBLEM 18   Remove grouping symbols and combine like terms.

(A) $3y + 2[y - 3(y - 5)]$

(B) $3m - \{m - [2m - (m - 1)]\}$

ANSWERS TO MATCHED PROBLEMS

13. (A) -7;   (B) 3

14. (A) 2;   (B) -1;   (C) -2

15. (A) $-x$;   (B) $-10x$;   (C) $2x - 2y$;   (D) $x^2 - 2x - 10$

16. (A) $8x - 13y$;   (B) $-x^2 - 13xy$;   (C) $5x - 2y$;   (D) $x + 6y$

17. (A) $3z - 4w$;   (B) $-3z + 4w$;   (C) $b - 3c$;   (D) $2v + 3w$

18. (A) $-y + 30$;   (B) $3m + 1$

EXERCISE 2.6   **A** *Evaluate:*

1. $3 - 2 + 4$
2. $3 + 4 - 2$

3. $4 - 8 - 9$
4. $-8 + 4 - 9$

5. $-4 + 7 - 6$
6. $7 - 6 - 4$

7. $2 - 3 - 6 + 5$
8. $5 + 2 - 6 - 3$

9. $-3 - 2 + 6 - 4$
10. $-8 + 9 - 4 - 3$

11. $-7 + 1 + 6 + 2 - 1$
12. $9 - 5 - 4 + 7 - 6 + 10$

13. $-5 - 3 - 8 + 15 - 1$
14. $1 - 12 + 5 + 7 - 1 + 6 - 8$

Remove symbols of grouping where present and combine like terms.

15. $7x - 3x$
16. $9x - 4x$

17. $2x - 5x - x$
18. $4t - 8t - 9t$

19. $-3y + 2y - 5y - 6y$
20. $2y - 3y - 6y + 5y$

21. $2x - 3y - 5x$
22. $4y - 3x - y$

23. $2x + 8y - 7x - 5y$
24. $5m + 3n - m - 9n$

25. $2(m + 3n) + 4(m - 2n)$
26. $3(u - 2v) + 2(3u + v)$

27. $2(x - y) - 3(x - 2y)$
28. $4(m - 3n) - 3(2m + 4n)$

29. $(x + 3y) + (2x - 5y)$
30. $(2u - v) + (3u - 5v)$

31. $x - (2x - y)$
32. $m - (3m + n)$

33. $(x + 3y) - (2x - 5y)$
34. $(2u - v) - (3u - 5v)$

35. $3(2x - 3y) - (2x - y)$ **36.** $2(3m - n) - (4m + 2n)$

B **37.** $3xy + 4xy - xy$ **38.** $3xy - xy + 4xy$

39. $-x^2y + 3x^2y - 5x^2y$ **40.** $-4r^3t^3 - 7r^3t^3 + 9r^3t^3$

41. $3x^2 - 2x + 5 - x^2 + 4x - 8$ **42.** $y^3 + 4y^2 - 10 + 2y^3 - y + 7$

43. $2x^2y + 3xy^2 - 5xy + 2xy^2 - xy$ **44.** $a^2 - 3ab + b^2 + 2a^2 + 3ab - 2b^2$

45. $x - 3(x + 2y) + 5y$ **46.** $y - 2(x - y) - 3x$

47. $-3(-t + 7) - (t - 1)$ **48.** $-2(-3x + 1) - (2x + 4)$

49. $-2(y - 7) - 3(2y + 1) - (-5y + 7)$

50. $2(x - 1) - 3(2x - 3) - (4x - 5)$

51. $3x(2x^2 - 4) - 2(3x^3 - x)$ **52.** $5y(2y - 3) + 3y(-2y + 4)$

53. $3x - 2[2x - (x - 7)]$ **54.** $y - [5 - 3(y - 2)]$

55. $2t - 3t[4 - 2(t - 1)]$ **56.** $2u - 3u[4 - (u - 3)]$

Replace each question mark with an appropriate algebraic expression.

57. $2 + 3x - y = 2 + (\ ?\ )$ **58.** $5 + m - 2n = 5 + (\ ?\ )$

59. $2 + 3x - y = 2 - (\ ?\ )$ **60.** $5 + m - 2n = 5 - (\ ?\ )$

61. $x - 4y - 8z = x - 4(\ ?\ )$ **62.** $2x - 3a + 12b = 2x - 3(\ ?\ )$

63. $w^2 - x + y - z = w^2 - (\ ?\ )$ **64.** $w^2 - x + y - z = w^2 + (\ ?\ )$

65. The width of a rectangle is 5 meters less than its length. If x is the length of the rectangle, write an algebraic expression that represents the perimeter of the rectangle, then multiply and combine like terms.

66. The length of a rectangle is 8 ft more than its width. If y is the width of the rectangle, write an algebraic expression that represents its area. Change the expression to a form without parentheses.

C *Remove symbols of grouping and combine like terms.*

67. $x - \{x - [x - (x - 1)]\}$

68. $2t - 3\{t + 2[t - (t + 5)] + 1\}$

69. $2x[3x - 2(2x + 1)] - 3x[8 + (2x - 4)]$

70. $-2t\{-2t(-t - 3) - [t^2 - t(2t + 3)]\}$

71. $3x^2 - 2\{x - x[x + 4(x - 3)] - 5\}$

72. $w - \{x - [z - (w - x) - z] - (x - w)\} + x$

73. A coin purse contains dimes and quarters only. There are 4 more dimes than quarters. If x represents the number of quarters, write an algebraic expression that represents the value of the money in the purse in cents. Clear grouping symbols and combine like terms.

74. A pile of coins consists of nickels, dimes, and quarters. There are twice as many dimes as nickels and 4 fewer quarters than dimes. If x represents the number of nickels, write an algebraic expression that represents the value of the pile of coins in cents. Remove grouping symbols and combine like terms.

2.7 EQUATIONS INVOLVING INTEGERS

We have reached the place where we can discuss methods of solving equations other than by guessing. For example, you would not likely guess the solution to

$$2x + 2(x - 6) = 52$$

an equation related to a practical problem that we will consider later.

A **solution** or **root** of an equation is a replacement of x that makes the left side equal to the right. The set of all solutions is called the **solution set.** To **solve an equation** is to find its solution set.

Knowing what we mean by the solution set of an equation is one thing, finding it is another. Our objective now is to develop a systematic method of solving equations that is free from guess work. We start by introducing the idea of equivalent equations. We say that **two equations are equivalent** if they both have the same solution set.

The basic idea in solving equations is to perform operations on equations that produce simpler equivalent equations and to continue the process until we reach an equation whose solution is obvious— generally, an equation such as

$$x = -3$$

With a little practice you will find the methods that we are going to develop very easy to use and very powerful. Before proceeding further, it is recommended that you briefly review the properties of equality discussed in Section 1.4. The following important theorem is a direct consequence of these properties.

THEOREM 6 (Further properties of equality.) For a, b, and c integers,

(A) If $a = b$, then $a + c = b + c$ addition property of equality

(B) If $a = b$, then $a - c = b - c$ subtraction property of equality

(C) If $a = b$, then $ca = cb$ multiplication property of equality

(D) If $a = b$ and $c \neq 0$, then $\dfrac{a}{c} = \dfrac{b}{c}$ division property of equality

In words, these properties say that if we start with two equal quantities, then the results obtained by adding to each, subtracting from each, multiplying each by, or dividing each by the same quantity (excluding division by zero) are equal.

The proofs of these properties are very easy. We will prove part **(A)** and leave the others as exercises:

$a + c = a + c$ reflexive property of equality

$\quad a = b$ given

$a + c = b + c$ substitution principle

The next theorem, which we will freely use but not prove, provides us with our final instructions for solving simple equations.

THEOREM 7 An equivalent equation will result if

(A) An equation is changed in any way by use of Theorem 6, except for multiplication or division by 0.

(B) Any algebraic expression in an equation is replaced by its equal (substitution principle).

We are now ready to solve equations! Several examples will illustrate the process.

EXAMPLE 19 Solve $x - 5 = -2$ and check.

Solution

$x - 5 = -2$	How can we eliminate the -5 from the left side?
$x - 5 + 5 = -2 + 5$	Add 5 to each side (addition property of equality).
$x = 3$	Solution is obvious.

CHECK

$x - 5 = 2$	Replace x with 3.
$3 - 5 \overset{?}{=} -2$	Evaluate both sides.
$-2 \overset{\checkmark}{=} -2$	We have a check since both sides are equal.

NOTE: Gradually the steps in the dotted box are done mentally. If in doubt, include them.

PROBLEM 19 Solve $x + 8 = -6$ and check.

EXAMPLE 20 Solve $-3x = 15$ and check.

Solution

$-3x = 15$	How can we make the coefficient of x plus 1?
$\dfrac{-3x}{-3} = \dfrac{15}{-3}$	Divide each side by -3 (division property of equality).
$x = -5$	Solution is obvious.

CHECK

$$-3x = 15 \qquad \text{Replace } x \text{ with } -5.$$

$$(-3)(-5) \overset{?}{=} 15 \qquad \text{Evaluate both sides.}$$

$$15 \overset{\checkmark}{=} 15 \qquad \text{We have a check since both sides are equal.}$$

PROBLEM 20   Solve $5x = -20$ and check.

The following examples are a little more difficult, but each can be converted into the types in Examples 19 and 20 by following the following strategy.

Strategy for Solving Equations

---

1. Simplify the left- and right-hand sides of the equations by removing grouping symbols and combining like terms.

2. Perform operations on the resulting equation using equality properties (Theorem 6) that will get all of the variable terms on one side (usually the left) and all of the constant terms on the other side (usually the right), then solve as in Examples 19 and 20.

EXAMPLE 21   Solve $2x - 8 = 5x + 4$ and check.

Solution

$$2x - 8 = 5x + 4$$

To remove -8 from the left side, add 8 to both sides (addition property of equality).

$$2x - 8 + 8 = 5x + 4 + 8$$

$$2x = 5x + 12$$

To remove $5x$ from the right side, subtract $5x$ from both sides (subtraction property of equality).

$$2x - 5x = 5x + 12 - 5x$$

$$-3x = 12$$

To isolate x on the left side with a coefficient of $+1$, divide both sides by -3 (division property of equality).

$$\frac{-3x}{-3} = \frac{12}{-3}$$

$$x = -4$$

We have solved the equation!

CHECK

$$2x - 8 = 5x + 4$$

Replace x with -4 and proceed as in Examples 19 and 20.

$$2(-4) - 8 \overset{?}{=} 5(-4) + 4$$

$$-8 - 8 \overset{?}{=} -20 + 4$$

$$-16 \overset{\checkmark}{=} -16$$

PROBLEM 21 Solve $3x - 9 = 7x + 3$ and check.

EXAMPLE 22 Solve $3x - 2(2x - 5) = 2(x + 3) - 8$ and check.

Solution This equation is not as difficult as it might at first appear; simplify the expressions on each side of the equal sign first, and then proceed as in the preceding example. (Note that some steps in the following solution are done mentally.)

$3x - 2(2x - 5) = 2(x + 3) - 8$	Remove grouping symbols.
$3x - 4x + 10 = 2x + 6 - 8$	Combine like terms.
$-x + 10 = 2x - 2$	Subtract 10 from each side.
$-x = 2x - 12$	Subtract $2x$ from each side.
$-3x = -12$	Divide each side by -3.
$x = 4$	We now have the solution.

CHECK

$$3x - 2(2x - 5) = 2(x + 3) - 8 \qquad \text{Replace } x \text{ with 4.}$$
$$3(4) - 2[2(4) - 5] \stackrel{?}{=} 2[(4) + 3] - 8$$
$$12 - 2(8 - 5) \stackrel{?}{=} 2(7) - 8$$
$$12 - 2(3) \stackrel{?}{=} 14 - 8$$
$$12 - 6 \stackrel{?}{=} 6$$
$$6 \stackrel{\checkmark}{=} 6$$

PROBLEM 22 Solve $8x - 3(x - 4) = 3(x - 4) + 6$ and check.

ANSWERS TO MATCHED PROBLEMS

19. $x = -14$ **20.** $x = -4$ **21.** $x = -3$ **22.** $x = -9$

EXERCISE 2.7

A *Solve and check.*

1. $x + 5 = 8$
2. $x + 2 = 7$
3. $x + 8 = 5$
4. $x + 7 = 2$
5. $x + 9 = -3$
6. $x + 4 = -6$
7. $x - 3 = 2$
8. $x - 4 = 3$
9. $x - 5 = -8$
10. $x - 7 = -9$
11. $y + 13 = 0$
12. $x - 5 = 0$
13. $4x = 32$
14. $9x = 36$

15. $6x = -24$
16. $7x = -21$
17. $-3x = 12$
18. $-2x = 18$
19. $-8x = -24$
20. $-9x = -27$
21. $3y = 0$
22. $-5m = 0$
23. $4x - 7 = 5$
24. $3y - 8 = 4$
25. $2y + 5 = 9$
26. $4x + 3 = 19$
27. $2y + 5 = -1$
28. $2w + 18 = -2$
29. $-3t + 8 = -13$
30. $-4m + 3 = -9$
31. $4m = 2m + 8$
32. $3x = x + 6$
33. $2x = 8 - 2x$
34. $3x = 10 - 7x$
35. $2n = 5n + 12$
36. $3y = 7y + 8$
37. $2x - 7 = x + 1$
38. $4x - 9 = 3x + 2$
39. $3x - 8 = x + 6$
40. $4y + 8 = 2y - 6$
41. $2t + 9 = 5t - 6$
42. $3x - 4 = 6x - 19$

B 43. $x - 3 = x + 7$
44. $2y + 8 = 2y - 6$
45. $2x + 2(x - 6) = 52$
46. $5x + 10(x + 7) = 100$
47. $x + (x + 2) + (x + 4) = 54$
48. $10x + 25(x - 3) = 275$
49. $2(x + 7) - 2 = x - 3$
50. $5 + 4(t - 2) = 2(t - 7) + 1$
51. $-3(4 - t) = 5 - (t + 1)$
52. $5x - (7x - 4) - 2 = 5 - (3x + 2)$
53. $x(x + 2) = x(x + 4) - 12$
54. $x(x - 1) + 5 = x^2 + x - 3$
55. $t(t - 6) + 8 = t^2 - 6t - 3$
56. $x(x - 4) - 2 = x^2 - 4(x + 3)$
57. Which of the following are equivalent to $3x - 6 = 6$: $3x = 12$, $3x = 0$, $x = 4$, $x = 0$?
58. Which of the following are equivalent to $2x + 5 = x - 3$: $2x = x - 8$, $2x = x + 2$, $3x = -8$, $x = -8$?

C *Which of the following are true (I is the set of integers)?*

59. $\{x \in I \mid 3x = 5\} = \emptyset$
60. $\{t \in I \mid 2t - 1 = 19\} = \{9, 10\}$
61. $\{t \in I \mid 2t - 1 = 19\} = \{9\}$
62. $\{t \in I \mid 2t - 1 = 19\} = \{10\}$
63. $\{y \in I \mid -2y = 8, y > 0\} = \emptyset$
64. $\{u \in I \mid 2u - 5 = 11\} = \{8\}$
65. $\{x \in I \mid 3x + 11 = 5, x > 0\} = \emptyset$
66. $\{x \in I \mid x^2 = 4\} = \{-2, 2\}$
67. Prove the subtraction property of equality in Theorem 6.
68. Prove the multiplication property of equality in Theorem 6.

2.8 WORD PROBLEMS AND APPLICATIONS

In preceding sections we considered the problem of translating words into algebraic expressions and equations. We are now ready to consider the more general problem of solving word problems and real world applications using algebra. We will consider a number of examples in detail and will state a strategy (method of attack) that will be useful in many problems.

Let us start by solving a fairly simple problem dealing with numbers. Through this problem you will learn some basic ideas about setting up and solving word problems in general.

EXAMPLE 23   Find three consecutive integers whose sum is 66.

Solution

Let

$x =$ the first integer

then

$x + 1 =$ the next integer

and

$x + 2 =$ the third integer

Identify one of the unknowns with a variable, say x, then write other unknowns in terms of this variable. In this case, x, $x + 1$, and $x + 2$ represent three consecutive integers starting with the integer x.

$$x + (x + 1) + (x + 2) = 66$$

Write an equation that relates the unknown quantities with other facts in the problem (sum of three consecutive integers is 66).

$$x + x + 1 + x + 2 = 66$$

Solve the equation.

$$3x + 3 = 66$$
$$3x = 63$$
$$x = 21$$
$$x + 1 = 22$$

Write all answers requested.

$$x + 2 = 23$$

CHECK

21
22
23
‾
66

Checking back in the equation is not enough since you might have made a mistake in setting up the equation; a final check is provided only if the conditions in the original problem are satisfied.

Thus we have found three consecutive integers whose sum is 66.

PROBLEM 23   Find three consecutive integers whose sum is 54.

EXAMPLE 24   Find three consecutive even numbers such that twice the second plus 3 times the third is 7 times the first.

Solution Let

x = the first even number

then

$x + 2$ = the second even number

and

$x + 4$ = the third even number

$$\begin{array}{ccc} \text{twice the second} \\ \text{even number} \end{array} + \begin{array}{c} \text{three times the} \\ \text{third even number} \end{array} = \begin{array}{c} \text{seven times the} \\ \text{first even number} \end{array}$$

$$2(x + 2) \quad + \quad 3(x + 4) \quad = \quad 7x$$
$$2x + 4 + 3x + 12 = 7x$$
$$5x + 16 = 7x$$
$$-2x = -16$$
$$x = 8$$
$$x + 2 = 10$$
$$x + 4 = 12$$

CHECK

8, 10, and 12 are three consecutive even numbers

$$2 \cdot 10 + 3 \cdot 12 \overset{?}{=} 7 \cdot 8$$
$$20 + 36 \overset{?}{=} 56$$
$$56 \overset{\checkmark}{=} 56$$

PROBLEM 24 Find three consecutive even numbers such that the second plus twice the third is 4 times the first.

EXAMPLE 25 Find the dimensions of a rectangle with a perimeter of 52 cm if its length is 5 cm more than twice its width.

Solution First draw a figure, then label parts using an appropriate variable. In this case, since the length is given in terms of the width, we let x represent the width, then the length will be given by $2x + 5$.

x

$2x + 5$

$$2(\text{length}) + 2(\text{width}) = \text{perimeter}$$

$$2(2x + 5) + 2x = 52$$

$$4x + 10 + 2x = 52$$

$$6x = 42$$

$$x = 7 \text{ cm} \qquad \text{width}$$

$$2x + 5 = 19 \text{ cm} \qquad \text{length}$$

CHECK

19 is 5 more than twice 7

$$2 \cdot 19 + 2 \cdot 7 \overset{?}{=} 52$$

$$38 + 14 \overset{?}{=} 52$$

$$52 \overset{\checkmark}{=} 52$$

PROBLEM 25 Find the dimensions of a rectangle, given that the perimeter of the rectangle is 30 meters and the length of the rectangle is 7 meters more than its width.

EXAMPLE 26 In a pile of coins which is composed of only dimes and nickels, there are 7 more dimes than nickels. If the total value of all of the coins in the pile is $1, how many of each type of coin is in the pile?

Solution Let

$x = $ the number of nickels in the pile

then

$x + 7 = $ the number of dimes in the pile

Do not confuse the number of nickels with the value of the nickels or the number of dimes with the value of the dimes.

$$\begin{pmatrix} \text{value of nickels} \\ \text{in cents} \end{pmatrix} + \begin{pmatrix} \text{value of dimes} \\ \text{in cents} \end{pmatrix} = \begin{pmatrix} \text{value of pile} \\ \text{in cents} \end{pmatrix}$$

$$5x + 10(x + 7) = 100$$

$$5x + 10x + 70 = 100$$

$$15x = 30$$

$$x = 2 \qquad \text{nickels}$$

$$x + 7 = 9 \qquad \text{dimes}$$

CHECK

Nine dimes is 7 more than two nickels

value of nickels in cents = 10
value of dimes in cents = 90
total value = 100

PROBLEM 26 A person has dimes and quarters worth $1.80 in a pocket. If there are twice as many dimes as quarters, how many of each does he or she have?

EXAMPLE 27 An airplane flew out to an island from the mainland and back in 5 hr. How far is the island from the mainland if the pilot averaged 600 mph going to the island and 400 mph returning?

Solution In this problem we will find it convenient to find the time out to the island first. The formula $d = rt$, of course, will be of great use to us here. Let

x = the time it took to get to the island

then

$5 - x$ = the time to return (since round trip time is 5 hr)

distance out = distance back
(rate out)(time out) = (rate back)(time back)

$$600x = 400(5 - x)$$
$$600x = 2{,}000 - 400x$$
$$1{,}000x = 2{,}000$$
$$x = 2 \text{ hr} \quad \text{time going}$$
$$5 - x = 3 \text{ hr} \quad \text{time returning}$$

Since distance = (rate) · (time), the distance to the island from the mainland is $600 \cdot 2 = 1{,}200$ miles.

CHECK

time going + time returning = 5   If $d = rt$, then $t = \dfrac{d}{r}$.

$$\frac{\text{distance out}}{\text{rate out}} + \frac{\text{distance back}}{\text{rate back}} = 5$$

$$\frac{1{,}200}{600} + \frac{1{,}200}{400} \overset{?}{=} 5$$

$$2 + 3 \overset{\checkmark}{=} 5$$

PROBLEM 27   An airplane flew from San Francisco to a distressed ship out at sea and back in 7 hr. How far was the ship from San Francisco if the pilot averaged 400 mph going and 300 mph returning?

Summary

You are now beginning to see the power of algebra. It was a historic occasion when it was realized that a solution to a problem that was difficult to obtain by arithmetical computation could be obtained instead by a deductive process involving a symbol that represented the solution.

There are many different types of algebraic applications, so many, in fact, that no single approach will apply to all. The following suggestions, however, may be of help to you:

Strategy for Solving Word Problems

---

1. Read the problem very carefully—several times if necessary.

2. Write down important facts and relationships on a piece of scratch paper. Draw figures if it is helpful. Write down any formulas that might be relevant.

3. Identify unknown quantities in terms of a single variable if possible.

4. Look for key words and relationships in a problem that will lead to an equation involving the variables introduced in Step **3**.

5. Solve the equation.

6. Write down all the solutions asked for in the original problem.

7. Check the solution(s) in the original problem.

Remember, mathematics is not a spectator sport! Just reading examples is not enough; you must set up and solve problems yourself.

ANSWERS TO MATCHED PROBLEMS

23. 17, 18, 19   **24.** 10, 12, 14   **25.** 4 meters by 11 meters
26. 4 quarters and 8 dimes   **27.** 1,200 miles

EXERCISE 2.8

A

1. Find three consecutive integers whose sum is 78.

2. Find three consecutive integers whose sum is 96.

3. Find three consecutive even numbers whose sum is 54.

4. Find three consecutive even numbers whose sum is 42.

5. How long would it take you to drive from San Francisco to Los Angeles, a distance of about 424 miles, if you could average 53 mph? (Use $d = rt$.)

6. If you drove from Berkeley to Lake Tahoe, a distance of 200 miles, in 4 hr, what is your average speed?

7. About 8 times the height of an iceberg of uniform cross section is under water as is above the water. If the total height of an iceberg from bottom to top is 117 ft, how much is above and how much is below the surface?

8. A chord called an octave can be produced by dividing a stretched string into two parts so that one part is twice as long as the other part. How long will each part of the string be if the total length of the string is 57 in?

9. The sun is about 390 times as far from the earth as the moon. If the sun is approximately 93,210,000 miles from the earth, how far is the moon from the earth?

10. You are asked to construct a triangle with two equal angles so that the third angle is twice the size of either of the two equal ones. How large should each angle be? NOTE: The sum of the three angles in any triangle is 180°.

B **11.** Find three consecutive odd numbers such that the sum of the first and second is 5 more than the third.

12. Find three consecutive odd numbers such that the sum of the second and third is 1 more than 3 times the first.

13. Find the dimensions of a rectangle with perimeter 66 ft if its length is 3 ft more than twice the width.

14. Find the dimensions of a rectangle with perimeter 128 in if its length is 6 in less than 4 times the width.

15. In a pile of coins containing only quarters and dimes, there are 3 less quarters than dimes. If the total value of the pile is $2.75, how many of each type of coin is in the pile?

16. If you have 20 dimes and nickels in your pocket worth $1.40, how many of each do you have?

17. A toy rocket shot vertically upward with an initial velocity of 160 ft per sec (fps), has at time t a velocity given by the equation $v = 160 - 32t$, where air resistance is neglected. In how many seconds will the rocket reach its highest point? HINT: Find t when $v = 0$.

18. In the preceding problem, when will the rocket's velocity be 32 fps?

19. Air temperature drops approximately 5°F per 1,000 ft in altitude above the surface of the Earth up to 30,000 ft. If T represents temperature and A represents altitude in thousands of feet, and if the temperature on the ground is 60°F, then we can write

$$T = 60 - 5A \qquad 0 \leq A \leq 30$$

If you were in a balloon, how high would you be if the thermometer registered $-50°$F?

20. A mechanic charges $6 per hr for his or her labor and $4 per hr for the assistant. On a repair job the bill was $190 with $92 for labor and $98 for parts. If the assistant worked 2 hr less than the mechanic, how many hours did each work?

21. In a recent election involving five candidates, the winner beat the opponents by 805, 413, 135, and 52, respectively. If the total number of votes cast was 10,250, how many votes did each receive?

22. If an adult with pure brown eyes marries an adult with blue eyes, their children, because of the dominance of brown, will all have brown eyes but will be carriers of the gene for blue. If the children marry others with the same type of parents, then according to Mendel's laws of heredity, we would expect the third generation (the children's children) to include 3 times as many with brown eyes as with blue. Out of a sample of 1,748 third-generation children with second-generation parents as described, how many brown-eyed children and blue-eyed children would you expect?

C **23.** A man in a canoe went up a river and back in 6 hr. If his rate up the river was 2 mph and back 4 mph, how far did he go up the river?

24. You are at a river resort and rent a motor boat for 5 hr at 7 A.M. You are told that the boat will travel at 8 mph upstream and 12 mph returning. You decide that you would like to go as far up the river as you can and still be back at noon. At what time should you turn back, and how far from the resort will you be at that time?

25. One ship leaves England and another leaves the United States at the same time. The distance between the two ports is 3,150 miles. The ship from the United States averages 25 mph and the one from England 20 mph. If they both travel the same route, how long will it take the ships to reach a rendezvous point, and how far from the United States will they be at that time?

26. At 8 A.M. your father left by car on a long trip. An hour later you find that he has left his wallet behind. You decide to take another car to try to catch up with him. From past experience you know that he averages about 48 mph. If you can average 60 mph, how long will it take you to catch him?

27. In a computer center two electronic card sorters are used to sort 52,000 IBM cards. If the first sorter operates at 225 cards per min and the second

66

66
2: INTEGERS

sorter operates at 175 cards per min, how long will it take both sorters together to sort all of the cards?

28. Find four consecutive even numbers so the sum of the first and last is the same as the sum of the second and third. (Be careful!)

2.9 CHAPTER REVIEW: IMPORTANT TERMS AND SYMBOLS, REVIEW EXERCISE, PRACTICE TEST

Important terms and symbols

positive integers (2.1)   negative integers (2.1)   zero (2.1)   integers (2.1)   opposite of (2.2) $-x$ (2.2)   absolute value (2.2) $|x|$ (2.2)   integer addition (2.3)   integer subtraction (2.4)   integer multiplication (2.5)   integer division (2.5) a/b (2.5)   equations (2.7)   properties of equality (2.7)   word-problem strategy (2.8)

Exercise 2.9 review exercise

All variables represent integers.

A *Evaluate as indicated.*

1. $-(+4)$
2. $|-(+3)|$
3. $(-8) + (+3)$
4. $(-9) + (-4)$
5. $(-3) - (-9)$
6. $(+4) - (+7)$
7. $(-7)(-4)$
8. $(+3)(-6)$
9. $(-16)/(+4)$
10. $(-12)/(-2)$
11. $0/(+2)$
12. $(-6)/0$
13. $-6 + 8 - 5 + 1$
14. $(-2)(+5) + (-6)$
15. $(-4)(-3) - (+20)$
16. $\frac{-8}{-2} - (-4)$
17. $(-6)(0) + \frac{0}{-3}$
18. $\frac{-9}{+3} - \frac{-12}{-4}$

Remove symbols of grouping, if present, and combine like terms.

19. $4x - 3 - 2x - 5$
20. $(2x - 3) + (3x + 1)$
21. $3(m + 2n) - (m - 3n)$
22. $2(x - 3y) - 4(2x + 3y)$

Solve and check.

23. $4x - 9 = x - 15$
24. $2x - 5 = 3x + 2$
25. Express each quantity by means of an appropriate integer.
(A) Salton Sea's surface at 245 ft below sea level
(B) Mount Whitney's height of 14,495 ft
26. Find three consecutive integers whose sum is 159.

B *Evaluate for $x = -12$, $y = -2$, and $z = +3$.*

27. $-x$
28. $-(-z)$
29. $-|-y|$
30. $x - y$
31. $|x + z|$
32. $(z - y) - x$
33. $(3y + x)/z$
34. $(x/y) - yz$
35. $(4z + x)/y$

36. $\left(yz - \dfrac{x}{z}\right) - xz$ **37.** $\dfrac{0}{x} + x(0)y$

Remove grouping symbols and combine like terms.

38. $(3x^2y^2 - xy) - (5x^2y^2 + 4xy)$

39. $3y(2y^2 - y + 4) - 2y(y^2 + 2y - 3)$

40. $7x - 3[(x + 7y) - (2x - y)]$

41. $2x^2y(3xy - 5) - 3xy^2(4x^2 - 1)$

Replace each question mark with an appropriate algebraic expression.

42. $3x - 6y + 9 = 3x - 3(\ ? \ )$

43. $3(x - 2y) - x + 2y = 3(x - 2y) - (\ ? \ )$

Solve and check.

44. $3(m - 2) - (m + 4) = 8$

45. $2x + 3(x - 1) = 8 - (x - 1)$

46. If the sum of four consecutive even numbers is 188, find the numbers.

47. A pile of coins consists of nickels and quarters. How many of each kind is there if the whole pile is worth \$1.45 and there are 3 fewer quarters than twice the number of nickels?

48. Express the net gain by means of an appropriate integer: A 15° rise in temperature followed by a 30° drop, another 15° drop, a 25° rise, and finally, a 40° drop.

C **49.** Show that subtraction is not associative by evaluating (A) $(x - y) - z$ and (B) $x - (y - z)$ for $x = +7$, $y = -3$, and $z = -5$.

50. Show that division is not associative by evaluating (A) $(x \div y) \div z$ and (B) $x \div (y \div z)$ for $x = +16$, $y = -8$, and $z = -2$.

51. Simplify: $3x - 2\{x - 2[x - (4x + 2)]\}$.

52. $\{x \in I \mid 2(x - 5) = 4x + 8\} = ?$

53. An unmanned space capsule passes over Cape Canaveral at 8 A.M. traveling at 17,000 mph. A manned capsule, attempting a rendezvous, passes over the same spot at 9 A.M. traveling at 18,000 mph. How long will it take the second capsule to catch up with the first?

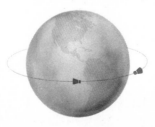

**Practice test
Chapter 2**

Evaluate:

1. (A) $-|-(+4)|$;   (B) $-[(+3) + (-8)]$

2. (A) $\dfrac{-15}{-5} + (-2)(+4)$;   (B) $(-3)[(-4) - (-10)]$

3. (A) $\dfrac{x}{y} - xyz$ and   (B) $(yz)/x$ for $x = 0$, $y = -3$, and $z = +2$

4. $w - \left(wx - \dfrac{y}{w}\right)$ for $w = -2$, $x = +4$, $y = +10$

Remove grouping symbols and combine like terms.

5. $5(x - 2y + 3) - (x - 8y + 10)$

6. $4y^2(3y - 2) - 3y(4y^2 - y + 3)$

Solve the next two equations:

7. $7x - 4 = 14 + 4x$

8. $3(x - 4) - 2 = 6(x + 1) + 4$

9. Replace each question mark with an appropriate algebraic expression:

(A) $5w - x + 3y = 5w - (\ \ ?\ \ )$

(B) $x + 2y + 3x + 6y = x + 2y + 3(\ \ ?\ \ )$

10. Find three consecutive odd numbers whose sum is 111. Set up an equation and solve.

11. A purse contains nickels and dimes. If the total value is $1.60 and there are 4 fewer dimes than twice the number of nickels, how many of each coin is in the purse? Set up an equation and solve.

12. A person can travel 4 km/hr up a river and 6 km/hr returning. If 5 hr are allotted for a round trip, how far can the person go up the river in order to be back by the end of the 5 hr? Set up an equation and solve.

3

RATIONAL NUMBERS

3.1 THE SET OF RATIONAL NUMBERS

In the last chapter we formed the set of integers by extending the natural numbers to include 0 and the negative integers. With this extension came more power to perform arithmetic operations and more power to solve equations. In spite of this added power, however, we are still not able to solve the simple equation

$$2x = 3$$

or find

$$-2 \div 5$$

You have no doubt guessed the direction of the next extension of the number system. We need fractions!

You will recall in the last chapter that we said that we would use

$$a \div b \qquad a/b \qquad \frac{a}{b}$$

interchangeably; hence,

$$4 \div 2 \qquad 4/2 \qquad \frac{4}{2}$$

are different names for the number 2. However, what does

$$\frac{3}{2}$$

name? Certainly not an integer. We are going to extend the set of integers so that $\frac{3}{2}$ will name a number and division will always be defined (except by 0). The extended number system will be called the set of rational numbers.

Any number that can be written in the form a/b where a and b are integers with $b \neq 0$ is called a **rational number.**

Thus

$$\frac{1}{3} \qquad \frac{3}{5} \qquad \frac{8}{1} \qquad \frac{-2}{7} \qquad \frac{10}{-5} \qquad \frac{-3}{-2}$$

all name rational numbers. It is important to note that the integers are a subset of the rational numbers since any integer can be expressed as the quotient of two integers. For example,

$$9 = \frac{9}{1} = \frac{18}{2}$$

$$23 = \frac{23}{1} = \frac{-46}{-2}$$

and so on. Thus, every integer is a rational number, but not every rational number is an integer—for example, $\frac{3}{2}$ is not an integer.

It would seem reasonable, from our experience with multiplying and dividing signed quantities in the preceding chapter, to define the quotient of any two integers with like signs as a **positive rational number** and the quotient of any two integers with unlike signs as a **negative rational number.** Thus,

$$\frac{+2}{+3} = \frac{2}{3} \qquad \frac{-2}{-3} = \frac{2}{3} \qquad \frac{-2}{+3} = -\frac{2}{3} \qquad \frac{+2}{-3} = -\frac{2}{3}$$

Usually, we will write $-\frac{2}{3}$ as $\frac{-2}{3}$ or $-2/3$.

Identifying rational numbers with points on a number line proceeds as one would expect: the positive numbers are located to the right of zero and the negative numbers to the left. Where do we locate a number such as $\frac{7}{4}$? We divide each unit into 4 segments and identify $\frac{7}{4}$ with the endpoint of the seventh segment to the right of zero. Where is $-\frac{3}{2}$ located? Halfway between -1 and -2.

Proceeding as described, every rational number can be associated with a point on a number line.

EXAMPLE 1   Locate $\frac{1}{2}$, $-\frac{3}{4}$, $\frac{5}{2}$, $-\frac{9}{4}$ on a number line.

Solution

PROBLEM 1   Locate $\frac{3}{4}$, $-\frac{1}{2}$, $\frac{7}{4}$, $-\frac{5}{2}$ on a number line.

The **opposite of (negative of) a rational number** and the **absolute value of a rational number** are defined as in the integers. Thus

$$-\left(\frac{2}{3}\right) = -\frac{2}{3} \qquad -\left(-\frac{2}{3}\right) = \frac{2}{3} \qquad \left|\frac{2}{3}\right| = \frac{2}{3} \qquad \left|-\frac{2}{3}\right| = \frac{2}{3}$$

ANSWERS TO
MATCHED
PROBLEMS

1.

3.2 MULTIPLICATION AND DIVISION

In this section we will consider multiplication and division of algebraic forms representing rational numbers, and in the next section we will consider addition and subtraction.

Multiplication

In arithmetic you learned to multiply fractions by multiplying their numerators (tops) and multiplying their denominators (bottoms). This is exactly what we do with rational numbers in general.

> ### Definition of Multiplication for Rational Numbers
> ---
> If a, b, c, and d are integers with b and d different from 0, then
>
> $$\frac{a}{b} \cdot \frac{c}{d} = \frac{a \cdot c}{b \cdot d}$$

EXAMPLE 2

(A) $\dfrac{2}{5} \cdot \dfrac{3}{7} = \dfrac{2 \cdot 3}{5 \cdot 7} = \dfrac{6}{35}$

(B) $(-8) \cdot \dfrac{9}{5} = \dfrac{-8}{1} \cdot \dfrac{9}{5} = \dfrac{(-8)(9)}{(1)(5)} = \dfrac{-72}{5}$ or $-\dfrac{72}{5}$

(C) $\dfrac{2x}{3y^2} \cdot \dfrac{x^2}{5y} = \dfrac{(2x)(x^2)}{(3y^2)(5y)} = \dfrac{2x^3}{15y^3}$

PROBLEM 2 Multiply:

(A) $\dfrac{3}{4} \cdot \dfrac{3}{5}$
(B) $(-5) \cdot \dfrac{3}{4}$
(C) $\dfrac{3x^2}{2y} \cdot \dfrac{x}{4y^2}$

An immediate consequence of the definition of multiplication is the fact that the rational numbers are associative and commutative relative to multiplication. Thus, we can continue to rearrange and regroup factors that represent rational numbers in the same way we rearrange and regroup factors that represent integers or natural numbers.

We now state an important theorem that is used very frequently when working with rational numbers.

THEOREM 1 For any nonzero integers b and k, and any integer a,

(A) $\dfrac{b}{b} = 1$

(B) $1 \cdot \dfrac{a}{b} = \dfrac{a}{b}$

(C) $\dfrac{ak}{bk} = \dfrac{a}{b}$, or, equivalently, $\dfrac{a}{b} = \dfrac{ka}{kb}$ **fundamental principle of fractions**

Thus, for x and y nonzero integers,

$$\frac{7}{7} = 1 \qquad \frac{3x^2}{3x^2} = 1 \qquad 1 \cdot \frac{2x}{3y} = \frac{2x}{3y}$$

$$\frac{8}{12} = \frac{2 \cdot 4}{3 \cdot 4} = \frac{2}{3} \qquad \frac{3}{4} = \frac{3 \cdot 3}{3 \cdot 4} = \frac{9}{12}$$

Parts (A) and (B) of Theorem 1 are rather obvious; part (C) states that we may cancel common factors from the numerator and denominator (that is, divide the numerator and denominator by the same nonzero quantity) or multiply the numerator and denominator by any nonzero quantity. Part **(C)** provides the basis for reducing fractions to lower terms and raising fractions to higher terms. To reduce a fraction to **lowest terms** we cancel *all* common factors from the numerator and denominator; to raise a fraction to **higher terms** we multiply the numerator and denominator by the same nonzero quantity.

EXAMPLE 3

(A) $\dfrac{27}{18} = \dfrac{3 \cdot \cancel{9}}{2 \cdot \cancel{9}} = \dfrac{3}{2}$ lowest terms

(B) $\dfrac{6x}{9x} = \dfrac{2 \cdot \cancel{3x}}{3 \cdot \cancel{3x}} = \dfrac{2}{3}$ lowest terms

(C) $\dfrac{3}{7} = \dfrac{6 \cdot 3}{6 \cdot 7} = \dfrac{18}{42}$ higher terms

(D) $\dfrac{5x}{3} = \dfrac{2x \cdot 5x}{2x \cdot 3} = \dfrac{10x^2}{6x}$ higher terms

PROBLEM 3   Replace question marks with appropriate symbols:

(A) $\dfrac{24}{32} = \dfrac{?}{4}$

(B) $\dfrac{8m}{12m} = \dfrac{2}{?}$

(C) $\dfrac{2}{3} = \dfrac{?}{12y}$

(D) $\dfrac{7}{4} = \dfrac{14y^2}{?}$

The next theorem, pertaining to rational numbers and signs, is used with great frequency in mathematics. Its misuse accounts for many algebraic errors.

THEOREM 2   For each integer a and nonzero integer b

(A) $\dfrac{-a}{-b} = \dfrac{a}{b}$

(B) $\dfrac{-a}{b} = \dfrac{a}{-b} = -\dfrac{a}{b}$

(C) $(-1)\dfrac{a}{b} = -\dfrac{a}{b}$

We are now ready to consider multiplication examples of a more general type.

EXAMPLE 4

(A) $\dfrac{5x^2}{9y^2} \cdot \dfrac{6y}{10x} = \dfrac{5xx \cdot 2 \cdot 3y}{3 \cdot 3yy \cdot 2 \cdot 5x}$ Factor numerator and denominator.

$= \dfrac{\cancel{5}x\cancel{x} \cdot \cancel{2} \cdot \cancel{3} \cdot \cancel{y}}{3 \cdot \cancel{3}y\cancel{y} \cdot \cancel{2} \cdot \cancel{5}\cancel{x}}$ Cancel common factors.

$= \dfrac{x}{3y}$ Answer is in lowest terms.

After a little experience you will probably proceed by repeated division of the numerator and denominator by common factors until all common factors are eliminated. That is, you will proceed something like this:

$$\frac{5x^2}{9y^2} \cdot \frac{6y}{10x} = \frac{(\overset{x}{\cancel{5x^2}})(\overset{1}{\cancel{6y}})}{(\underset{3y}{\cancel{9y^2}})(\underset{1}{\cancel{10x}})} = \frac{x}{3y}$$

(B) $\dfrac{-3x}{2y} \cdot \dfrac{6y^2}{9x^2} = \dfrac{(\cancel{-3x})(\overset{y}{\cancel{6y^2}})}{(\underset{1}{\cancel{2y}})(\underset{3x}{\cancel{9x^2}})} = \dfrac{-y}{x}$ or $-\dfrac{y}{x}$

We could also proceed as in the first part of part **A** by factoring the numerator and denominator and canceling common factors.

PROBLEM 4   Multiply and write answer in lowest terms:

(A) $\dfrac{10x}{6y^2} \cdot \dfrac{12y}{5x^2}$ **(B)** $\dfrac{-7x^2}{3y^2} \cdot \dfrac{12y}{14x}$

Division of fractions

In arithmetic courses you were probably told: "To divide one fraction by another, invert the divisor and multiply." It is not difficult to see why this mechanical rule is valid. To start, we define division for rational numbers as in the integers.

Definition of Division

If a/b and c/d are any two rational numbers, then

$$\frac{a}{b} \div \frac{c}{d} = Q \quad \text{if and only if} \quad \frac{a}{b} = \frac{c}{d} \cdot Q \text{ and } Q \text{ is unique}$$

That is, the quotient Q is the number that c/d must be multiplied by to produce a/b.

As a result of this definition we can establish the following mechanical rule for carrying out division.

Mechanical Rule for Division

To divide one rational number by another rational number different from zero, invert the divisor and multiply. That is,

$$\frac{a}{b} \div \frac{c}{d} = \frac{a}{b} \cdot \frac{d}{c} \qquad \text{divisor} \quad \text{inverted divisor}$$

To establish this rule one has to show that the product of the divisor (c/d) and the quotient $[(a/b) \cdot (d/c)]$ is equal to the dividend (a/b):

$$\frac{c}{d}\left(\frac{a}{b} \cdot \frac{d}{c}\right) = \frac{c}{d} \cdot \frac{ad}{bc}$$

$$= \frac{\overset{1}{\cancel{c}}a\overset{1}{\cancel{d}}}{\underset{1}{\cancel{d}}b\underset{1}{\cancel{c}}} = \frac{a}{b}$$

EXAMPLE 5 **(A)** $\dfrac{6}{14} \div \dfrac{21}{2} = \dfrac{\overset{}{\cancel{6}}}{\underset{7}{\cancel{14}}} \cdot \dfrac{\overset{1}{\cancel{2}}}{\underset{7}{\cancel{21}}} = \dfrac{2}{49}$

Do not cancel in division before inverting the divisor.

(B) $\dfrac{12x}{5y} \div \dfrac{9y}{8x} = \dfrac{\overset{4x}{\cancel{12x}}}{5y} \cdot \dfrac{8x}{\underset{3y}{\cancel{9y}}} = \dfrac{32x^2}{15y^2}$

(C) $\dfrac{18a^2b}{15c} \div \dfrac{12ab^2}{5c} = \dfrac{\overset{\overset{a}{\cancel{3a}}}{\cancel{18a^2b}}}{\underset{\underset{1}{\cancel{3}}}{\cancel{15c}}} \cdot \dfrac{\overset{1}{\cancel{5c}}}{\underset{2b}{\cancel{12ab^2}}} = \dfrac{a}{2b}$

(D) $\dfrac{-3x}{yz} \div 12x = \dfrac{-3x}{yz} \div \dfrac{12x}{1} = \dfrac{\overset{-1}{\cancel{-3x}}}{yz} \cdot \dfrac{1}{\underset{4}{\cancel{12x}}} = \dfrac{-1}{4yz} \quad \text{or} \quad -\dfrac{1}{4yz}$

PROBLEM 5 Divide and reduce to lowest terms:

(A) $\dfrac{8}{9} \div \dfrac{4}{3}$

(B) $\dfrac{8x}{3y} \div \dfrac{6x}{9y}$

(C) $\dfrac{15mn^2}{12x} \div \dfrac{9m^2n}{8x}$ **(D)** $\dfrac{6x}{wz} \div (-3x)$

2. (A) $\dfrac{9}{20}$; (B) $\dfrac{-15}{4}$ or $-\dfrac{15}{4}$; (C) $\dfrac{3x^3}{8y^3}$

3. (A) 3; (B) 3; (C) $8y$; (D) $8y^2$

4. (A) $\dfrac{4}{xy}$; (B) $\dfrac{-2x}{y}$ or $-\dfrac{2x}{y}$

5. (A) $\dfrac{2}{3}$; (B) 4; (C) $\dfrac{10n}{9m}$; (D) $\dfrac{-2}{wz}$ or $-\dfrac{2}{wz}$

EXERCISE 3.2

In answers do not change improper fractions to mixed fractions; that is, write $\frac{7}{2}$, not $3\frac{1}{2}$. All variables represent integers.

A **1.** What rational numbers are associated with points *a*, *b*, and *c*?

2. What rational numbers are associated with points *c*, *d*, and *e*?

Multiply.

3. $\dfrac{2}{5} \cdot \dfrac{3}{7}$ **4.** $\dfrac{3}{8} \cdot \dfrac{3}{5}$ **5.** $\dfrac{4}{5} \cdot \dfrac{7x}{3y}$

6. $\dfrac{5}{7} \cdot \dfrac{2x}{3y}$ **7.** $\dfrac{x}{2y} \cdot \dfrac{3x}{y^2}$ **8.** $\dfrac{2m}{n^2} \cdot \dfrac{3m^2}{5n^2}$

9. $\dfrac{3}{7} \cdot \dfrac{-2}{11}$ **10.** $\dfrac{-2}{5} \cdot \dfrac{4}{3}$ **11.** $\dfrac{-5}{3} \cdot \dfrac{2}{-7}$

12. $\dfrac{2}{-5} \cdot \dfrac{-3}{7}$

Divide.

13. $\dfrac{3}{5} \div \dfrac{5}{7}$ **14.** $\dfrac{3}{4} \div \dfrac{4}{5}$ **15.** $\dfrac{2x}{3} \div \dfrac{5}{7y}$

16. $\dfrac{3}{2u} \div \dfrac{5v}{11}$ **17.** $\dfrac{3}{7} \div \dfrac{-2}{3}$ **18.** $\dfrac{-4}{5} \div \dfrac{3}{7}$

Replace question marks with appropriate symbols.

19. $\dfrac{8}{12} = \dfrac{?}{3}$ **20.** $\dfrac{12}{16} = \dfrac{?}{4}$ **21.** $\dfrac{1}{5} = \dfrac{3}{?}$

22. $\dfrac{3}{4} = \dfrac{?}{20}$ **23.** $\dfrac{21x}{28x} = \dfrac{?}{4}$ **24.** $\dfrac{36y}{54y} = \dfrac{2}{?}$

25. $\dfrac{3}{7} = \dfrac{?}{21x^2}$ **26.** $\dfrac{4}{5} = \dfrac{28m^3}{?}$ **27.** $\dfrac{6x^3}{4xy} = \dfrac{?}{2y}$

28. $\dfrac{9xy}{12y^2} = \dfrac{3x}{?}$

B   *Reduce to lowest terms.*

29. $\dfrac{9x}{6x}$ **30.** $\dfrac{27y}{15y}$ **31.** $\dfrac{-3}{12}$

32. $\dfrac{18}{-8}$ **33.** $\dfrac{2y^2}{8y^3}$ **34.** $\dfrac{6x^3}{15x}$

35. $\dfrac{12a^2b}{3ab^2}$ **36.** $\dfrac{21x^2y^3}{35x^3y}$ **37.** $\dfrac{-2xy^2}{8x^2}$

38. $\dfrac{25mn^3}{-15m^2n^2}$

Multiply or divide as indicated and reduce to lowest terms.

39. $\dfrac{3}{4} \cdot \dfrac{8}{9}$ **40.** $\dfrac{2}{9} \cdot \dfrac{3}{10}$

41. $\dfrac{1}{25} \div \dfrac{15}{4}$ **42.** $\dfrac{7}{3} \div \dfrac{2}{3}$

43. $3 \cdot \dfrac{5}{3}$ **44.** $\dfrac{5}{7} \cdot 7$

45. $\dfrac{2}{3} \div \dfrac{4}{9}$ **46.** $\dfrac{5}{11} \div \dfrac{55}{44}$

47. $\dfrac{4}{-5} \cdot \dfrac{15}{16}$ **48.** $\dfrac{8}{3} \cdot \dfrac{-12}{24}$

49. $\dfrac{2x}{3yz} \cdot \dfrac{6y}{4x}$ **50.** $\dfrac{2a}{3bc} \cdot \dfrac{9c}{a}$

51. $\dfrac{6x}{5y} \div \dfrac{3x}{10y}$ **52.** $\dfrac{9m}{8n} \div \dfrac{3m}{4n}$

53. $2xy \div \dfrac{x}{y}$ **54.** $\dfrac{x}{3y} \div 3y$

55. $\dfrac{2x^2}{3y^2} \cdot \dfrac{9y}{4x}$ **56.** $\dfrac{3x^2}{4} \cdot \dfrac{16y}{12x^3}$

57. $\dfrac{2x}{3y} \div \dfrac{4x}{6y^2}$ **58.** $\dfrac{a}{4c} \div \dfrac{a^2}{12c^2}$

59. $\dfrac{6a^2}{7c} \cdot \dfrac{21cd}{12ac}$ **60.** $\dfrac{8x^2}{3xy} \cdot \dfrac{12y^3}{6y}$

61. $\dfrac{3uv^2}{5w} \div \dfrac{6u^2v}{15w}$ **62.** $\dfrac{21x^2y^2}{12cd} \div \dfrac{14xy}{9d}$

63. $\dfrac{-6x^3}{5y^2} \div \dfrac{18x}{10y}$ **64.** $\dfrac{9u^4}{4v^3} \div \dfrac{-12u^2}{15v}$

C *Perform the operations as indicated and reduce to lowest terms.*

65. $\left(\dfrac{9}{10} \div \dfrac{4}{6}\right) \cdot \dfrac{3}{5}$

66. $\dfrac{9}{10} \div \left(\dfrac{4}{6} \cdot \dfrac{3}{5}\right)$

67. $\dfrac{-21}{16} \cdot \dfrac{12}{-14} \cdot \dfrac{8}{9}$

68. $\dfrac{18}{15} \cdot \dfrac{-10}{21} \cdot \dfrac{3}{-1}$

69. $\dfrac{2x^2}{3y^2} \cdot \dfrac{6yz}{2x} \cdot \dfrac{y}{-xz}$

70. $\dfrac{-a}{-b} \cdot \dfrac{12b^2}{15ac} \cdot \dfrac{-10}{4b}$

71. $\left(\dfrac{a}{b} \div \dfrac{c}{d}\right) \div \dfrac{e}{f}$

72. $\dfrac{a}{b} \div \left(\dfrac{c}{d} \div \dfrac{e}{f}\right)$

3.3 ADDITION AND SUBTRACTION

As in the preceding sections, we will again generalize from arithmetic. In adding $\frac{1}{2}$ and $\frac{2}{3}$ you probably proceed somewhat as follows:

$$\frac{1}{2} = \frac{3}{6}$$
$$\frac{2}{3} = \frac{4}{6}$$
$$\overline{\phantom{\frac{2}{3}} = \frac{7}{6}}$$

That is, you changed each fraction to an equivalent form having a common denominator, then added the numerators and placed the sum over the common denominator. We will proceed in the same way with rational numbers, but we will find it more convenient to work horizontally. We start by defining addition and subtraction of rational numbers with common denominators.

Addition and Subtraction of Rational Numbers

If a, b, and c are integers with $b \neq 0$, then

$$\frac{a}{b} + \frac{c}{b} = \frac{a+c}{b} \qquad \frac{a}{b} - \frac{c}{b} = \frac{a-c}{b} \qquad (1)$$

Thus, two rational numbers with common denominators are added or subtracted by adding or subtracting the numerators and placing the result over the common denominator.

An immediate consequence of this definition (and material in the preceding sections) is: the rational numbers are commutative and associative relative to addition, and multiplication distributes over addition.

EXAMPLE 6†   **(A)** $\dfrac{1}{8} + \dfrac{3}{8} = \dfrac{1+3}{8} = \dfrac{4}{8} = \dfrac{1}{2}$

(B) $\dfrac{5x}{5x} - \dfrac{2}{5x} = \dfrac{5x-2}{5x}$ $\dfrac{5x-2}{5x} = -2$ is wrong. Only common

                                            factors cancel. $5x$ is a term in the

(C) $\dfrac{2x}{6x^2} + \dfrac{5}{6x^2} = \dfrac{2x+5}{6x^2}$ numerator, not a factor.

PROBLEM 6   Combine into single fractions:

(A) $\dfrac{7}{3} - \dfrac{5}{3}$ **(B)** $\dfrac{3}{2u} + \dfrac{2u}{2u}$ **(C)** $\dfrac{3m^2}{12m^3} - \dfrac{2m}{12m^3}$

How do we add or subtract rational numbers, or algebraic expressions representing rational numbers, when the denominators are not the same? We use the fundamental principle of fractions

$$\frac{a}{b} = \frac{ka}{kb} \qquad k, b \neq 0 \qquad \text{fundamental principle of fractions} \qquad (2)$$

(which states that we can multiply the numerator and denominator of a fraction by any nonzero quantity) to obtain equivalent forms having common denominators. The common denominator that generally results in the least amount of computation is the least common multiple (LCM) of all the denominators. The LCM (see Section 1.2) of the denominators is the "smallest" quantity exactly divisible by each denominator and is called the **least common denominator** (LCD).

Finding the LCD

1. Determine the LCD (the "smallest" quantity exactly divisible by each denominator) by inspection, if possible.
2. If the LCD is not obvious, then it can always be found as follows:
   (A) Factor each denominator completely, representing multiple factors as powers.
   (B) The LCD must contain each *different* factor from these factorizations to the highest power it occurs in any one.

EXAMPLE 7   Add or subtract and reduce to lowest terms.

(A) $\dfrac{1}{3} + \dfrac{5}{12}$

---

†All variables in this section represent nonzero integers.

Solution   *Step 1* Find the LCD. The smallest number exactly divisible by 3 and 12 is 12. Therefore,

$$LCD = 12$$

Step 2 Use the fundamental principle of fractions (2) to convert each fraction into a form having the LCD as a denominator, then combine into a single fraction and reduce to lowest terms.

$$\frac{1}{3} + \frac{5}{12} = \frac{4 \cdot 1}{4 \cdot 3} + \frac{5}{12}$$ Convert to forms having the same LCD.

$$= \frac{4}{12} + \frac{5}{12}$$ Combine into a single fraction.

$$= \frac{4 + 5}{12}$$

$$= \frac{9}{12}$$ Reduce to lowest terms.

$$= \frac{3}{4}$$

(B) $\dfrac{5}{24y} - \dfrac{7}{10y}$

Solution   In this case the LCD is not obvious, so we factor each denominator completely:

$$24y = 8 \cdot 3y = 2 \cdot 2 \cdot 2 \cdot 3 \cdot y = 2^3 \cdot 3 \cdot y$$

$$10y = 2 \cdot 5 \cdot y$$

Since the LCD must contain each *different* factor from these factorizations to the highest power it occurs in any one, we first write down all the different factors:

2, 3, 5, *y*

then observe the highest power each occurs in any one factorization to obtain

$$LCD = 2^3 \cdot 3 \cdot 5 \cdot y = 120y$$

Now multiply numerators and denominators in the original problem by appropriate quantities to obtain 120*y* as a common denominator.

$$\frac{5}{24y} - \frac{7}{10y} = \frac{5 \cdot 5}{5 \cdot 24y} - \frac{12 \cdot 7}{12 \cdot 10y}$$

$$= \frac{25}{120y} - \frac{84}{120y}$$

$$= \frac{25 - 84}{120y}$$

$$= \frac{-59}{120y}$$

(C) $\quad \dfrac{3}{8x} + \dfrac{7}{12x^2}$

Solution   Find the LCD:

$$8x = 2^3 \cdot x$$

$$12x^2 = 2^2 \cdot 3 \cdot x^2$$

$$\text{LCD} = 2^3 \cdot 3 \cdot x^2 = 24x^2$$

$$\frac{3}{8x} + \frac{7}{12x^2} = \frac{(3x)(3)}{(3x)(8x)} + \frac{2 \cdot 7}{2(12x^2)}$$

$$= \frac{9x}{24x^2} + \frac{14}{24x^2}$$

$$= \frac{9x + 14}{24x^2}$$

PROBLEM 7   Add or subtract and reduce to lowest terms:

(A) $\quad \dfrac{4}{5} - \dfrac{2}{15}$ **(B)** $\quad \dfrac{5}{18xy} + \dfrac{3}{4xy}$ **(C)** $\quad \dfrac{2}{9x^2y} - \dfrac{7}{12xy^2}$

EXAMPLE 8   **(A)** $\quad \dfrac{-3}{4} - \dfrac{-1}{3} + \dfrac{5}{6} = \dfrac{3(-3)}{3(4)} - \dfrac{4(-1)}{4(3)} + \dfrac{2(5)}{2(6)} = \dfrac{-9}{12} - \dfrac{-4}{12} + \dfrac{10}{12}$

$$= \frac{-9 - (-4) + 10}{12}$$

NOTE: LCD $= 12$

$$= \frac{-9 + 4 + 10}{12}$$

$$= \frac{5}{12}$$

(B) $\quad \left(\dfrac{3}{2x^2} - \dfrac{-5}{x} + 1 \right) = \dfrac{3}{2x^2} - \dfrac{2x(-5)}{2x(x)} + \dfrac{2x^2}{2x^2} = \dfrac{3}{2x^2} - \dfrac{-10x}{2x^2} + \dfrac{2x^2}{2x^2}$

$$= \frac{3 - (-10x) + 2x^2}{2x^2}$$

NOTE: LCD $= 2x^2$

$$= \frac{3 + 10x + 2x^2}{2x^2}$$

PROBLEM 8 Combine into one fraction:

(A) $\dfrac{-1}{2} - \dfrac{-3}{4} + \dfrac{2}{3}$

(B) $2 - \dfrac{-3}{x} + \dfrac{4}{3x^2}$

ANSWERS TO
MATCHED
PROBLEMS

6. (A) $\dfrac{2}{3}$; (B) $\dfrac{3 + 2u}{2u}$; (C) $\dfrac{3m^2 - 2m}{12m^3}$

7. (A) $\dfrac{2}{3}$; (B) $\dfrac{37}{36xy}$; (C) $\dfrac{8y - 21x}{36x^2y^2}$

8. (A) $\dfrac{11}{12}$; (B) $\dfrac{6x^2 + 9x + 4}{3x^2}$

EXERCISE 3.3

Combine into single fractions and reduce to lowest terms (work horizontally).

A

1. $\dfrac{2}{3} + \dfrac{4}{3}$
2. $\dfrac{3}{4} + \dfrac{5}{4}$
3. $\dfrac{-3}{5} + \dfrac{7}{5}$

4. $\dfrac{2}{3} + \dfrac{-5}{3}$
5. $\dfrac{3}{8} + \dfrac{1}{2}$
6. $\dfrac{2}{5} + \dfrac{3}{10}$

7. $\dfrac{2}{3} + \dfrac{3}{5}$
8. $\dfrac{1}{2} + \dfrac{4}{7}$
9. $\dfrac{7}{11} - \dfrac{3}{11}$

10. $\dfrac{5}{3} - \dfrac{2}{3}$
11. $\dfrac{7}{11} - \dfrac{-3}{11}$
12. $\dfrac{5}{3} - \dfrac{-2}{3}$

13. $\dfrac{1}{2} - \dfrac{3}{8}$
14. $\dfrac{2}{5} - \dfrac{3}{10}$
15. $\dfrac{3}{5} - \dfrac{2}{3}$

16. $\dfrac{1}{2} - \dfrac{4}{7}$
17. $\dfrac{3}{5xy} + \dfrac{-6}{5xy}$
18. $\dfrac{-6}{5x^2} + \dfrac{4}{5x^2}$

19. $\dfrac{3y}{x} + \dfrac{2y}{x}$
20. $\dfrac{x}{5y} + \dfrac{2x}{5y}$
21. $\dfrac{3}{7y} - \dfrac{-3}{7y}$

22. $\dfrac{-2}{3x} - \dfrac{2}{3x}$
23. $\dfrac{1}{2x} + \dfrac{2}{3x}$
24. $\dfrac{3}{5m} + \dfrac{5}{2m}$

25. $\dfrac{3x}{2} + \dfrac{2x}{3}$
26. $\dfrac{4m}{3} + \dfrac{m}{7}$
27. $\dfrac{3}{5x} - \dfrac{2}{3}$

28. $\dfrac{2}{3} - \dfrac{3}{4y}$

B

29. $\dfrac{x}{y} - \dfrac{y}{x}$
30. $\dfrac{a}{b} + \dfrac{b}{a}$
31. $\dfrac{x}{y} - 2$

32. $1 - \dfrac{1}{x}$
33. $5 - \dfrac{-3}{x}$
34. $\dfrac{-2}{m} - 4$

35. $\dfrac{1}{xy} - \dfrac{3}{y}$
36. $\dfrac{2a}{b} + \dfrac{-1}{ab}$
37. $\dfrac{3}{2x^2} + \dfrac{4}{3x}$

38. $\dfrac{5}{3y} + \dfrac{3}{4y^2}$
39. $\dfrac{5}{8m^3} - \dfrac{1}{12m}$
40. $\dfrac{2}{9n^2} - \dfrac{5}{12n^4}$

41. $\dfrac{1}{3} - \dfrac{-1}{2} + \dfrac{5}{6}$
42. $\dfrac{-3}{4} + \dfrac{2}{5} - \dfrac{-3}{2}$

43. $\dfrac{x^2}{4} - \dfrac{x}{3} + \dfrac{-1}{2}$

44. $\dfrac{2}{5} - \dfrac{x}{2} - \dfrac{-x^2}{3}$

45. $\dfrac{3}{4x} - \dfrac{2}{3y} + \dfrac{1}{8xy}$

46. $\dfrac{1}{xy} - \dfrac{1}{yz} + \dfrac{-1}{xz}$

47. $\dfrac{3}{y^3} - \dfrac{-2}{3y^2} + \dfrac{1}{2y} - 3$

48. $\dfrac{1}{5x^3} + \dfrac{-3}{2x^2} - \dfrac{-2}{3x} - 1$

C 49. $\dfrac{y}{9} - \dfrac{-1}{28} - \dfrac{y}{42}$

50. $\dfrac{5x}{6} - \dfrac{3}{8} + \dfrac{x}{15} - \dfrac{3}{20}$

51. $\dfrac{x^2}{12} + \dfrac{x}{18} - \dfrac{1}{30}$

52. $\dfrac{3x}{50} - \dfrac{x}{15} - \dfrac{-2}{6}$

3.4 EQUATIONS INVOLVING FRACTIONS

Most of the equations we solved earlier had integer coefficients. In most practical applications, rational number coefficients occur more frequently than integers. We now have the tools to convert any equation with rational coefficients into one with integer coefficients, thus placing it in a position to be solved by earlier methods. A couple of examples will make the process clear.

EXAMPLE 9   Solve

$$\frac{x}{4} = \frac{3}{8}$$

Solution   To isolate x on the left side with a coefficient of $+1$, we multiply both sides by 4 (that is, $\frac{4}{1}$) using the multiplication property of equality (Section 2.7).

$$\frac{4}{1} \cdot \frac{x}{4} = \frac{4}{1} \cdot \frac{3}{8} \qquad \text{Usually done mentally after some practice.}$$

$$x = \frac{\overset{1}{\cancel{4}}}{1} \cdot \frac{3}{\underset{2}{\cancel{8}}} = \frac{3}{2}$$

PROBLEM 9   Solve $\dfrac{x}{9} = \dfrac{2}{3}$.

EXAMPLE 10   Solve

$$\frac{x}{3} - \frac{1}{2} = \frac{5}{6}$$

Solution | What operation can we perform on the equation to eliminate the denominators? If we could find a number exactly divisible by each denominator, then we would be able to use the multiplication property of equality to clear the denominators. The LCM (see Section 1.2) of the denominators is exactly what we are looking for. The LCM (the smallest number exactly divisible by 3, 2, and 6) is 6. Thus, we multiply both sides of the equation by 6 (or $\frac{6}{1}$).

$$\frac{6}{1} \cdot \left(\frac{x}{3} - \frac{1}{2} \right) = \frac{6}{1} \cdot \frac{5}{6}$$ Clear ( ) before canceling.

$$\overset{2}{\cancel{\frac{6}{1}}} \cdot \frac{x}{\underset{1}{\cancel{3}}} - \overset{3}{\cancel{\frac{6}{1}}} \cdot \frac{1}{\underset{1}{\cancel{2}}} = \frac{6}{1} \cdot \frac{5}{\underset{1}{\cancel{6}}}$$ All denominators should cancel, resulting in an equation with integer coefficients.

$$2x - 3 = 5$$ Add 3 to each side.

$$2x = 8$$ Divide each side by 2.

$$x = 4$$

With a little experience, the dotted box steps are done mentally. Notice each term on each side of the equation is multiplied by 6, and since 6 is the LCM of the denominators, each denominator will divide into 6 exactly.

Before you try the matched problem, observe that

$$\frac{5}{12}x$$

can also be written in the form

$$\frac{5x}{12}$$

since

$$\frac{5}{12}x = \frac{5}{12} \cdot \frac{x}{1} = \frac{5x}{12}$$

You should be able to shift easily from one form to the other.

PROBLEM 10 | Solve $\frac{1}{4}x - \frac{2}{3} = \frac{5}{12}x$

(First, write each term as a single fraction; that is, $\frac{1}{4}x = \frac{x}{4}$, and $\frac{5}{12}x = \frac{5x}{12}$.)

EXAMPLE 11   Solve

$$5 - \frac{2x - 1}{4} = \frac{x + 2}{3}$$

Solution   It is a good idea to first enclose any numerator with more than one term in parentheses before multiplying both sides by the LCM of the denominators. The LCM of 4 and 3 is 12.

$$5 - \frac{(2x - 1)}{4} = \frac{(x + 2)}{3}$$ Multiply both sides by 12.

$$12 \cdot 5 - \overset{3}{\cancel{12}} \cdot \frac{(2x - 1)}{\underset{1}{\cancel{4}}} = 12 \cdot \overset{4}{} \frac{(x + 2)}{\underset{1}{\cancel{3}}}$$ Cancel denominators.

$$60 - 3(2x - 1) = 4(x + 2)$$ Clear ( ).

$$60 - 6x + 3 = 4x + 8$$ Combine like terms.

$$-6x + 63 = 4x + 8$$ Subtract 63 from each side.

$$-6x = 4x - 55$$ Subtract 4x from each side.

$$-10x = -55$$ Divide both sides by -10.

$$x = \frac{-55}{-10}$$ Reduce answer.

$$x = \frac{11}{2} \quad \text{or} \quad 5.5$$

PROBLEM 11   Solve $\dfrac{x + 3}{4} - \dfrac{x - 4}{2} = \dfrac{3}{8}$

A COMMON ERROR

A very common error occurs about now—students tend to confuse expressions involving fractions with equations involving fractions. Consider the two problems:

(A)   Solve: $\frac{x}{2} + \frac{x}{3} = 10$ **(B)**   Add: $\frac{x}{2} + \frac{x}{3} + 10$

The problems look very much alike, but are actually quite different. To solve the equation in part **A** we multiply both sides by 6 to clear the fractions. This works so well for equations, students want to do the same thing for problems like **B**. The only catch is that part **B** is not an equation and the multiplication property of equality does not apply. If we multiply the expressions in part **B** by 6 we obtain an expression 6 times as large as the original. To add in part **B** we find the LCD and proceed as in Section 3.3.

Compare the following:

(A) $\frac{x}{2} + \frac{x}{3} = 10$

$6 \cdot \frac{x}{2} + 6 \cdot \frac{x}{3} = 6 \cdot 10$

$3x + 2x = 60$

$5x = 60$

$x = 12$

(B) $\frac{x}{2} + \frac{x}{3} + 10$

$= \frac{3 \cdot x}{3 \cdot 2} + \frac{2 \cdot x}{2 \cdot 3} + \frac{6 \cdot 10}{6 \cdot 1}$

$= \frac{3x}{6} + \frac{2x}{6} + \frac{60}{6}$

$= \frac{5x + 60}{6}$

ANSWERS TO MATCHED PROBLEMS

9. 6   **10.** -4 **11.** $\frac{19}{2}$ or 9.5

EXERCISE 3.4

Solve. (*In Problems 13–16 multiply both sides of the equation by a number that will clear the decimals. For example, if we start with 0.7x = 3.5 we could multiply both sides by 10 to obtain 7x = 35, an equation free of decimals.*)

A

1. $\frac{x}{7} - 1 = \frac{1}{7}$ **2.** $\frac{x}{5} - 2 = \frac{3}{5}$ **3.** $\frac{y}{4} + \frac{y}{2} = 9$

4. $\frac{x}{3} + \frac{x}{6} = 4$ **5.** $\frac{x}{2} + \frac{x}{3} = 5$ **6.** $\frac{y}{4} + \frac{y}{3} = 7$

7. $\frac{x}{3} - \frac{1}{4} = \frac{3x}{8}$ **8.** $\frac{y}{5} - \frac{1}{3} = \frac{2y}{15}$ **9.** $\frac{n}{5} - \frac{n}{6} = \frac{6}{5}$

10. $\frac{m}{4} - \frac{m}{3} = \frac{1}{2}$ **11.** $\frac{2}{3} - \frac{x}{8} = \frac{5}{6}$ **12.** $\frac{5}{12} - \frac{m}{3} = \frac{4}{9}$

13. $0.8x = 16$ **14.** $0.5x = 35$

15. $0.3x + 0.5x = 24$ **16.** $0.7x + 0.9x = 32$

B

17. $\frac{x+3}{2} - \frac{x}{3} = 4$ **18.** $\frac{x-2}{3} + 1 = \frac{x}{7}$

19. $\frac{2x+1}{4} = \frac{3x+2}{3}$ **20.** $\frac{4x+3}{9} = \frac{3x+5}{7}$

21. $3 - \frac{x-1}{2} = \frac{x-3}{3}$ **22.** $4 - \frac{x-3}{4} = \frac{x-1}{8}$

23. $3 - \frac{2x-3}{3} = \frac{5-x}{2}$ **24.** $1 - \frac{3x-1}{6} = \frac{2-x}{3}$

25. $0.4(x + 5) - 0.3x = 17$ **26.** $0.1(x - 7) + 0.05x = 0.8$

C **27.** $\dfrac{3x - 1}{8} - \dfrac{2x + 1}{3} = \dfrac{1 - x}{12} - 1$

28. $\dfrac{2x - 3}{9} - \dfrac{x + 5}{6} = \dfrac{3 - x}{2} + 1$

3.5 APPLICATIONS: NUMBER AND GEOMETRIC PROBLEMS

We are now ready to consider a variety of word problems and significant applications. This section deals with relatively easy number and geometric problems to give you more practice in translating words into symbolic forms. The sections following include problems of a slightly more difficult nature from a large variety of fields. You may be surprised at the number and variety of applications that are now within your reach after having had less than three chapters of algebra.

To start our discussion, we restate a strategy for solving word problems:

A Strategy for Solving Word Problems

1. Read the problem carefully—several times if necessary, that is, until you understand the problem and know what is to be found.

2. Draw figures or diagrams and label given and unknown parts.

3. Look for formulas connecting the given with the unknown.

4. Let one of the unknown quantities be represented by a variable, say x, and try to represent all other unknown quantities in terms of x. This is an important step and must be done carefully.

5. Form an equation relating the unknown quantities with known quantities.

6. Solve the equation and write answers for all parts of the problem requested.

7. Check all solutions back in the original problem.

Number problems

Recall that if x is a number, then two-thirds x can be written

$$\frac{2}{3}x \qquad \text{or} \qquad \frac{2x}{3}$$

The latter form will be more convenient for our purposes.

EXAMPLE 12 Find a number such that 10 less than two-thirds the number is one-fourth the number.

Solution   Let $x =$ the number. Symbolize each part of the problem:

$$\frac{2x}{3} - 10 \qquad \text{10 less than two-thirds the number}$$

$$= \qquad \text{is}$$

$$\frac{x}{4} \qquad \text{one-fourth the number}$$

Write an equation involving the symbolic forms and solve:

$$\frac{2x}{3} - 10 = \frac{x}{4}$$

$$12 \cdot \frac{2x}{3} - 12 \cdot 10 = 12 \cdot \frac{x}{4} \qquad \text{Multiply by 12, the LCM of 3 and 4.}$$

$$8x - 120 = 3x \qquad \text{Cancel denominators and simplify.}$$

$$5x = 120$$

$$x = 24$$

CHECK

$$\tfrac{2}{3}(24) - 10 = 16 - 10 = 6 \qquad \text{left side}$$

$$\tfrac{1}{4}(24) = 6 \qquad \text{right side}$$

PROBLEM 12   Find a number such that 6 more than one-half the number is two-thirds the number.

Geometric problems   Recall that the **perimeter** of a triangle or rectangle is the distance around the figure. Symbolically:

$$P = a + b + c \qquad\qquad P = 2a + 2b$$

EXAMPLE 13   If one side of a triangle is one-third the perimeter, the second side is 7 cm, and the third side is one-fifth the perimeter, what is the perimeter of the triangle?

Solution   Let P = the perimeter. Draw a triangle and label sides:

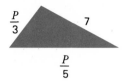

Thus,

$$P = \frac{P}{3} + 7 + \frac{P}{5}$$

$$15 \cdot P = 15 \cdot \frac{P}{3} + 15 \cdot 7 + 15 \cdot \frac{P}{5} \qquad \text{15 is LCM of 3 and 5}$$

$$15P = 5P + 105 + 3P$$

$$7P = 105$$

$$P = 15 \, \text{cm}$$

CHECK

$$\frac{15}{3} + 7 + \frac{15}{5} = 5 + 7 + 3 = 15 \text{ (the perimeter)}$$

PROBLEM 13   If one side of a triangle is one-fourth the perimeter, the second side is 7 meters, and the third side is two-fifths the perimeter, what is the perimeter?

EXAMPLE 14   Find the dimensions of a rectangle with perimeter 176 cm if its width is three-eighths its length.

Solution   Draw a rectangle and label sides:

Let x = length

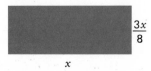

Thus,

$$2x + 2 \cdot \frac{3x}{8} = 176 \qquad \text{Use perimeter formula:} \quad 2a + 2b = P$$

$$2x + \frac{3x}{4} = 176$$

$$4 \cdot 2x + 4 \cdot \frac{3x}{4} = 4 \cdot 176$$

$$8x + 3x = 704$$

$$11x = 704$$

$$x = 64 \text{ cm} \qquad \text{length}$$

$$\frac{3x}{8} = 24 \text{ cm} \qquad \text{width}$$

CHECK

$2 \cdot 64 + 2 \cdot 24 = 128 + 48 = 176$ cm (the perimeter)

PROBLEM 14 Find the dimensions of a rectangle with perimeter 84 cm if its width is two-fifths its length.

ANSWERS TO **12.** 36   **13.** 20 meters   **14.** 30 cm by 12 cm
MATCHED
PROBLEMS

EXERCISE 3.5

A *If x represents a number, write an algebraic expression for each of the following numbers:*

1. one-half x
2. one-sixth x
3. two-thirds x
4. three-fourths x
5. 2 more than one-third x
6. 5 more than one-fourth x
7. 8 less than two-thirds x
8. 6 less than three-fourths x
9. one-half the number that is 3 less than twice x
10. one-third the number that is 5 less than four times x

Find numbers meeting each of the indicated conditions. (A) Write an equation using x and (B) solve the equation.

11. 2 more than one-fourth the number is $\frac{1}{2}$.
12. 3 more than one-sixth the number is $\frac{2}{3}$.
13. 2 less than one-half the number is one-third the number.
14. 3 less than one-third the number is one-fourth the number.

B 15. 5 less than half the number is 3 more than one-third the number.
16. 2 less than one-sixth the number is 1 more than one-fourth the number.

17. 5 more than two-thirds the number is 10 less than one-fourth the number.

18. 4 less than three-fifths the number is 8 more than one-third the number.

Solve.

19. If one side of a triangle is one-fourth the perimeter, the second side is 3 meters, and the third side is one-third the perimeter, what is the perimeter?

20. If one side of a triangle is two-fifths the perimeter, the second side is 70 cm, and the third side is one-fourth the perimeter, what is the perimeter?

21. An electrical transmission tower is located in a lake. If one-fifth of the tower is in the sand, 10 meters in the water, and two-thirds of it in the air, what is the total height of the tower from the rock foundation to the top (see Figure 1)?

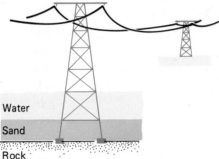

Figure 1

22. On a safari in Africa a group traveled one-half the distance by Land Rover, 55 km by horse, and the last one-third of the distance by boat. How long was the trip?

23. Find the dimensions of a rectangle with perimeter 72 cm, if its width is one-third its length.

24. Find the dimensions of a rectangle with perimeter 84 meters, if its width is one-sixth its length.

25. Find the dimensions of a rectangle with perimeter 216 meters, if its width is two-sevenths its length.

26. Find the dimensions of a rectangle with perimeter 100 cm, if its width is two-thirds its length.

C **27.** Find the dimensions of a rectangle with perimeter 112 cm, if its width is 7 cm less than two-fifths its length.

28. Find the dimensions of a rectangle with perimeter 264 cm, if its width is 11 cm less than three-eighths its length.

3.6 APPLICATIONS: RATIO AND PROPORTION

One of the first applications of algebra in elementary science and technology courses you are likely to encounter deals with ratio and proportion. Many problems in these courses can be solved using the methods developed in this section. In addition, we will find a convenient way to convert metric units into English units and English units into metric units.

Ratios

The use of ratios is very common. You have no doubt been using ratios for many years and will recall that the ratio of two quantities is the first divided by the second; that is

---
Ratio of a to b

The ratio of a to b, assuming $b \neq 0$, is $\dfrac{a}{b}$.

EXAMPLE 15   If there are 10 men and 20 women in a class, then the ratio of men to women is $\frac{10}{20}$ or $\frac{1}{2}$ (which is also written 1:2 and is read 1 to 2).

PROBLEM 15   If there are 500 men and 400 women in a school:

(A) What is the ratio of men to women;   (B) Women to men?

Proportion

In addition to comparing known quantities, another reason we want to know something about ratios is that they often lead to a simple way of finding unknown quantities.

EXAMPLE 16   Suppose you are told that the ratio of women to men in a college is 3:5 and that there are 1,450 men. How many women are in the college?

Solution   Let $x =$ the number of women; then the ratio of women to men is $x/1,450$. Thus,

$$\frac{x}{1,450} = \frac{3}{5} \qquad \text{To isolate } x, \text{ multiply both sides by 1,450.}$$

$$x = 1,450 \cdot \frac{3}{5}$$

$$= 870 \text{ women}$$

PROBLEM 16   If in a college the ratio of men to women is 2:3, and there are 1,200 women, how many men are in the school?

A statement of equality between two ratios is called a proportion. That is,

---

Proportion

$$\frac{a}{b} = \frac{c}{d} \qquad b, d \neq 0$$

EXAMPLE 17   If a car can travel 192 miles on 8 gal of gas, how far will it go on 15 gal?

Solution   Let $x =$ distance traveled on 15 gal. Thus,

$$\frac{x}{15} = \frac{192}{8}$$

Both ratios represent miles per gallon.

$$x = 15 \cdot \frac{192}{8}$$

We can isolate x by multiplying both sides by 15—we do not need to use the LCM of 15 and 8.

$$= 360 \text{ miles}$$

PROBLEM 17   If there are 24 grams of hydrochloric acid in 64 grams of solution, how many grams of acid will be in 48 grams of the same solution? Set up a proportion and solve.

Metric conversion   A summary of metric units is located inside the back cover of the text. Here we show how proportion can be used to convert from one system to the other.

EXAMPLE 18   If there are 2.2 pounds in 1 kilogram how many kilograms are in 100 pounds?

Solution   Let $x =$ number of kilograms in 100 lb. Set up a proportion with x in a numerator; that is, a proportion of the form:

$$\frac{\text{kilograms}}{\text{pounds}} = \frac{\text{kilograms}}{\text{pounds}}$$

Each ratio represents kilograms per pound.

Thus,

$$\frac{x}{100} = \frac{1}{2.2}$$

$$x = \frac{100}{2.2}$$

$$= 45.45 \text{ kg}$$

PROBLEM 18 If there is 0.45 kilogram in 1 pound, how many pounds are in 90 kilograms? Set up a proportion (with the variable in the numerator) and solve.

EXAMPLE 19 If there are 1.09 yards in 1 meter, how many meters are in 80 yards?

Solution Let x = number of meters in 80 yd. Set up a proportion of the form:

$$\frac{\text{meters}}{\text{yards}} = \frac{\text{meters}}{\text{yards}}$$

Thus,

$$\frac{x}{80} = \frac{1}{1.09}$$

$$x = \frac{80}{1.09}$$

$$= 73.39 \text{ meters}$$

PROBLEM 19 If there is 0.94 liter in 1 quart, how many quarts are in 50 liters? Set up a proportion (with the variable in a numerator) and solve.

ANSWERS TO MATCHED PROBLEMS

15. (A) $\frac{5}{4}$ or 5:4; (B) $\frac{4}{5}$ or 4:5

16. 800 men

17. $\frac{x}{48} = \frac{24}{64}$, $x = 18$ grams

18. $\frac{x}{90} = \frac{1}{0.45}$, $x = 200$ lb

19. $\frac{x}{50} = \frac{1}{0.94}$, $x = 53.19$ qt

EXERCISE 3.6

A *Write as a ratio.*

1. 16 men to 64 women
2. 64 women to 16 men
3. 25 cm to 5 cm
4. 30 meters to 10 meters
5. 30 km² to 90 km²
6. 25 square meters to 100 square meters

Solve each proportion.

7. $\frac{x}{12} = \frac{2}{3}$

8. $\frac{y}{16} = \frac{5}{4}$

9. $\frac{d}{12} = \frac{27}{18}$

10. $\frac{y}{13} = \frac{21}{39}$

11. $\frac{18}{27} = \frac{h}{6}$

12. $\frac{35}{56} = \frac{x}{32}$

Set up proportions and solve.

13. If in a college the ratio of men to women is $\frac{5}{7}$ and there are 840 women, how many men are there?

14. If the ratio of women to men is $\frac{7}{9}$ and there are 630 men, how many women are there?

15. If the ratio of the length of a rectangle to its width is $\frac{3}{2}$ and the width is 24 cm, how long is the rectangle?

16. If the ratio of the width of a rectangle to its length is $\frac{3}{5}$ and its length is 30 meters, how wide is it?

17. If the ratio of grade points to the total number of units for a student is $\frac{7}{2}$ and the student has completed 60 units, how many grade points does the student have?

18. If the ratio of grade points to the total number of units for a student is $\frac{14}{5}$ and the student has completed 90 units, how many grade points does the student have?

19. If a car can travel 100 km on 12 liters of gas, how far will it go on 15 liters?

20. If an airplane can fly 2,400 miles in 9 hr, how far would it fly in 6 hr?

B **21.** MIXTURE. If there are 8 grams of sulfuric acid in 70 grams of solution, how many grams of sulfuric acid are in 21 grams of the same solution?

22. MIXTURE. If 1.5 cups of sugar are needed in a recipe for 4 people, how many cups of sugar are needed for 6 people?

23. PHOTOGRAPHY. If you enlarge a 3- by 6-in picture so that the longer side is 8 in, how wide will the enlargement be?

24. SCALE DRAWINGS. An architect wishes to make a scale drawing of a 12- by 8-meter building. If she uses 9 cm for the length, what should she use for the width?

25. PRICE-EARNING RATIO. If the price-earning ratio for a common stock is $\frac{5}{2}$ and the stock earns $36 per share, what is the price of the stock?

26. COMMISSIONS. If you were charged a commission of $57 on the purchase of 300 shares of stock, what would be the proportionate commission on 500 shares of the same stock?

27. ENGINEERING. If an engineer knows that a 1.5-meter piece of steel rod weighs 12 kg, how much would a 5-meter piece of the same rod weigh?

28. ENGINEERING. If in Figure 2 the area a of the small pipe is 2 cm^2 and

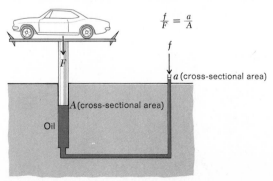

$$\frac{f}{F} = \frac{a}{A}$$

Figure 2

the area A of the larger pipe is 136 cm², how much force f in kilograms is required to lift a car weighing 1,700 kg (3,740 lb)?

29. METRIC CONVERSION. If there is 1 kilogram in 2.2 pounds, how many kilograms are in 12 pounds?

30. METRIC CONVERSION. If there is 1 liter in 1.06 quarts, how many liters are in 1 gallon (4 quarts)?

31. METRIC CONVERSION. If there are 1.61 kilometers in 1 mile, how many miles are in 40 kilometers?

32. METRIC CONVERSION. If there are 1.09 yards in 1 meter, how many meters are in 100 yards?

33. METRIC CONVERSION. If there are 28.57 grams in 1 ounce, how many ounces are in 1 kilogram (1,000 grams)?

34. METRIC CONVERSION. If there are 39.37 inches in 1 meter, how many meters are in 10 feet?

35. METRIC CONVERSION. If there is 0.92 meter in 1 yard, how many yards are in 50 meters?

36. METRIC CONVERSION. If there is 0.94 liter in 1 quart, how many quarts are in 20 liters?

C **37.** WILDLIFE MANAGEMENT. Zoologists Green and Evans (1940) estimated the total population of snowshoe hares in the Lake Alexander area of Minnesota as follows. They captured and banded 948 hares, then released them. After an appropriate period for mixing, they again captured a sample of 421 and found 167 of these marked. Assuming the ratio of marked hares to the total number captured in the second sample is the same as the ratio of those banded in the first sample to the total population, set up a proportion and estimate the total hare population in the region.

38. WILDLIFE MANAGEMENT. Estimate the total number of trout in a lake if a sample of 500 are netted, marked, and released, and after a period for mixing, a second sample of 375 produces 25 marked ones. (See Problem 37.)

39. ASTRONOMY. Do you have any idea how one might measure the circumference of the earth? In 240 B.C. Eratosthenes measured the size of the earth from its curvature. At Syene, Egypt (lying on the Tropic of Cancer), the sun was directly overhead at noon on June 21. At the same time in Alexandria, a town 500 miles directly north, the sun's rays fell at an angle of 7.5° to

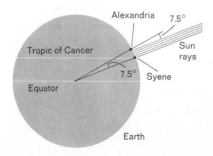

Figure 3

the vertical. Using this information and a little knowledge of geometry (see Figure 3), Eratosthenes was able to approximate the circumference of the earth using the following proportion: 360° is to 7.5° as the circumference of the earth is to 500 miles. Compute Eratosthenes' estimate.

3.7 APPLICATIONS: RATE-TIME PROBLEMS

If a car travels 400 km in 5 hr, then the ratio

$$\frac{400 \text{ km}}{5 \text{ hr}} \qquad \text{or} \qquad 80 \text{ km/hr}$$

(read 80 kilometers per hour) is called the rate of motion, or speed. It is the number of miles produced in each unit of time. Similarly, if a person types 420 words in 6 min, the ratio

$$\frac{420 \text{ words}}{6 \text{ min}} \qquad \text{or} \qquad 70 \text{ words per min}$$

is the rate of typing. It is the number of words produced in each unit of time.

In general, if q is the quantity produced in t units of time, then

$$\frac{\text{quantity}}{\text{time}} = \text{rate}$$

or

$$\frac{q}{t} = r \tag{3}$$

Thus, the **rate** r is the amount of q produced in each unit of time. If both sides of (3) are multiplied by t, we obtain the more familiar form

$$q = rt \tag{4}$$

If q is distance d, then

$$d = rt \tag{5}$$

a special form of (4) with which you are likely familiar. Formulas (4) and (5) enter into the solutions of many rate-time problems.

Important Rate-Time Formulas

---

quantity = (rate)(time)

$$q = rt \qquad (4)$$

distance = (rate)(time)

$$d = rt \qquad (5)$$

EXAMPLE 20   If a person jogs 9 miles in 2 hr, what is his or her rate of motion (in miles per hour)? Set up an equation and solve.

Solution   We write (4) in the form

$$rt = d$$

and let $t = 2$ and $d = 9$, then solve for r:

$$r \cdot 2 = 9$$

$$r = \frac{9}{2} \quad \text{or} \quad 4.5 \text{ mph}$$

PROBLEM 20   If a gas station pump pumps 10 gal in 4 min, what is its rate of pumping (in gallons per minute)? Set up an equation and solve.

EXAMPLE 21   If an IBM card sorting machine can sort 350 cards per min, how long will it take the machine to sort 3,850 cards? Set up an equation and solve.

Solution   We write (4) in the form

$$rt = q$$

and let $r = 350$ and $q = 3,850$, then solve for t:

$$350t = 3,850$$

$$t = \frac{3,850}{350} = 11 \text{ min}$$

PROBLEM 21   If an airplane flies at 800 km/hr, how long will it take to fly 2,600 km? Set up an equation and solve.

EXAMPLE 22   A car leaves town A and travels at 55 mph toward town B at the same time a car leaves town B and travels 45 mph toward town A. If the towns are 450 miles apart, how long will it take the two cars to meet? Set up an equation and solve.

Solution   Let $t = $ number of hours until both cars meet; then draw a diagram and label known and unknown parts:

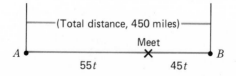

$$\begin{pmatrix} \text{distance car} \\ \text{from } A \text{ travels} \end{pmatrix} + \begin{pmatrix} \text{distance car} \\ \text{from } B \text{ travels} \end{pmatrix} = \begin{pmatrix} \text{total} \\ \text{distance} \end{pmatrix}$$

$$55t \quad + \quad 45t \quad = \quad 450$$

$$100t = 450$$

$$t = \frac{450}{100}$$

$$t = 4.5 \, \text{hr}$$

PROBLEM 22 If an older printing press can print 50 handbills per min and a newer press can print 75, how long will it take both together to print 1,500 handbills? Set up an equation and solve.

EXAMPLE 23 A car leaves a town traveling at 60 km/hr. How long will it take a second car traveling at 80 km/hr to catch up to the first car if it leaves 2 hr later? Set up an equation and solve.

Solution Let t = number of hours for second car to catch up, then draw a diagram and label known and unknown parts.

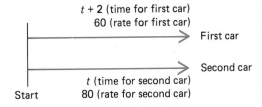

When the second car catches up to the first car, both cars will have traveled the same distance; hence,

$$\begin{pmatrix} \text{distance first} \\ \text{car travels} \end{pmatrix} = \begin{pmatrix} \text{distance second} \\ \text{car travels} \end{pmatrix}$$

$$\begin{pmatrix} \text{rate of} \\ \text{first car} \end{pmatrix}\begin{pmatrix} \text{time for} \\ \text{first car} \end{pmatrix} = \begin{pmatrix} \text{rate of} \\ \text{second car} \end{pmatrix}\begin{pmatrix} \text{time for} \\ \text{second car} \end{pmatrix}$$

$$60 \quad (t + 2) \quad = \quad 80 \quad t$$

$$60t + 120 = 80t$$

$$-20t = -120$$

$$t = \frac{-120}{-20} = 6 \, \text{hr}$$

PROBLEM 23 Find the total amount of time to complete the job in Problem 22 above if the newer press is brought on the job 5 min after the first press starts and both continue until the job is finished.

ANSWERS TO
MATCHED
PROBLEMS

20. $r \cdot 4 = 10$, $r = 2.5$ gal per min

21. $800t = 2,600$, $t = 3.25$ hr

22. $50t + 75t = 1,500$, $t = 12$ min

23. $50t + 75(t - 5) = 1,500$, $t = 15$ min

EXERCISE 3.7

In each problem set up an equation and solve.

A   **1.**   A car traveling at an average rate of 48 mph will take how long to go 156 miles?

2.   If a typist can type 76 words per min, how long will it take to type 1,520 words?

3.   If a water pump on a boat can pump at the rate of 12 gal per min, how long will it take to pump 30 gal?

4.   If the labor costs for a painter are $480, how long did the painter spend on the job if he or she receives $8 per hour?

5.   If a person earns $220 for a 40-hr week, what is the rate per hour?

6.   If a typist can type 2,400 words in 50 min, what is the rate per minute?

7.   If a car travels 550 km in 5.5 hr, what is its rate?

8.   If a printing press can print the evening run of 12,000 papers in 2.5 hr, what is its rate?

9.   Two cars leave New York at the same time and travel in opposite directions. If one travels at 55 mph and the other at 50 mph, how long will it take them to be 630 miles apart?

10.   Two airplanes leave Atlanta at the same time and fly in opposite directions. If one travels at 600 km/hr and the other at 500 km/hr, how long will it take them to be 3,850 km apart?

B   **11.**   If one machine can fill and cap 20 bottles per min, and another machine can do 30, how long will it take both together to complete a 30,000-bottle order?

12.   Two people volunteer to fold and stuff envelopes for a political campaign. If one person can produce 6 per minute and the other 8, how long will it take both together to prepare 1,267 envelopes?

13.   The distance between towns A and B is 630 km. If a passenger train leaves town A and travels toward B at 100 km/hr at the same time a freight train leaves town B and travels toward A at 40 km/hr, how long will it take the two trains to meet?

14. Repeat Problem 13 using 750 km for the distance between the two towns, 90 km/hr as the rate for the passenger train, and 35 km/hr as the rate for the freight train.

15. A car leaves town traveling at 45 mph. How long will it take a second car traveling at 50 mph to catch up to the first car if it leaves 1 hr later?

16. Repeat Problem 15 if the second car leaves 2 hr later and travels at 55 mph.

17. Find the total amount of time to complete the job in Problem 11 if the second (faster) machine is brought on the job 1 hr after the first machine starts and both continue until the job is finished.

18. Find the total amount of time to complete the job in Problem 12 if the second (faster) person is brought on the job 28 min after the first person starts and both continue until the job is finished.

19. A research chemist charges $20 per hr for his services and $8 per hr for his assistant. On a given job a customer received a bill for $1,040. If the chemist worked 10 hr more on the job than his assistant, how much time did each spend?

20. A contractor has just finished constructing a swimming pool in your back yard, and you have turned on the large water valve to fill it. The large valve lets water into the pool at a rate of 60 gal per min. After 2 hr you get impatient and turn on the garden hose which lets water in at 15 gal per min. If the swimming pool holds 30,000 gal of water, what will be the total time required to fill it?

C **21.** An earthquake emits a primary wave and a secondary wave. Near the surface of the earth the primary wave travels at about 5 miles per sec, and the secondary wave at about 3 miles per sec. From the time lag between the two waves arriving at a given seismic station, it is possible to estimate the distance to the quake. (The "epicenter" can be located by getting distance bearings at three or more stations.) Suppose a station measured a time difference of 12 sec between the arrival of the two waves. How far would the earthquake be from the station?

22. A skydiver free falls (because of air resistance) at about 176 fps or 120 mph (Figure 4); with parachute open the rate of fall is about 22 fps or 15 mph. If the skydiver opened the chute halfway down and the total time for the descent was 6 min, how high was the plane when the skydiver jumped?

Figure 4

**3.8 APPLICATIONS:
MIXTURE PROBLEMS**

A variety of applications can be classified as mixture problems. Even though the problems come from different areas, their mathematical treatment is essentially the same. Let us start with an example you may recall from the last chapter.

EXAMPLE 24 Suppose you have 40 coins consisting only of nickels and quarters in your pocket that are worth $4. How many of each type of coin do you have?

Solution Let

x = the *number* of nickels

then

$40 - x$ = the *number* of quarters

$$\underbrace{\begin{pmatrix} \text{value of} \\ \text{nickels} \\ \text{in cents} \end{pmatrix} + \begin{pmatrix} \text{value of} \\ \text{quarters} \\ \text{in cents} \end{pmatrix}}_{\textit{Value before mixing}} = \underbrace{\begin{pmatrix} \text{total value} \\ \text{of mixture} \\ \text{in cents} \end{pmatrix}}_{\textit{Value after mixing}}$$

$$5x + 25(40 - x) = 400 \quad \text{(not 4.00)}$$
$$5x + 1{,}000 - 25x = 400$$
$$-20x = -600$$
$$x = 30 \quad \text{(nickels)}$$
$$40 - x = 10 \quad \text{(quarters)}$$

PROBLEM 24 A noon concert brought in $2,000 on the sale of 1,500 tickets. If tickets sold for $1 and $2, how many of each were sold? Set up an equation and solve.

We now consider mixture problems involving percent. Recall from arithmetic that 12 percent in decimal form is 0.12, and 3.5 percent is 0.035, and so on. (Appendix A.3 has a brief review of percent for those who need it.)

EXAMPLE 25 How many centiliters (cl) of pure alcohol must be added to 35 cl of a 20% solution to obtain a 30% solution?

Solution Let x = the number of centiliters of pure alcohol to be added. Let us illustrate the situation before and after mixing, as we did in Example 24. The amount of alcohol present before mixing must equal the amount of alcohol present after mixing.

Before Mixing			After Mixing
20% solution	Pure alcohol		30% solution

35 cl + x cl = $(x+35)$ cl

| (Amount of pure alcohol in first solution) | + | (Amount of pure alcohol in second solution) | = | (Amount of pure alcohol in mixture) |

$$0.2(35) \quad + \quad x \quad = \quad 0.3(x+35)$$
$$7 + x \quad = \quad 0.3x + 10.5$$
$$0.7x \quad = \quad 3.5$$
$$x \quad = \quad 5 \text{ cl}$$

PROBLEM 25 How many centiliters of distilled water must be added to 80 cl of a 60% acid solution to obtain a 50% acid solution? Set up an equation and solve.

EXAMPLE 26 A chemical storeroom has a 40% acid solution and an 80% solution. How many deciliters (dl) must be used from each to obtain 12 dl of a 50% solution?

Solution We proceed as in Example 25.

Let $\qquad x =$ amount of 40% solution used

Then, $\qquad 12 - x =$ amount of 80% solution used

Before Mixing			After Mixing
40% solution	80% solution		50% solution

x dl + $(12-x)$dl = 12 dl

| (Amount of pure acid in first solution) | + | (Amount of pure acid in second solution) | = | (Amount of pure acid in mixture) |

$$0.4x \quad + \quad 0.8(12-x) \quad = \quad 0.5(12)$$
$$0.4x + 9.6 - 0.8x \quad = \quad 6$$
$$-0.4x \quad = \quad -3.6$$
$$x \quad = \quad 9 \text{ dl of 40\% solution}$$
$$12 - x \quad = \quad 3 \text{ dl of 80\% solution}$$

PROBLEM 26   A coffee shop wishes to blend coffee that sells for \$3 per pound with coffee that sells for \$4.25 per pound to produce a blend selling for \$3.50 per pound. How much of each should be used to produce 50 lb of the new blend? Set up an equation and solve. (This problem is mathematically very close to Example 26.)

ANSWERS TO MATCHED PROBLEMS

24. $1x + 2(1,500 - x) = 2,000$; $x = 1,000$ (\$1 tickets) and $1,500 - x = 500$ (\$2 tickets)

25. $0.6(80) + 0.0x = 0.5(x + 80)$; $x = 16$ cl

26. $3x + 4.25(50 - x) = 3.5(50)$; $x = 30$ lb (\$3 coffee) and $50 - x = 20$ lb (\$4.25 coffee)

EXERCISE 3.8

A

1. A parking meter takes only nickels and dimes. If it contains 50 coins at a total value of \$3.50, how many of each type of coin is in the meter?

2. An all-day parking meter takes only dimes and quarters. If it contains 100 coins at a total value of \$14.50, how many of each type of coin is in the meter?

3. A jazz concert brought in \$60,000 on the sale of 8,000 tickets. If the tickets sold for \$6 and \$10 each, how many of each type was sold?

4. A student production brought in \$7,000 on the sale of 3,000 tickets. If tickets sold for \$2 and \$3 each, how many of each type was sold?

B

5. How many deciliters (dl) of pure alcohol must be added to 12 dl of a 30% solution to obtain a 40% solution?

6. How many milliliters (ml) of pure acid must be added to 100 ml of a 40% acid solution to obtain a 50% acid solution?

7. How many milliliters of distilled water must be added to 500 ml of a solution of 60% acid to obtain a 50% solution?

8. How many centiliters (cl) of distilled water must be added to 140 cl of a 80% acid solution to obtain a 70% acid solution?

9. A chemist has two solutions in a stockroom—one is 30% acid and the other is 70% acid. How many milliliters of each should be mixed to obtain 100 ml of a 40% solution?

10. A chemical stockroom has a 20% acid solution and a 50% acid solution. How many centiliters must be taken from each to obtain 90 cl of a 30% solution?

11. A coffee and tea shop wishes to blend a \$3.50-per-pound coffee with a \$4.75-per-pound coffee in order to produce a blend selling for \$4 per pound. How much of each would have to be used to produce 100 lb of the new blend?

12. A coffee and tea shop wishes to blend a $2.50-per-pound tea with a $3.25-per-pound tea to produce a blend selling for $3 per pound. How much of each should be used to produce 75 lb of the new blend?

13. You have just inherited $10,000 and wish to invest part at 8 percent interest and the rest at 12 percent interest. How much should be invested at each rate in order to produce the same yield as if you had invested it all at 9 percent?

14. An investor has $20,000 to invest. If part is to be invested at 8 percent and the rest at 12 percent, how much should be invested at each rate to yield the same amount as if all had been invested at 11 percent?

C   **15.** A 10-liter radiator contains a 60% solution of antifreeze in distilled water. How much should be drained and replaced with pure antifreeze to obtain an 80% solution?

16. A 3-gal radiatior contains a 50% solution of antifreeze in distilled water. How much should be drained and replaced with pure antifreeze to obtain a 70% solution?

17. It is known that a carton contains 100 packages and that some of the packages weigh $\frac{1}{2}$ lb each and the rest weigh $\frac{1}{3}$ lb each. To save time counting each type of package in the carton, you can weigh the whole contents of the box (45 lb) and determine the number of each kind of package by use of algebra. How many are there of each kind?

3.9 SUPPLEMENTAL APPLICATIONS (OPTIONAL)

Now that you have had experience in solving several specific types of word problems, we present a supplemental exercise set with a wide variety of real-world applications from several fields. Some are easy and others are more difficult to solve. The more difficult problems are starred with either a double or single star.

It is again worthwhile noting that with less than three chapters of algebra behind you, you are in a position to solve a fair number of practical problems. In the following chapters we will increase our problem-solving power even more.

If you are having difficulty in solving word problems (most people do at first), do not become discouraged. Keep working on the easier problems, then gradually work up to the more difficult problems.

EXERCISE 3.9

This supplemental set of exercises contains a variety of applications arranged according to subject area. The more difficult problems are marked with two stars (**), the moderately difficult problems with one star (*) and the easier problems are not marked.

BUSINESS

1. If an IBM electronic card sorter can sort 1,250 cards in 5 min, how long will it take the card sorter to sort 11,250 cards?

2. A person borrowed a sum of money from a lending group at 12 percent simple interest. At the end of 2.5 years the loan was cleared by paying $520. How much was originally borrowed? HINT: $A = P + Prt$, where A is the amount repaid, P the amount borrowed, r is the interest rate expressed as a decimal, and t is time in years.

3. If you paid $168 for a camera after receiving a 20 percent discount, what was the original price of the camera?

4. The retail price of a record is $2.80. The markup on the cost is 40 percent. What did the store pay for the record? HINT: Cost + markup = retail. If C = cost, then markup = $0.4C$.

DOMESTIC

5. A father, in order to encourage his daughter to do better in algebra, agrees to pay her 25 cents for each problem she gets right on a test and to fine her 10 cents for each problem she misses. On a test with 20 problems she received $3.60. How many problems did she get right?

6. A car rental company charges $5 per day and 5 cents per mile. If a car was rented for two days, how far was it driven if the total rental bill was $30?

**7. The cruising speed of an airplane is 150 mph (relative to ground). You wish to hire the plane for a 3-hr sightseeing trip. You instruct the pilot to fly north as far as possible and still return to the airport at the end of the allotted time.

(A) How far north should the pilot fly if there is a 30-mph wind blowing from the north?

(B) How far north should the pilot fly if there is no wind blowing?

CHEMISTRY

8. Two-thirds of a standard bar of silver (Ag) balances exactly with one-half of a standard bar of silver and a $\frac{1}{4}$-kg weight (Figure 5). How much does one whole standard bar of silver weigh?

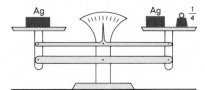

Figure 5

9. In the study of gases there is a simple law called Boyle's law that expresses a relationship between volume and pressure. It states that the product of the pressure and volume, as these quantities change and all other variables are held fixed, remains constant. Stated as a formula, $P_1V_1 = P_2V_2$. If 500 cc of air at 70-cm pressure were converted to 100-cm pressure, what volume would it have?

LIFE SCIENCE

10. A fairly good approximation for the normal weight of a person over 60 in (5 ft) tall is given by the formula $w = 5.5h - 220$, where h is height in inches and w is weight in pounds. How tall should a 121-lb person be?

*11. A wildlife management team estimated the number of bears in a national forest by the popular capture-mark-recapture technique. Using live traps (Figure 6), they captured and marked 30 bears, then released them. After a period for mixing, they captured another 30 and found 5 marked among them. Assuming the ratio of the total bear population to the bears marked in the first sample is the same as the ratio of bears in the second sample to those found marked in the second sample, estimate the bear population in the forest.

Figure 6

EARTH SCIENCE

12. About $\frac{1}{9}$ of the height of an iceberg is above water. If 20 meters are observed above water, what is the total height of the iceberg?

13. Pressure in sea water increases by 1 atmosphere (15 lb per sq in) for each 33 ft of depth; it is 15 lb per sq in at the surface. Thus, $p = 15 + 15(d/33)$ where p is the pressure in pounds per square foot at a depth of d ft below the surface. How deep is a diver if he or she observes that the pressure is 165 lb per sq in?

**14. As dry air moves upward, it expands and in so doing cools at the rate of about 5.5°F for each 1,000 ft in rise. This ascent is known as the "adiabatic process." If the ground temperature is 80°F, write an equation that relates temperature T with altitude h (in feet). How high is an airplane if the pilot observes that the temperature is 25°F?

MUSIC

*15. Starting with a string tuned to a given note, one can move up and down the scale simply by decreasing or increasing its length (while maintaining the same tension) according to simple whole number ratios (see Figure 7). Chords can also be formed by taking two strings whose lengths form ratios involving certain whole numbers. Find the lengths of 7 strings (each less than 30 in) that will produce the following seven chords when paired with a 30-in string:

(A) Octave   1:2   (B) Fifth   2:3
(C) Fourth   3:4   (D) Major third   4:5
(E) Minor third   5:6   (F) Major sixth   3:5
(G) Minor sixth   5:8

	C	D	E	F	G	A	B	C	D	E	F	G	A	B	C
Relative string length	2	$\frac{16}{9}$	$\frac{8}{5}$	$\frac{3}{2}$	$\frac{4}{3}$	$\frac{6}{5}$	$\frac{16}{15}$	1	$\frac{8}{9}$	$\frac{4}{5}$	$\frac{3}{4}$	$\frac{2}{3}$	$\frac{3}{5}$	$\frac{8}{15}$	$\frac{1}{2}$
Scale ratios (proportional to frequencies)	$\frac{1}{2}$	$\frac{9}{16}$	$\frac{5}{8}$	$\frac{2}{3}$	$\frac{3}{4}$	$\frac{5}{6}$	$\frac{15}{16}$	1	$\frac{9}{8}$	$\frac{5}{4}$	$\frac{4}{3}$	$\frac{3}{2}$	$\frac{5}{3}$	$\frac{15}{8}$	2
Frequencies	132	149	165	176	198	220	248	264	297	330	352	396	440	495	528

Figure 7
Diatonic scale.

PHYSICS—ENGINEERING

16. An important problem in physics and engineering is the lever problem (Figure 8). In order for the system to be balanced (not move), the product of the force and distance on one side must equal the product of the force and distance on the other. If a person has a 1-meter wrecking bar and places a fulcrum 10 cm from one end, how much can be lifted with a force of 25 kg on the long end?

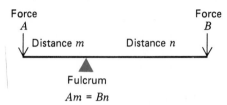

Force A   Force B

Distance m   Distance n

Fulcrum

$Am = Bn$

Figure 8
Lever system.

17. How heavy a rock can be moved with an 8-ft steel rod if the fulcrum is placed 1 ft from the rock end and a force of 100 lb is exerted on the other end (see Figure 9).

Figure 9

***18.** How far would a fulcrum have to be placed from an end with a 65-kg weight to balance 85 kg on the other end (see Figure 8), if the bar is 3 meters long?

***19.** In 1849, during a celebrated experiment, the Frenchman Fizeau made the first accurate approximation of the speed of light. By using a rotating disc with notches equally spaced on the circumference and a reflecting mirror 5 miles away (Figure 10), he was able to measure the elapsed time for the light traveling to the mirror and back. Calculate his estimate for the speed of light (in miles per second) if his measurement for the elapsed time was $\frac{1}{20,000}$ sec?

Figure 10

PSYCHOLOGY

20. Psychologists define IQ (Intelligence Quotient) as

$$IQ = \frac{\text{mental age}}{\text{chronological age}} \times 100$$

$$= \frac{MA}{CA} \times 100$$

If a person has an IQ of 120 and a mental age of 18 years, what is the person's chronological (actual) age?

21. In 1948 Professor Brown, a psychologist, trained a group of rats (in an experiment on motivation) to run down a narrow passage in a cage to receive food in a goal box. He then put a harness on each rat and connected it to an overhead wire that was attached to a scale (Figure 11). In this way he could place the rat at different distances from the food and measure the pull (in grams) of the rat toward the food. He found that a relation between motivation (pull) and position was given approximately by the equation

$$p = -\tfrac{1}{5}d + 70 \qquad 30 \leq d \leq 175$$

where pull p is measured in grams and distance d is measured in centimeters. If the pull registered was 40 grams, how far was the rat from the goal box?

Figure 11

PUZZLES

*22. Diophantus, an early Greek algebraist (A.D. 280), was the subject for a famous ancient puzzle. See if you can find Diophantus' age at death from the following information: Diophantus was a boy for one-sixth of his life; after one-twelfth more he grew a beard; after one-seventh more he married, and after 5 years of marriage he was granted a son; the son lived one-half as long as his father; and Diophantus died 4 years after his son's death.

**23. A classic problem is the courier problem. If a column of soldiers 3 miles long is marching at 5 mph, how long will it take a courier on a motorcycle traveling at 25 mph to deliver a message from the end of the column to the front and then return?

**24. After 12:00 noon exactly, what time will the hands of a clock be together again?

3.10 Chapter Review: Important Terms and Symbols, Review Exercise, Practice Tests

Important terms and symbols

rational number (*3.1*)   rational number multiplication (*3.2*)   lowest terms (*3.2*)   higher terms (*3.2*)   rational number division (*3.2*)   rational number addition (*3.3*) least common denominator, LCD (*3.3*)   rational number subtraction (*3.3*)   word-problem strategy (*3.5*)   ratio (*3.6*)   proportion (*3.6*)   metric conversion (*3.6*)

Exercise 3.10
Review exercise

All variables represent nonzero integers.

A   1.   Graph $\{-\frac{7}{4}, -\frac{3}{4}, \frac{3}{2}\}$ on a number line.

Perform the indicated operations and reduce to lowest terms.

2. $\dfrac{3}{2y} \cdot \dfrac{5x}{4}$

3. $\dfrac{3}{2y} \div \dfrac{5x}{4}$

4. $\dfrac{y}{2} + \dfrac{y}{3}$

5. $\dfrac{3}{2y} - \dfrac{5x}{4}$

Solve.

6. $6x = 5$

7. $3x - 5 = 5x - 8$

8. $\dfrac{y}{8} = \dfrac{3}{4}$

9. $0.7x = 4.2$

10. $\dfrac{x}{2} - \dfrac{1}{3} = \dfrac{x}{6}$

11.   Three-tenths of what number is $\frac{2}{5}$? (A) Write an equation and (B) solve.

12.   If the width of a rectangle with perimeter 80 cm is three-fifths the length, what are the dimensions of the rectangle?

13.   How much pure alcohol must be added to 30 ml of a 60% solution to obtain a 70% solution? Set up an equation and solve.

B   *Perform the indicated operations and reduce to lowest terms.*

14. $\dfrac{3y}{5xz} \cdot \dfrac{10z}{15xy}$

15. $\dfrac{3y}{5xy} \div \dfrac{10z}{15xy}$

16. $\dfrac{3}{4xy^2} - \dfrac{1}{3x^2y}$

17. $\dfrac{3}{x^2} - \dfrac{2}{x} + 1$

18. $\dfrac{1}{4y} + \dfrac{3}{2z} - \dfrac{1}{3x} - 2$

19. $\dfrac{-4}{9} - \dfrac{35}{18} - \dfrac{-10}{3}$

Solve.

20. $-\dfrac{3}{5}y = \dfrac{2}{3}$

21. $\dfrac{x}{4} - \dfrac{x-3}{3} = 2$

22. $0.4x - 0.3(x - 3) = 5$

23. If the ratio of all the trout in a lake to the ones that had been captured, marked, and released is $20:3$ and there are 450 marked trout, how many trout are in the lake? Set up an equation and solve.

24. If there are 2.54 centimeters in 1 inch, how many inches are in 40 centimeters? Set up a proportion and solve.

25. Two boats leave from opposite ports along the same shipping route, which is 2,800 miles long. If one boat travels at 22 mph and the other at 13 mph, how long will it take them to meet? Set up an equation and solve.

26. If one car leaves town traveling 48 mph, how long will it take a second car traveling at 54 mph to catch up to the first car, if the second car leaves 1 hr later? Set up an equation and solve.

27. A 40% acid solution and a 70% acid solution are in a stockroom. How many deciliters (dl) of each must be used to obtain 100 dl of a 49% solution? Set up an equation and solve.

C *Perform the indicated operations and reduce to lowest terms.*

28. $\dfrac{10x}{9y} \div \left(\dfrac{15xy}{2z} \div \dfrac{3y^2}{-z} \right)$

29. $\dfrac{3}{10x^2} - \dfrac{2}{15xy} + \dfrac{5}{18y^2}$

30. Solve $\dfrac{x + 3}{10} - \dfrac{x - 2}{15} = \dfrac{3 - x}{6} - 1$

31. If one printing press can print 90 leaflets per min and a newer press can print 110, how long will it take to print 6,000 leaflets if the newer press is brought on the job 20 min after the first press starts and both continue until finished? Set up an equation and solve.

32. A radiator with a capacity of 12 liters contains a 40% solution of antifreeze in distilled water. How much should be drained and replaced with pure antifreeze to bring the level up to 50 percent? Set up an equation and solve.

Practice test
Chapter 3

Perform the indicated operations and reduce to lowest terms.

1. (A) $\dfrac{6x^2}{9y^2} \cdot \dfrac{3y}{4x}$; (B) $\dfrac{6x^2}{9y^2} \div \dfrac{3y}{4x}$

2. $\dfrac{1}{2} - \dfrac{2}{3} + \dfrac{-5}{6}$

3. $\dfrac{2}{9xy^2} - \dfrac{5}{12x^2}$

4. $\dfrac{-6xz}{2y} \div \left(\dfrac{3z^2}{2y^2} \cdot \dfrac{-xy}{z} \right)$

Solve.

5. $\dfrac{x}{6} = \dfrac{5}{9}$

6. $\dfrac{x}{5} - 2 = \dfrac{x}{7}$

7. $\dfrac{x}{3} - \dfrac{3x - 3}{2} = -9$

8. $0.2(x + 5) - 0.3x = 12$

9. Find a number such that one-half of the number is 5 less than two-thirds of the number. Set up an equation and solve.

10. A stockroom has a 60% alcohol solution and an 80% alcohol solution. How many milliliters of each should be used to obtain 40 ml of a 75% solution? Set up an equation and solve.

11. If there are 1.61 kilometers in 1 mile, how many miles are in 30 kilometers? Set up a proportion and solve.

12. One pipe can fill a tank at 8 gal per min and a larger pipe at 12 gal per min. If the tank holds 1,000 gal, and the second pipe is opened 50 min after the first is opened, how long will it take to fill the tank? Set up an equation and solve.

4

GRAPHING AND LINEAR FORMS

4.1 REAL NUMBERS, INEQUALITY STATEMENTS, AND LINE GRAPHS

In this section we will extend the rational numbers to the real numbers and introduce inequality forms and their graphs on number lines.

The set of real numbers

By extending the integers to the rational numbers, we significantly increased our ability to perform certain operations, to solve equations, and to attack practical problems. It appears that the rational numbers are capable of satisfying all of our number needs. Can we stop here or do we need to go further?

Suppose we wish to find the length of the side of a square with area 2. We then have

and

$$x^2 = 2$$

We ask, are there any rational numbers whose square is 2? It turns out that one can prove that there is no rational number whose square is 2. If a square of area 2 is to have a number that represents the length of a side, then we must invent a new kind of number. This new kind of number is called an **irrational number.** In this case a number whose square is 2 is called a square root of 2 and is symbolized by $\sqrt{2}$.

When associating rational numbers with points on a number line it may appear, since between any two points with a rational number as a label one can always find another point with a rational number as a label, that the rational numbers name all of the points. This, of course, is false, since the side of the square above corresponds to a point on the number line that cannot be labeled with a rational number. It may surprise you to learn that in a certain sense there are more points that have not been named by rational numbers than have been named.

To make a long story short, all the points on a number line that have not been named by rational numbers are named by irrational numbers; that is, using the two sets of numbers, every point on the line will be named and no numbers will be left over.

The set of rational and irrational numbers form the **real number system,** and this is the number system in which most of you will

TABLE 1
The set of real numbers.

Symbol	Number set	Description	Examples
N	Natural numbers	Counting numbers (positive integers)	1, 3, 3,525
I	Integers	Set of counting numbers, their opposites, and 0	-31, -1, 0, 4, 702
Q	Rationals	Any number that can be represented in the form $\frac{a}{b}$, $b \neq 0$, where a and b are integers	-4, $-\frac{3}{5}$, 0, $\frac{2}{3}$, 3.57
R	Reals	Set of all rational and irrational numbers	$-\sqrt{2}$, $\frac{-4}{7}$, 0, 62.48, $\sqrt{5}$, π, 3,407

NOTE: Each set in the left-hand column is a subset of every set below it.

operate most of the time. Table 1 compares the various sets of numbers we have discussed, and Figure 1 shows some of these numbers on a **real number line.**

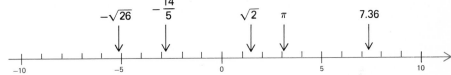

Figure 1
A real number line

Every rational number has a repeating decimal representation. For example,

$$5 = 5.000 \cdots = 5.\overline{0}$$

$$3.14 = 3.1400 \cdots = 3.14\overline{0}$$

$$\tfrac{4}{3} = 1.33 \cdots = 1.\overline{3}$$

$$\tfrac{5}{7} = 0.\overline{714285}$$

$$\tfrac{71}{330} = 0.21515 \cdots = 0.2\overline{15}$$

where the overbar indicates the block of numbers that continues to repeat indefinitely.

Every irrational number, when represented in decimal form, has an infinite nonrepeating decimal representation. For example,

$$\sqrt{2} = 1.4142135 \cdots$$

$$\pi = 3.1415926 \cdots$$

$$-\sqrt{6} = -2.4494897 \cdots$$

and no block of numbers will continue to repeat.

If we write

$$1.414 \approx \sqrt{2}$$

$$3.412 \approx \pi$$

where $\approx$ means approximately equal to, then we are using rational number approximations for the irrational numbers $\sqrt{2}$ and π.

This is enough on irrational numbers for now; we will return to them in more detail in Chapter 7. Basic properties of the real number system can be found inside the front cover of the text. Let us now turn to inequality statements and their graphs on a real number line.

Inequality symbols

In Chapter 1 we discussed the equal sign and some of its important properties. In this and the next section we will introduce and discuss the concept of inequality. This relation involves "less than" and "greater than."

Just as we use "=" to replace the words "is equal to," we will use the **inequality symbols** "<" and ">" to represent "is less than" and "is greater than," respectively. Thus, we can write the following equivalent forms:

$a < b$	a is less than b
$a > b$	a is greater than b
$a \leq b$	a is less than or equal to b
$a \geq b$	a is greater than or equal to b

Note that the small end (the point) of the inequality symbol is directed toward the smaller of the two numbers.

It no doubt seems obvious to you that

$$3 < 7$$

is a true statement. But does it seem equally obvious that

$$-8 < -1 \qquad -15 < 1 \qquad -2 > -10,000$$

are also true statements? To make the inequality relation precise so that we can interpret it relative to *all* real numbers, we need a careful definition of the concept.

Definition of $<$ and $>$

If a and b are real numbers, then we write

$$a < b$$

if there exists a positive real number p such that $a + p = b$. We write

$$c > d$$

if there exists a positive real number q such that $c - q = d$.

Certainly, one would expect that if a positive number were added to *any* real number it would make it larger and if it were subtracted from *any* real number it would make it smaller. That is essentially what the definition states. Note that if $a > b$, then it follows that $b < a$, and vice versa.

EXAMPLE 1
 (**A**) $2 < 3$ since $2 + 1 = 3$

(**B**) $-8 < -1$ since $-8 + 7 = -1$

(**C**) $0 > -5$ since $0 - 5 = -5$

(**D**) $-2 > -10{,}000$ since $-2 - 9{,}998 = -10{,}000$

PROBLEM 1
Replace each question mark with $<$ or $>$:

(**A**) $4\,?\,6$ (**B**) $6\,?\,4$ (**C**) $-6\,?\,-4$

(**D**) $-9\,?\,9$ (**E**) $3\,?\,-9$ (**F**) $0\,?\,-14$

Inequality statements and the real number line

The inequality symbols have a very clear geometric interpretation on the real number line. If $a < b$, then a is to the left of b; if $c > d$, then c is to the right of d (Figure 2).

Figure 2
$a < b,\ c > d$

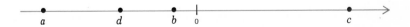

Now let us turn to simple inequality statements of the form

$$x \geq -3 \qquad -2 < x \leq 3$$

To solve such inequality statements is to find the set of all replacements of the variable x (from some specified set of numbers) that makes the inequality true. This set is called the **solution set** for the inequality. To **graph** an inequality statement on a real number line is to graph its solution set.

EXAMPLE 2   Graph each inequality statement on a real number line.

(A) $x \geq -3$ x an integer
(B) $x \geq -3$ x a real number

Solution   (A)   The solution set for

$x \geq -3$ x an integer

is the set of all integers greater than or equal to -3. Graphically, the general solution can be represented as follows:

(B)   The solution set for

$x \geq -3$ x a real number

is the set of *all* real numbers greater than or equal to -3. Graphically, this includes *all* the points from -3 to the right—a solid line:

PROBLEM 2   Graph each inequality statement on a real number line.

(A) $x \leq 2$ x an integer
(B) $x \leq 2$ x a real number

EXAMPLE 3   Graph each inequality statement on a real number line.

(A) $-2 < x \leq 3$ x an integer
(B) $-2 < x \leq 3$ x a real number

Solution   (A)   The double inequality

$-2 < x \leq 3$ x an integer

is a short way of writing

$-2 < x$ and $x \leq 3$ x an integer

which means that x is greater than -2 and at the same time x is less than or equal to 3. In other words, x can be any integer between -2 and 3, excluding

−2, but including 3.† This is the set {−1, 0, 1, 2, 3}, which is graphed as follows:

(B) The solution set for

$$-2 < x \leq 3 \qquad x \text{ a real number}$$

is the set of *all* real numbers between −2 and 3, excluding −2, but including 3. The graph is a solid line excluding the endpoint −2 and including the endpoint 3. Note that a hollow circle indicates that an endpoint is not included and a solid circle indicates that an endpoint is included.

PROBLEM 3   Graph each inequality statement on a real number line.

    (A) $-4 \leq x < 2$ x an integer
    (B) $-4 \leq x < 2$ x a real number

ANSWERS TO   **1.** (A) <;   (B) >;   (C) <;   (D) <;   (E) >;   (F) >
MATCHED   **2.** (A)
PROBLEMS

    (B)

    3. (A)

    (B)

EXERCISE 4.1   **A** *Indicate whether true (T) or false (F).*

    1. 5 is a natural number.
    2. $\frac{2}{3}$ is a rational number.
    3. −3 is an integer.
    4. $\frac{3}{4}$ is a natural number.
    5. 0 is an integer.
    6. $-\frac{2}{3}$ is an integer.
    7. $\sqrt{2}$ is an irrational number.
    8. π is an irrational number.

†We do not use double inequality forms where the inequality symbols point away from each other or toward each other. Study the following forms carefully to see why:

$$2 < x > -1 \qquad -2 < x > 5 \qquad 5 > x < 8$$

9. $\sqrt{5}$ is a real number. **10.** 7 is a real number.

11. -5 is a real number. **12.** $\frac{3}{7}$ is a real number.

Replace each question mark with $<$ or $>$.

13. 7 ? 5	**14.** 3 ? 6	**15.** 5 ? 7
16. 6 ? 3	**17.** -7 ? -5	**18.** -3 ? -6
19. -5 ? -7	**20.** -6 ? -3	**21.** 0 ? 8
22. 5 ? 0	**23.** 0 ? -8	**24.** -5 ? 0
25. -7 ? 5	**26.** -6 ? 3	**27.** -842 ? 0
28. -905 ? -10	**29.** 900 ? $-1,000$	**30.** 505 ? -55

Referring to

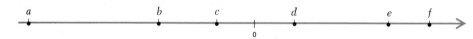

replace each question mark in Problems 31–36 with either $<$ or $>$.

31. a ? d	**32.** e ? a	**33.** b ? a
34. 0 ? d	**35.** e ? f	**36.** d ? e

B **37.** If we add a positive number to any real number, will the sum be to the right or left of the original number on a real number line?

38. If we subtract a positive number from any real number, will the difference be to the right or left of the original number on a real number line?

Graph each inequality statement on a real number line for (A) x an integer and (B) x a real number.

39. $x > -4$	**40.** $x < 1$	**41.** $x \leq -1$
42. $x \geq -1$	**43.** $-3 < x \leq 2$	**44.** $-5 \leq x < 0$
45. $-3 < x < -1$	**46.** $-5 < x < -3$	**47.** $-3 \leq x \leq 3$
48. $-1 \leq x \leq 4$		

C *Express each rational number as a repeating decimal. (Divide each denominator into the numerator, and continue until a repeating pattern is observed.)*

49. $\frac{1}{4}$	**50.** $\frac{7}{4}$	**51.** $\frac{23}{9}$
52. $\frac{10}{6}$	**53.** $\frac{7}{13}$	**54.** $\frac{15}{21}$

Graph each set on a real number line. (Recall, I is the set of integers and R is the set of real numbers.)

55. $\{x \in I | -3 < x < 3\}$ **56.** $\{x \in I | -10 < x \leq -5\}$

57. $\{x \in R | -3 < x < 3\}$ **58.** $\{x \in R | -6 \leq x < -2\}$

59. $\{x \in R | \frac{-9}{4} \leq x < \frac{3}{2}\}$ **60.** $\{x \in R | \frac{-15}{8} \leq x \leq \frac{-1}{8}\}$

4.2 SOLVING INEQUALITIES

In Section 4.1 we introduced simple inequality statements of the form

$$x < 5 \qquad -3 \leq x < 4 \qquad x \geq -4$$

with obvious solutions. In this section we will consider inequality statements that do not have obvious solutions. Can you guess the real number solutions for

$$3x - 4 > x - 2$$

By the end of this section you will be able to solve this type of inequality almost as easily as you solved equations of the corresponding type.

When solving equations we made considerable use of the addition, subtraction, multiplication, and division properties of equality. We can use similar properties to help us solve inequalities. We will start with several numerical examples and generalize from these.

1. *Add the same quantity to each side of an inequality:*

<table>
<tr><td align="center">add a positive quantity
to each side</td><td align="center">add a negative quantity
to each side</td></tr>
<tr><td align="center">$-2 < 4$</td><td align="center">$-2 < 4$</td></tr>
<tr><td align="center">$-2 + 3 \qquad 4 + 3$</td><td align="center">$-2 + (-3) \qquad 4 + (-3)$</td></tr>
<tr><td align="center">$1 < 7$</td><td align="center">$-5 < 1$</td></tr>
<tr><td align="center">sense of inequality
remains the same</td><td align="center">sense of inequality
remains the same</td></tr>
</table>

2. *Subtract the same quantity from each side of an inequality:*

<table>
<tr><td align="center">subtract a positive
quantity from each side</td><td align="center">subtract a negative
quantity from each side</td></tr>
<tr><td align="center">$-2 < 4$</td><td align="center">$-2 < 4$</td></tr>
<tr><td align="center">$-2 - 5 \qquad 4 - 5$</td><td align="center">$-2 - (-7) \qquad 4 - (-7)$</td></tr>
<tr><td align="center">$-7 < -1$</td><td align="center">$5 < 11$</td></tr>
<tr><td align="center">sense of inequality
remains the same</td><td align="center">sense of inequality
remains the same</td></tr>
</table>

3. *Multiply each side of an inequality by the same nonzero quantity:*

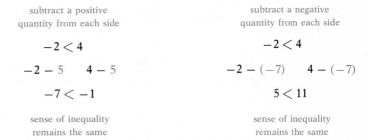

<table>
<tr><td align="center">multiply both sides by
a positive quantity</td><td align="center">multiply both sides by
a negative quantity</td></tr>
<tr><td align="center">$-2 < 4$</td><td align="center">$-2 < 4$</td></tr>
<tr><td align="center">$2(-2) \qquad 2(4)$</td><td align="center">$(-2)(-2) \qquad (-2)(4)$</td></tr>
</table>

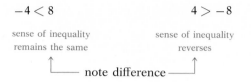

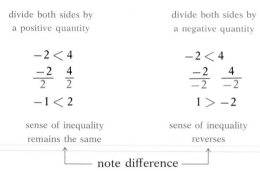

4. *Divide each side of an inequality by the same nonzero quantity:*

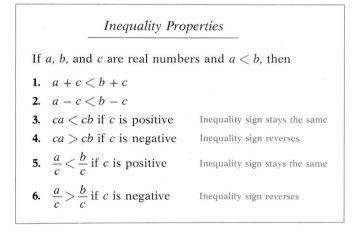

The descriptions above generalize completely, and are summarized as follows without proof:

Inequality Properties

If a, b, and c are real numbers and $a < b$, then

1. $a + c < b + c$

2. $a - c < b - c$

3. $ca < cb$ if c is positive   Inequality sign stays the same

4. $ca > cb$ if c is negative   Inequality sign reverses

5. $\dfrac{a}{c} < \dfrac{b}{c}$ if c is positive   Inequality sign stays the same

6. $\dfrac{a}{c} > \dfrac{b}{c}$ if c is negative   Inequality sign reverses

---

Similar properties hold if each inequality sign is reversed, or if $>$ is replaced with $\geq$ and $<$ is replaced with $\leq$. Thus, we find that we can perform essentially the same operations on inequality statements to produce equivalent statements that we perform on equations to produce equivalent equations with the exception that

The inequality sign reverses if we multiply or divide both sides of an inequality by a negative number.

We will now solve some inequalities using the properties above. Unless otherwise stated, in **solving an inequality** we find *all* real number solutions.

EXAMPLE 4   Solve each inequality.

(A) $x + 3 < -2$

(B) $5x \geq 10$

(C) $-3x - 2 < 7$

(D) $-\dfrac{x}{2} > -3$

Solution   (A) $x + 3 < -2$ Subtract 3 from each side.

$x + 3 - 3 < -2 - 3$ Inequality sign does not reverse.

$x < -5$ Solution set is set of all real numbers less than -5.

(B) $5x \geq 10$ Divide each side by 5.

$\dfrac{5x}{5} \geq \dfrac{10}{5}$ Inequality sign does not reverse, since we divided by a positive number.

$x \geq 2$ Solution set is set of all real numbers greater than or equal to 2.

(C) $-3x - 2 < 7$ Add 2 to each side.

$-3x - 2 + 2 < 7 + 2$ Inequality sign does not reverse.

$-3x < 9$ Divide each side by -3.

$\dfrac{-3x}{-3} > \dfrac{9}{-3}$ Inequality sign reverses, since we divided by a negative number.

$x > -3$ Solution set is set of all real numbers greater than -3.

(D) $-\dfrac{x}{2} > -3$ Multiply each side by -2.

$(-2)\left(-\dfrac{x}{2}\right) < (-2)(-3)$ Inequality sign reverses, since we multiplied by a negative number.

$x < 6$ Solution set is set of all real numbers less than 6.

PROBLEM 4   Solve each inequality

    (A) $x - 3 \geq -5$ **(B)** $-6x < 18$

    (C) $3x + 5 \leq -1$ **(D)** $-\dfrac{x}{3} > -4$

EXAMPLE 5   Solve and graph.

    (A) $3(x - 1) + 5 \leq 5(x + 2)$

    (B) $\dfrac{3x - 2}{2} - 5 > 1 - \dfrac{x}{4}$

Solution   **(A)** $3(x - 1) + 5 \leq 5(x + 2)$ Simplify left and right sides.

               $3x - 3 + 5 \leq 5x + 10$

                 $3x + 2 \leq 5x + 10$ Isolate x on the left side.

                     $3x \leq 5x + 8$

                     $-2x \leq 8$

                     $x \geq -4$ Inequality sign reverses (why?).

    (B) $\dfrac{3x - 2}{2} - 5 > 1 - \dfrac{x}{4}$ Multiply both sides by 4, the LCM of the denominators.

           $4\left[\dfrac{(3x - 2)}{2} - 5\right] > 4\left(1 - \dfrac{x}{4}\right)$ Inequality sign does not reverse (why?).

             $2(3x - 2) - 20 > 4 - x$ Simplify left side.

               $6x - 4 - 20 > 4 - x$

                 $6x - 24 > 4 - x$ Isolate x on the left side.

                    $6x > 28 - x$

                    $7x > 28$

                     $x > 4$ Inequality sign does not reverse (why?).

PROBLEM 5   Solve and graph.

    (A) $2(2x + 3) \geq 6(x - 2) + 10$

    (B) $\dfrac{2x - 3}{3} - 2 > \dfrac{x}{6} - 1$

EXAMPLE 6   Solve and graph: $-8 \leq 3x - 5 < 7$.

Solution   We proceed as above, except we try to isolate x in the middle.

$$-8 \leq 3x - 5 < 7 \qquad \text{Add 5 to each member.}$$

$$-8 + 5 \leq 3x - 5 + 5 < 7 + 5$$

$$-3 \leq 3x < 12 \qquad \text{Divide each member by 3.}$$

$$\frac{-3}{3} \leq \frac{3x}{3} < \frac{12}{3} \qquad \text{Inequality signs do not reverse.}$$

$$-1 \leq x < 4$$

PROBLEM 6   Solve and graph: $-3 < 2x + 3 \leq 9$.

We conclude this section with a word problem and an application.

EXAMPLE 7   What numbers satisfy the condition, "4 more than twice a number is less than or equal to that number"?

Solution   Let $x =$ the number, then

$$2x + 4 \leq x$$

$$x \leq -4$$

PROBLEM 7   What numbers satisfy the condition, "6 less than 3 times a number is greater than or equal to 9"?

EXAMPLE 8   If the temperature for a 24-hr period in Antarctica ranged between $-49°F$ and $14°F$ (that is, $-49 \leq F \leq 14$), what was the range in Celsius degrees? (Recall $F = \frac{9}{5}C + 32$.)

Solution   Since $F = \frac{9}{5}C + 32$, we replace F in $-49 \leq F \leq 14$ with $\frac{9}{5}C + 32$ and solve the double inequality:

$$-49 \leq \tfrac{9}{5}C + 32 \leq 14$$

$$-49 - 32 \leq \tfrac{9}{5}C + 32 - 32 \leq 14 - 32$$

$$-81 \leq \tfrac{9}{5}C \leq -18$$

$$(\tfrac{5}{9})(-81) \le (\tfrac{5}{9})(\tfrac{9}{5}C) \le (\tfrac{5}{9})(-18)$$

$$-45 \le C \le -10$$

PROBLEM 8 Repeat Example 8 for $-31 \le F \le 5$.

ANSWERS TO MATCHED PROBLEMS

4. (A) $x \ge -2$; (B) $x > -3$; (C) $x \le -2$; (D) $x < 12$

5. (A) $x \le 4$

(B) $x > 4$

6. $-3 < x \le 3$

7. $x \ge 5$

8. $-35 \le C \le -15$

EXERCISE 4.2

A *Solve.*

1. $x - 2 > 5$ **2.** $x - 4 < -1$ **3.** $x + 5 < -2$

4. $x + 3 > -4$ **5.** $2x > 8$ **6.** $3x < 6$

7. $-2x \ge 8$ **8.** $-3x \le 6$ **9.** $\dfrac{x}{3} < -7$

10. $\dfrac{x}{5} > -2$ **11.** $\dfrac{x}{-3} \le -7$ **12.** $\dfrac{x}{-5} \ge -2$

13. $3x + 7 < 13$ **14.** $2x - 3 > 5$

15. $-2x + 8 < 4$ **16.** $-4x - 7 > 5$

17. $7x - 8 \le 4x + 7$ **18.** $6m + 2 \le 4m + 6$

19. $4y - 7 \ge 9y + 3$ **20.** $x - 1 \le 4x + 8$

B *Solve and graph.*

21. $3 - (2 + x) > -9$ **22.** $2(1 - x) \ge 5x$

23. $3 - x \ge 5(3 - x)$ **24.** $2(x - 3) + 5 < 5 - x$

25. $3(u - 5) - 2(u + 1) \ge 2(u - 3)$

26. $4(2u - 3) < 2(3u + 1) - (5 - 3u)$

27. $\dfrac{m}{6} - \dfrac{1}{2} > \dfrac{2}{3} + m$ **28.** $\dfrac{x}{5} - 3 < \dfrac{3}{5} - x$

29. $-2 - \dfrac{1 + x}{3} < \dfrac{x}{4}$ **30.** $\dfrac{x}{4} - \dfrac{2x + 1}{6} < -1$

31. $2 < x + 3 < 5$ **32.** $-3 \le x - 5 \le 8$

33. $-4 \le 5x + 6 \le 21$ **34.** $2 < 3x - 7 < 14$

35. $-4 \le \tfrac{9}{5}C + 32 \le 68$ **36.** $-1 \le \tfrac{2}{3}m + 5 \le 11$

C   **37.** $-10 \le \tfrac{5}{9}(F - 32) \le 25$ **38.** $-5 \le \tfrac{5}{9}(F - 32) \le 10$

39. $-3 \le 3 - 2x < 7$ **40.** $-5 < 7 - 4x \le 15$

Applications

In Problems 41 to 48 set up appropriate inequality statements and solve.

41. NUMBER. What numbers satisfy the condition, "3 less than twice the number is greater than or equal to −6"?

42. NUMBER. What numbers satisfy the condition, "5 less than 3 times the number is less than or equal to 4 times the number"?

43. GEOMETRY. If the perimeter of a rectangle with a length of 10 cm must be smaller than 30 cm, how small must the width be?

44. GEOMETRY. If the area of a rectangle of length 10 in must be greater than 65 in² (square inches), how large must the width be?

*45. PHOTOGRAPHY. A photographic developer is to be kept so that its temperature is not below 68°F or above 77°F. What is the range of temperature in Celsius degrees? ($F = \frac{9}{5}C + 32$)

*46. CHEMISTRY. In a chemistry experiment the solution of hydrochloric acid is to be kept so that its temperature is not below 30°C or above 35°C. What would the range of temperature be in Fahrenheit degrees? [$C = \frac{5}{9}(F - 32)$]

*47. PSYCHOLOGY. A person's IQ is found by dividing mental age, as indicated by standard tests, by chronological age and then multiplying this ratio by 100. In terms of a formula,

$$IQ = \frac{MA \cdot 100}{CA}$$

If the IQ range of a group of 12-year-olds is $70 \leq IQ \leq 120$, what is the mental-age range of this group?

*48. BUSINESS. For a business to make a profit it is clear that revenue R must be greater than costs C; in short, a profit will result only if $R > C$. If a company manufactures records and its cost equation for a week is $C = 300 + 1.5x$, where x is the number of records manufactured in a week, and its revenue equation is $R = 2x$, where x is the number of records sold in a week, how many records must be sold for the company to realize a profit?

4.3 CARTESIAN COORDINATE SYSTEM

In Section 4.1 we introduced the real number line where each real number is associated with a point on the line and each point with a real number. In the last section we considered graphs on the real number line. We now move to a plane and develop a system, called a cartesian coordinate system, that will enable us to graph equations and inequalities with two variables instead of just one.

To form a cartesian coordinate system we start with two sets of objects, the set of all points in a plane (a set of geometric objects) and the set of all ordered pairs of real numbers. In a plane select two number lines, one vertical and one horizontal, and let them cross at their respective origins (Figure 3). Up and to the right are the usual choices for the positive directions. These two number lines are called the **vertical axis** and the **horizontal axis** or (together) the **coordinate axes.** The coordinate axes divide the plane into four parts called **quadrants.** The quadrants are numbered counterclockwise from I to IV.

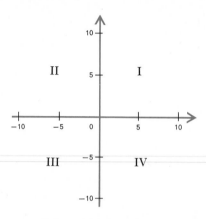

Figure 3

Pick a point P in the plane at random (see Figure 4). Pass horizontal and vertical lines through the point. The vertical line will intersect the horizontal axis at a point with coordinate a, and the horizontal line will intersect the vertical axis at a point with coordinate b. The coordinate of each point of intersection, a and b, respectively, form the **coordinates**

(a, b)

of the point P in the plane. Thus, point Q has coordinates $(5, 10)$ and point R coordinates $(-10, 0)$.

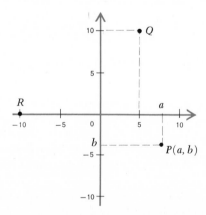

Figure 4

The first number a (called the **abscissa** of P) in the ordered pair (a, b) is the directed distance of the point from the vertical axis (measured on the horizontal scale); the second number b (called the **ordinate** of P) is the directed distance of the point from the horizontal axis (measured on the vertical scale). The abscissa of Q is 5 and the ordinate of Q is 10.

We know from Section 4.1 that coordinates a and b exist for each point in the plane since every point on each axis has a real number associated with it. Hence, by the procedure described, every point in the plane can be labeled with a pair of real numbers. Conversely, by reversing the process, every pair of real numbers can be associated with a point in the plane.

The system that we have just defined to produce this correspondence is called a **cartesian coordinate system** (sometimes referred to as a **rectangular coordinate system**).

EXAMPLE 9   Find the coordinates of each of the points A, B, C, and D.

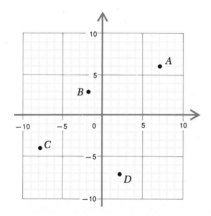

Solution $A(7, 6)$ $B(-2, 3)$ $C(-8, -4)$ $D(2, -7)$

PROBLEM 9   Find the coordinates, using the figure in Example 9, for each of the following points:

(A)   2 units to the right and 1 unit up from A
(B)   2 units to the left and 2 units down from C
(C)   1 unit up and 1 unit to the left of D
(D)   2 units to the right of B.

EXAMPLE 10   Graph (associate each ordered pair of numbers with a point in the cartesian coordinate system):

$(2, 7)$, $(7, 2)$, $(-8, 4)$, $(4, -8)$, $(-8, -4)$, $(-4, -8)$

Solution

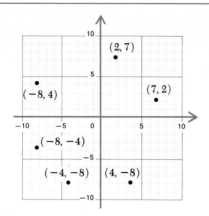

It is very important to note that the ordered pair (2, 7) and the set {2, 7} are not the same thing; {2, 7} = {7, 2}, but (2, 7) ≠ (7, 2).

PROBLEM 10   Graph in the same coordinate system: (3, 4), (−3, 2), (−2, −2), (4, −2), (0, 1), and (−4, 0)

ANSWERS TO   **9.**   (A) (9, 7);   (B) (−10, −6);   (C) (1, −6);   (D) (0, 3)
MATCHED
PROBLEMS   **10.**

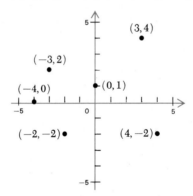

EXERCISE 4.3   **A**   *Write down the coordinates for each labeled point.*

1.

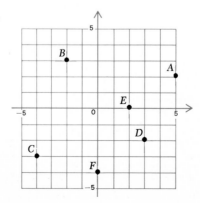

2.

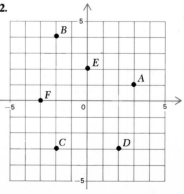

3.

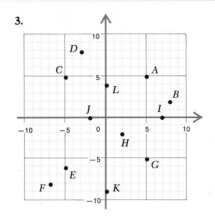

4.

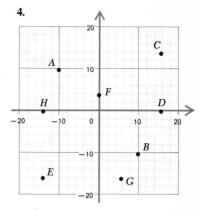

Graph each set of ordered pairs of numbers on the same coordinate system.

5. $(3, 1)$, $(-2, 3)$, $(-5, -1)$, $(2, -1)$, $(4, 0)$, $(0, -5)$

6. $(4, 4)$, $(-4, 1)$, $(-3, -3)$, $(5, -1)$, $(0, 2)$, $(-2, 0)$

7. $(-9, 8)$, $(8, -9)$, $(0, 5)$, $(4, -8)$, $(-3, 0)$, $(7, 7)$, $(-6, -6)$

8. $(2, 7)$, $(7, 2)$, $(-6, 3)$, $(-4, -7)$, $(2, 3)$, $(0, -8)$, $(9, 0)$

B *Write down the coordinates of each labeled point to the nearest quarter of a unit.*

9.

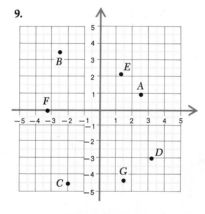

10.

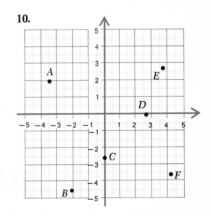

11. Graph the following ordered pairs of numbers on the same coordinate system: $A(1\frac{1}{2}, 3\frac{1}{2})$, $B(-3\frac{1}{4}, 0)$, $C(3, -2\frac{1}{2})$, $D(-4\frac{1}{2}, 1\frac{3}{4})$, and $E(-2\frac{1}{2}, -4\frac{1}{4})$.

12. Graph the following ordered pairs of numbers on the same coordinate system: $A(3\frac{1}{2}, 2\frac{1}{2})$, $B(-4\frac{1}{2}, 3)$, $C(0, -3\frac{3}{4})$, $D(-2\frac{3}{4}, -3\frac{3}{4})$, and $E(4\frac{1}{4}, -3\frac{3}{4})$.

13. Without graphing, tell which quadrants contain the graph of each of the following ordered pairs (see Figure 3):
(A) $(-20, -4)$ (B) $(-3, 22\frac{3}{4})$
(C) $(4, 35,000)$ (D) $(\sqrt{2}, -3)$

14. Without graphing, tell which quadrants contain the graph of each of the following ordered pairs (see Figure 3):

(A) $(-23, 403)$ (B) $(32\frac{1}{2}, -430)$
(C) $(201, 25)$ (D) $(-0.008, -3.2)$

C *What is the abscissa of point A in*

15. Exercise 1 16. Exercise 2
17. Exercise 3 18. Exercise 4

What is the ordinate of point B in

19. Exercise 1 20. Exercise 2
21. Exercise 3 22. Exercise 4

4.4 EQUATIONS AND STRAIGHT LINES

The development of the cartesian coordinate system represented a very important advance in mathematics. It was through the use of this system that René Descartes (1596–1650), a French philosopher-mathematician, was able to transform geometric problems requiring long tedious reasoning into algebraic problems that could be solved almost mechanically. This joining of algebra and geometry has now become known as **analytic geometry.**

Two fundamental problems of analytic geometry are the following:

1. Given an equation, find its graph.
2. Given a geometric figure, such as a straight line, circle, or ellipse, find its equation.

In this course we will be mainly interested in the first problem, with particular emphasis on equations whose graphs are straight lines.
Let us start by trying to find the solution set for

$$y = 2x - 4$$

A **solution** of an equation in two variables is an ordered pair of real numbers that satisfies the equation. If we agree that the first element in the ordered pair will replace x and the second y, then

$$(0, -4)$$

is a solution of $y = 2x - 4$, as can easily be checked. How do we find other solutions? The answer is easy: We simply assign to x in $y = 2x - 4$ any convenient value and solve for y. For example, if $x = 3$, then

$$y = 2(3) - 4$$

$$= 2$$

Hence,

(3, 2)

is another solution of $y = 2x - 4$. It is clear that by proceeding in this manner, we can get solutions to this equation without end. Thus, the solution set is infinite. Let us make up a table of some solutions and graph them in a cartesian coordinate system, identifying the horizontal axis with x and the vertical axis with y.

Choose x	Compute 2x − 4 = y	Write ordered pair (x, y)
−4	2(−4) − 4 = −12	(−4, −12)
−2	2(−2) − 4 = −8	(−2, −8)
0	2(0) − 4 = −4	(0, −4)
2	2(2) − 4 = 0	(2, 0)
4	2(4) − 4 = 4	(4, 4)
6	2(6) − 4 = 8	(6, 8)
8	2(8) − 4 = 12	(8, 12)

It appears that the graph of the equation is a straight line. If we knew this for a fact, then graphing $y = 2x - 4$ would be easy. We would simply find two solutions of the equation, plot them, then plot as much of $y = 2x - 4$ as we like by drawing a line through the two points using a ruler. It turns out that it is true that the graph of $y = 2x - 4$ is a straight line. In fact, we have the following general theorem, which we state without proof.

THEOREM 1   The graph of any equation of the form

$$Ax + By = C$$

where A, B, and C are constants (A and B both not 0) and x and y are variables, is a straight line. Every straight line in a cartesian coordinate system is the graph of an equation of this form.

It immediately follows that any equation of the form

$$y = mx + b$$

where m and b are constants, is also a straight line since it can be written in the form $-mx + y = b$, which is a special case of $Ax + By = C$.

Thus, to graph any equation of the form

$$Ax + By = C$$

or

$$y = mx + b$$

we plot any two points of the solution set and use a ruler to draw a line through these two points. It is sometimes wise to find a third point as a check point.

It should be obvious that we cannot draw a straight line extending indefinitely in either direction. We will settle for the part of the line in which we are interested—usually the part close to the origin unless otherwise stated.

EXAMPLE 11   Graph:

(A) $y = 2x - 4$

(B) $x + 3y = 6$

Solution   (A)   Make up a table of at least two solutions (ordered pairs of numbers that satisfy the equation), plot these, then join the points with a straight line using a ruler.

x	2x − 4 = y	(x, y)
0	2(0) − 4 = −4	(0, −4)
2	2(2) − 4 = 0	(2, 0)
4	2(4) − 4 = 4	(4, 4)

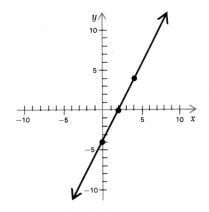

(B)   To graph $x + 3y = 6$, assign to either x or y any convenient value and solve for the other variable. If we let $x = 0$, a convenient value, then

$$0 + 3y = 6$$

$$3y = 6$$

$$y = 2$$

Thus, (0, 2) is a solution.

If we let $y = 0$, another convenient choice, then

$$x + 3(0) = 6$$

$$x + 0 = 6$$

$$x = 6$$

Thus, (6, 0) is a solution.

To find a check point, choose another value for x or y, say $x = -6$, then

$$-6 + 3y = 6$$

$$3y = 12$$

$$y = 4$$

Thus, $(-6, 4)$ is also a solution.

We summarize the above results in a table, then draw the graph. The first two solutions indicate where the graph crosses the coordinate axes and are called the **y and x intercepts**, respectively. The intercepts are often the easiest to find—we let $x = 0$ and solve for y, then let $y = 0$ and solve for x.

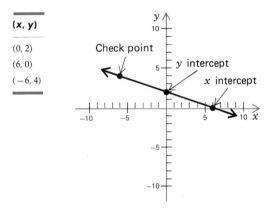

(x, y)
(0, 2)
(6, 0)
(−6, 4)

If a straight line does not pass through all three points, then you have made a mistake and must go back and check your work.

PROBLEM 11   Graph:

(A) $y = 2x - 6$ **(B)** $3x + y = 6$

Vertical and horizontal lines

Vertical and horizontal lines in rectangular coordinate systems have particularly simple equations.

EXAMPLE 12   Graph $y = 4$ and $x = 3$ in a rectangular coordinate system.

Solution   To graph $y = 4$ or $x = 3$ in a rectangular coordinate system, each equation must be provided with the missing variable (usually done mentally) as follows:

$y = 4$	is equivalent to	$0x + y = 4$
$x = 3$	is equivalent to	$x + 0y = 3$

In the first case, we see that no matter what value is assigned to x, $0x = 0$; thus, as long as $y = 4$, x can assume any value, and the graph of $y = 4$ is a horizontal line crossing the y axis at 4. Similarly, in the second case y can assume any value as long as $x = 3$, and the graph of $x = 3$ is a vertical line crossing the x axis at 3. Thus,

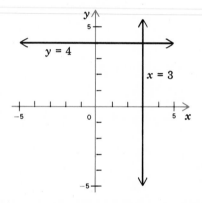

PROBLEM 12   Graph $y = -3$ and $x = -4$ in a rectangular coordinate system.

It should now be clear why equations of the form $Ax + By = C$ are called **linear equations.** Their graphs are straight lines!

ANSWERS TO
MATCHED
PROBLEMS

11. (A)

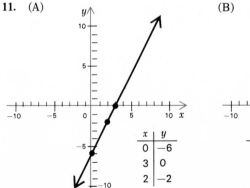

x	y
0	−6
3	0
2	−2

(B)

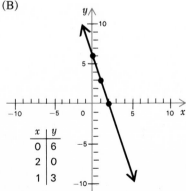

x	y
0	6
2	0
1	3

12.

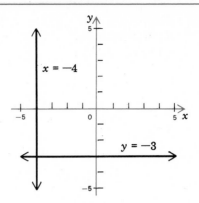

EXERCISE 4.4

Graph in a rectangular coordinate system.

A

1. $y = x$
2. $y = 2x$
3. $y = x - 1$

4. $y = 2x - 3$
5. $y = \dfrac{x}{2}$
6. $y = \dfrac{x}{3}$

7. $y = \dfrac{x}{2} + 1$
8. $y = \dfrac{x}{3} + 2$
9. $x + y = 6$

10. $x + y = -4$
11. $x - y = 5$
12. $x - y = 3$

13. $2x + 3y = 12$
14. $3x + 4y = 12$
15. $3x - 5y = 15$

16. $8x - 3y = 24$
17. $x = 2$
18. $y = 3$

19. $y = -3$
20. $x = -4$

B

21. $y = \tfrac{1}{4}x$
22. $y = \tfrac{1}{2}x$
23. $y = \tfrac{1}{4}x + 1$

24. $y = \tfrac{1}{2}x - 1$
25. $y = -2x + 6$
26. $y = -x + 2$

27. $y = -\tfrac{1}{2}x + 2$
28. $y = -\tfrac{1}{3}x - 1$
29. $2x + y = 7$

30. $3x + 2y = 10$
31. $7x - 4y = 21$
32. $5x - 6y = 15$

33. $x = 0$
34. $y = 0$

Write in the form $y = mx + b$ and graph.

35. $y - x - 2 = x + 1$
36. $x + 6 = 3x + 2 - y$

Write in the form $Ax + By = C$, $A > 0$, and graph.

37. $6x - 3 + y = 2y + 4x + 5$
38. $y + 8 = 2 - x - y$

Use a different scale on the vertical axis to keep the size of the graph within reason.

39. $d = 60t,\ 0 \le t \le 10$
40. $I = 6t,\ 0 \le t \le 10$
41. $A = 100 + 10t,\ 0 \le t \le 10$
42. $v = 10 + 32t,\ 0 \le t \le 5$
43. Graph $x + y = 3$ and $x - 2y = 0$ on the same coordinate system. Determine by inspection the coordinates of the point where the two graphs

160
4: GRAPHING AND LINEAR FORMS

cross. Show that the coordinates of the point of intersection satisfy both equations.

44. Repeat the preceding problem with the equations

$$2x - 3y = -6 \quad \text{and} \quad x + 2y = 11$$

C **45.** Graph $y = -\frac{1}{2}x + b$ for $b = -6$, $b = 0$, and $b = 6$, all on the same coordinate system.

46. Graph $y = mx - 2$ for $m = 2$, $m = \frac{1}{2}$, $m = 0$, $m = -\frac{1}{2}$, and $m = -2$, all on the same coordinate system.

47. Graph $y = |x|$. HINT: Graph $y = x$ for $x \geq 0$ and $y = -x$ for $x < 0$.

48. Graph $y = |2x|$ and $y = |\frac{1}{2}x|$ on the same coordinate system.

APPLICATIONS

Choose horizontal and vertical scales to produce maximum clarity in graphs.

49. BIOLOGY. In biology there is an approximate rule, called the bioclimatic rule, for temperate climates that states that in spring and early summer periodic phenomena such as blossoming for a given species, appearance of certain insects, and ripening of fruit usually come about 4 days later for each 500 ft of altitude. Stated as a formula,

$$d = 4\left(\frac{h}{500}\right)$$

where d = change in days and h = change in altitude in feet. Graph the equation for $0 \leq h \leq 4{,}000$.

50. PSYCHOLOGY. In 1948 Professor Brown, a psychologist, trained a group of rats (in an experiment on motivation) to run down a narrow passage in a cage to receive food in a goal box. He then put a harness on each rat and connected it to an overhead wire that was attached to a scale. In this way he could place the rat at different distances (in centimeters) from the food and measure the pull (in grams) of the rat toward the food. He found that a relation between motivation (pull) and position was given approximately by the equation $p = -\frac{1}{5}d + 70$, $30 \leq d \leq 175$. Graph this equation for the indicated values of d.

51. ELECTRONICS. If the resistance is 30 ohms in a simple electric circuit, such as that found in a flashlight, the current in the circuit I (in amperes) and the electromotive force E (in volts) are related by the equation $E = 30I$. Graph this equation for $0 \leq I \leq 1$.

4.5 SOLVING SYSTEMS OF LINEAR EQUATIONS BY GRAPHING

Many practical problems can be solved conveniently using two-equation–two-unknown methods. For example, if a 12-ft board is cut in two pieces so that one piece is 4 ft longer than the other piece, how long is each piece? We could solve this problem using one-equation–one-unknown methods studied earlier, but we can also proceed as follows using two variables:

Let x = the length of the longer piece
y = the length of the shorter piece

then

$$x + y = 12$$

$$x - y = 4$$

To solve this system is to find all the ordered pairs of real numbers that satisfy both equations at the same time. In general, we are interested in solving linear systems of the type

$$ax + by = m$$

$$cx + dy = n$$

where a, b, c, d, m, and n are real constants and x and y are variables. There are several methods of solving systems of this type. We will consider two that are widely used—a graphing method in this section and an elimination method in the next section.

**Solving
by graphing**

To solve a system of linear equations by graphing, we proceed by graphing both equations on the same coordinate system. Then the coordinates of any points that the graphs have in common must be solutions to the system since they must satisfy both equations.

EXAMPLE 13 Solve by graphing:

$$x + y = 12$$

$$x - y = 4$$

Solution Graph $x + y = 12$ and $x - y = 4$ in the same coordinate system. The coordinates of the point of intersection—if it exists—is the solution.

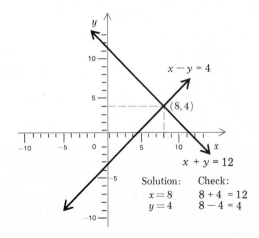

$x - y = 4$

$(8, 4)$

$x + y = 12$

Solution:
$x = 8$
$y = 4$

Check:
$8 + 4 = 12$
$8 - 4 = 4$

PROBLEM 13 Solve by graphing:

$$x + y = 10$$

$$x - y = 6$$

It is clear that Example 13 and Problem 13 each has exactly one solution since the two lines in each case intersect in exactly one point. Let us look at three typical cases that illustrate the three possible ways two lines can be related to each other in a rectangular coordinate system. Solving the following three systems graphically

(A) $2x - 3y = 2$
  $x + 2y = 8$

(B) $4x + 6y = 12$
  $2x + 3y = -6$

(C) $2x - 3y = -6$
  $-x + \frac{3}{2}y = 3$

we obtain

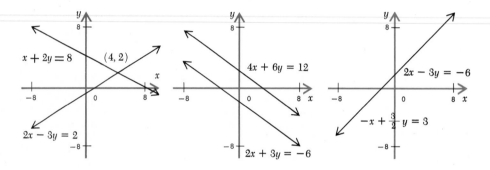

Lines intersect at one point only:

Exactly one solution
$x = 4, y = 2$

Lines are parallel:

No solution

Lines coincide:

Infinite number of solutions

Now we know exactly what to expect when solving a system of two linear equations in two unknowns:

Exactly one pair of numbers as a solution.

No solutions.

An infinite number of solutions.

In most applications the first case prevails. If we find a pair of numbers that satisfies the system of equations and the graphs of the equations meet at only one point, then that pair of numbers is the only solution of the system, and we need not look further for others.

The graphical method of solving systems of equations yields considerable information as to what to expect in the way of solutions to a system of two linear equations in two unknowns. In addition, graphs frequently reveal relationships in problems that would otherwise be hidden. On the other hand, if one is interested in solutions with several-decimal-place accuracy, the graphical method is often not practical. The method of elimination, to be considered in the next section, will take care of this deficiency.

ANSWERS TO MATCHED PROBLEMS

13.

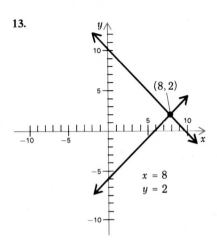

EXERCISE 4.5

Solve by graphing and check.

A

1. $x + y = 5$
$x - y = 1$

2. $x + y = 6$
$x - y = 2$

3. $x + y = 5$
$2x - y = 4$

4. $2x + y = 6$
$x - y = -3$

5. $x - 3y = 3$
$x - y = 7$

6. $x - 2y = -4$
$2x - y = 10$

B

7. $3x - y = 2$
$x + 2y = 10$

8. $x - 2y = 2$
$2x + y = 9$

9. $-2x + 3y = 12$
$2x - y = 4$

10. $3x - 2y = 12$
$7x + 2y = 8$

11. $-3x + y = 9$
$3x + 4y = -24$

12. $x + 5y = -10$
$-x + 2y = 3$

13. $x + 2y = 4$
$2x + 4y = -8$

14. $3x + 5y = 15$
$6x + 10y = -30$

15. $\frac{1}{2}x - y = -3$
$-x + 2y = 6$

16. $3x - 5y = 15$
$x - \frac{5}{3}y = 5$

C *It can be shown that if two linear equations are written in the form*

$y = mx + b$

and the coefficient of x is the same for both, then their graphs are parallel.

17. Show that the lines in Exercise 13 are parallel.

18. Show that the lines in Exercise 14 are parallel.

19. How can you use the idea above to show that the two lines in Exercise 15 coincide?

20. Show that the lines in Exercise 16 coincide.

4.6 SOLVING SYSTEMS OF LINEAR EQUATIONS BY ELIMINATION

The elimination method we are now going to develop can produce answers to any decimal accuracy desired. This method has to do with the replacement of systems of equations with simpler equivalent systems (by performing appropriate operations) until we get a system whose solution is obvious. **Equivalent systems** are, as you would expect, systems with the same solution set. What operations on a system produce equivalent systems? The following theorem, stated but not proved, answers this question.

THEOREM 2   Equivalent systems result if

(A) Both sides of the equations are multiplied by a nonzero constant.

(B) The two equations are combined by addition or subtraction and the result is paired with either equation.

Solving a system of equations by use of this theorem is best illustrated by examples.

EXAMPLE 14   Solve the system:

$$3x + 2y = 13$$
$$2x - y = 4$$

Solution   We use Theorem 2 to eliminate one of the variables, and thus obtain a system whose solution is obvious:

$$3x + 2y = 13$$
$$2x - y = 4$$

If we multiply the bottom equation by 2 and add this to the top equation, we can eliminate y.

$$3x + 2y = 13$$
$$\underline{4x - 2y = 8}$$
$$7x \qquad = 21$$

Now solve for x.

$$\boxed{x = 3}$$

$$2 \cdot 3 - y = 4$$
$$-y = -2$$

Substitute $x = 3$ back into either of the two original equations, the simpler of the two, and solve for y. We choose the second equation.

$$\boxed{y = 2}$$

CHECK

$$3x + 2y = 13 \qquad\qquad 2x - y = 4$$
$$3 \cdot 3 + 2 \cdot 2 \overset{?}{=} 13 \qquad\qquad 2 \cdot 3 - 2 \overset{?}{=} 4$$
$$9 + 4 \overset{\checkmark}{=} 13 \qquad\qquad 6 - 2 \overset{\checkmark}{=} 4$$

PROBLEM 14 Solve the system:

$$2x + 3y = 7$$
$$3x - y = 5$$

EXAMPLE 15 Solve the system:

$$2x + 3y = 1$$
$$5x - 2y = 12$$

Solution

$$2x + 3y = 1$$
$$5x - 2y = 12$$

If we multiply the top equation by 2 and the bottom equation by 3 and add, we can eliminate y.

$$4x + 6y = 2$$
$$\underline{15x - 6y = 36}$$
$$19x = 38$$
$$\boxed{x = 2}$$

$$2 \cdot 2 + 3y = 1$$
$$3y = -3$$
$$\boxed{y = -1}$$

Substitute $x = 2$ back into either of the two original equations, then solve for y.

CHECK

$$2x + 3y = 1 \qquad\qquad 5x - 2y = 12$$
$$2 \cdot 2 + 3(-1) \overset{?}{=} 1 \qquad\qquad 5 \cdot 2 - 2(-1) \overset{?}{=} 12$$
$$4 - 3 \overset{\checkmark}{=} 1 \qquad\qquad 10 + 2 \overset{\checkmark}{=} 12$$

PROBLEM 15 Solve the system:

$$3x - 2y = 8$$
$$2x + 5y = -1$$

EXAMPLE 16 Solve the system:

$$x + 3y = 2$$
$$2x + 6y = -3$$

Solution $\quad x + 3y = 2 \qquad$ Multiply the top equation by 2 and subtract.
$2x + 6y = -3$

$\quad\quad\quad 2x + 6y = 4$
$\quad\quad\quad \underline{2x + 6y = -3}$
$\quad\quad\quad\quad\quad\quad 0 = 7 \qquad$ A contradiction!

Hence, no solution.

Our assumption that there are values for x and y that satisfy both equations simultaneously must be false (otherwise, we have proved that $0 = 7$); thus, the system has no solutions. Systems of this type are said to be **inconsistent**—conditions have been placed on the unknowns x and y that are impossible to meet. Geometrically, the graphs of the two equations must be nonintersecting parallel lines.

PROBLEM 16 Solve the system:

$$2x - y = 2$$
$$-4x + 2y = 1$$

EXAMPLE 17 Solve the system:

$$-2x + y = -8$$
$$x - \tfrac{1}{2}y = 4$$

Solution $\quad -2x + y = -8 \qquad$ Multiply the bottom equation by 2 and add.
$\quad\quad\quad x - \tfrac{1}{2}y = 4$

$\quad\quad\quad -2x + y = -8$
$\quad\quad\quad \underline{2x - y = 8}$
$\quad\quad\quad\quad\quad\quad 0 = 0$

Both unknowns have been eliminated! Actually, if we had multiplied the bottom equation by -2, we would have obtained the top equation. When one equation is a constant multiple of the other, the system is said to be **dependent,** and their graphs will coincide. There are infinitely many solutions to the system—any solution of one equation will be a solution of the other.

PROBLEM 17 Solve the system:

$$4x - 2y = 3$$
$$-2x + y = -\tfrac{3}{2}$$

14. $x = 2$, $y = 1$ **15.** $x = 2$, $y = -1$ **16.** No solution
17. Infinitely many solutions—any solution of one equation is a solution of the other.

EXERCISE 4.6

Solve by the elimination method and check.

A   **1.** $x + y = 5$
$x - y = 1$

2. $x - y = 6$
$x + y = 10$

3. $x + 3y = 13$
$-x + y = 3$

4. $-x + y = 1$
$x - 2y = -5$

5. $2x + y = 0$
$3x + y = 2$

6. $x + 5y = 16$
$x - 2y = 2$

7. $2x + 3y = 1$
$3x - y = 7$

8. $3x - y = -3$
$5x + 3y = -19$

9. $3x - 4y = 1$
$-x + 3y = 3$

10. $-x + 5y = -3$
$2x - 3y = -1$

11. $2x + 4y = 6$
$-3x + y = 5$

12. $3x + y = -8$
$-5x + 3y = 4$

B   **13.** $3x + 2y = -2$
$4x + 5y = 2$

14. $5x + 7y = 8$
$3x + 2y = 7$

15. $6x + 5y = 4$
$7x + 2y = -3$

16. $9x + 4y = 1$
$4x + 3y = -2$

17. $11x + 2y = 1$
$9x - 3y = 24$

18. $3x - 11y = -7$
$4x + 3y = 26$

19. $3p + 8q = 4$
$15p + 10q = -10$

20. $5m - 3n = 7$
$7m + 12n = -1$

21. $4m + 6n = 2$
$6m - 9n = 15$

22. $5a - 4b = 1$
$3a - 6b = 6$

23. $3x + 5y = 15$
$6x + 10y = -5$

24. $x + 2y = 4$
$2x + 4y = -9$

25. $3x - 5y = 15$
$x - \frac{5}{3}y = 5$

26. $\frac{1}{2}x - y = -3$
$-x + 2y = 6$

Write in standard form

$ax + by = c$

$dx + ey = f$

and solve.

27. $y = 3x - 3$
$6x = 8 + 3y$

28. $3x = 2y$
$y = -7 - 2x$

29. $3m + 2n = 2m + 2$
$2m + 3n = 2n - 2$

30. $2x - 3y = 1 - 3x$
$4y = 7x - 2$

31. If 3 limes and 12 lemons cost 81 cents, and 2 limes and 5 lemons cost 42 cents, what is the cost of 1 lime and 1 lemon?

32. Find the capacity of each of two trucks if 3 trips of the larger and 4 trips of the smaller result in a total haul of 41 tons, and if 4 trips of the larger and 3 trips of the smaller result in a total haul of 43 tons.

C *Solve by the elimination method.*

33. $0.3x - 0.6y = 0.18$
$0.5x + 0.2y = 0.54$

34. $0.8x - 0.3y = 0.79$
$0.2x - 0.5y = 0.07$

35. $\dfrac{x}{3} + \dfrac{y}{2} = 4$

$\dfrac{x}{3} - \dfrac{y}{2} = 0$

36. $\dfrac{x}{4} + \dfrac{y}{3} = 0$

$-\dfrac{x}{4} + \dfrac{y}{3} = -4$

37. $\dfrac{x}{2} + \dfrac{y}{3} = 1$

$\dfrac{2x}{3} + \dfrac{y}{2} = 2$

38. $\dfrac{a}{4} - \dfrac{2b}{3} = -2$

$\dfrac{a}{2} - b = -2$

4.7 APPLICATIONS INVOLVING LINEAR SYSTEMS

Many problems that can be solved using the one-equation–one-unknown methods discussed in Chapter 3 can also be solved using two-equation–two-unknown methods. In fact, many practical problems are more naturally set up using two variables rather than one.

EXAMPLE 18

If you have 25 dimes and quarters in your pocket worth $4, how many of each do you have?

Solution

Let x = the number of dimes
y = the number of quarters

then

$x + y = 25$ number of coins

$10x + 25y = 400$ value of coins in cents

Multiply the top equation by -10 and add:

$$\begin{array}{r} -10x - 10y = -250 \\ \underline{10x + 25y = 400} \\ 15y = 150 \end{array}$$

$$y = 10 \text{ quarters}$$

$x + 10 = 25$ Now solve for x using the top equation.

$$x = 15 \text{ dimes}$$

CHECK

$10 + 15 = 25$ coins; $10(25) + 15(10) = 250 + 150 = 400$ cents or $4

PROBLEM 18   If you have 25 nickels and quarters in your pocket worth $2.25, how many of each do you have? Set up two equations with two unknowns and solve.

EXAMPLE 19   A chemical storeroom has a 40% acid solution and an 80% acid solution. How many deciliters (dl) must be taken from each to obtain 12 dl of a 50% solution?

Solution   Let x = amount of 40% solution used
y = amount of 80% solution used

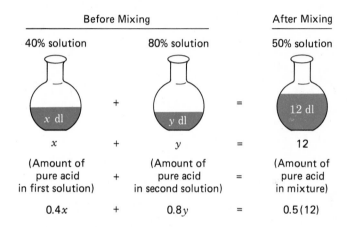

	Before Mixing		After Mixing
40% solution		80% solution	50% solution
x dl	+	y dl	12 dl
x	+	y =	12
(Amount of pure acid in first solution)	+	(Amount of pure acid in second solution) =	(Amount of pure acid in mixture)
$0.4x$	+	$0.8y$ =	$0.5(12)$

Thus, we obtain two equations and two unknowns:

$x +\quad y = 12$

$0.4x + 0.8y = 6$ Multiply by 10 to clear decimals.

$x +\quad y = 12$

$4x +\quad 8y = 60$ Divide by -4 to simplify.

$\begin{aligned} x +\quad y &= 12 \\ -x -\quad 2y &= -15 \\ \hline -y &= -3 \end{aligned}$ Add to eliminate x.

$y = 3$ dl of 80% solution

$x + y = 12$ Now solve for x using the top equation.

$x + 3 = 12$

$x = 9$ dl of 40% solution

PROBLEM 19   A coffee shop wishes to blend $3-per-pound coffee with $4.25-per-pound coffee to produce a blend selling for $3.50 per pound.

How much of each should be used to produce 50 lb of the new blend? Set up two equations with two unknowns and solve.

EXAMPLE 20   Two architects contract to design a building and agree to share the fee in the ratio of $3:5$. If the total fee to the client is $7,200, how much does each receive?

Solution   Let x = amount one receives
y = amount other receives

then

$$x + y = 7,200$$

and

$$\frac{x}{y} = \frac{3}{5} \qquad \text{Multiply both sides by } 5y \text{ to clear fractions.}$$

$$x + y = 7,200$$

$$5x = 3y$$

$$x + y = 7,200 \qquad \text{Write in standard form: } ax + by = c$$
$$\qquad\qquad\qquad\qquad\qquad\qquad dx + ey = f$$
$$5x - 3y = 0$$

$$\begin{array}{l} 3x + 3y = 21,600 \qquad \text{Multiply the top equation by 3 and add.} \\ \underline{5x - 3y = 0} \\ 8x \qquad\quad = 21,600 \end{array}$$

$$x = \$2,700 \quad \text{to the first architect}$$

$$x + y = 7,200 \qquad \text{Solve for } y.$$

$$2,700 + y = 7,200$$

$$y = \$4,500 \quad \text{to the second architect}$$

PROBLEM 20   If a guitar string is divided in the ratio of $5:6$, a minor third chord will result when the string is plucked. How would you divide a 44-cm string to produce this chord? Set up two equations and two unknowns and solve.

ANSWERS TO MATCHED PROBLEMS

18. $x + y = 25$, $5x + 25y = 225$; 20 nickels, 5 quarters
19. $x + y = 50$, $3x + 4.25y = (3.50)(50)$;
     $x = 30$ lb ($3 coffee), $y = 20$ lb ($4.25 coffee)
20. $x + y = 44$, $\dfrac{x}{y} = \dfrac{5}{6}$ (or $6x - 5y = 0$); 20 cm and 24 cm

EXERCISE 4.7

The problems in this exercise are grouped according to subject area. The most difficult problems are double-starred (**), moderately difficult problems are single-starred (*), and the easier problems are not starred. Solve all problems using two-equation–two-unknown methods.

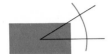

GEOMETRY

1. An 18-ft board is cut into two pieces so that one piece is 4 ft longer than the other piece. How long is each piece?

2. If the sum of two angles in a right triangle is 90° and their difference is 14°, find the two angles.

* **3.** Find the dimensions of a rectangle with perimeter 72 cm if its length is 1.25 times its width.

* **4.** Find the dimensions of a rectangle with perimeter 168 cm if its length is 1.8 times its width.

BUSINESS

5. A jazz concert brought in $60,000 on the sale of 8,000 tickets. If the tickets sold for $6 and $10 each, how many of each type were sold?

6. A student production brought in $7,000 on the sale of 3,000 tickets. If tickets sold for $2 and $3 each, how many of each type was sold?

* **7.** A person has $8,000 to invest. If part is invested at 6 percent and the rest at 8 percent, how much should be invested at each rate to have a total return of $520 per year?

* **8.** You have just inherited $10,000 and wish to invest part at 8 percent and the rest at 12 percent. How much should be invested at each rate to produce the same return as if you had invested it all at 9 percent?

* **9.** A coffee shop wishes to blend a $3.50-per-pound coffee with a $4.75-per-pound coffee to produce a blend selling for $4 per pound. How much of each should be used to produce 100 lb of the new blend?

***10.** A coffee and tea shop wishes to blend a $2.50-per-pound tea with a $3.25-per-pound tea to produce a blend selling for $3 per pound. How much of each should be used to produce 75 lb of the new blend?

CHEMISTRY

11. Alcohol and distilled water were mixed to produce 120 centiliters (cl) of solution. If 20 cl more water was used than alcohol, how many centiliters of each was mixed?

12. Hydrochloric acid and distilled water were mixed to produce 96 milliliters (ml) of solution. If 36 ml more water was used than acid how much of each was mixed?

*13.   A chemist has two solutions in a stockroom, one is a 30% acid solution and the other is a 70% acid solution. How many milliliters (ml) of each should be mixed to obtain 100 ml of a 40% solution?

*14.   A chemical storeroom has a 20% acid solution and a 50% acid solution. How many centiliters (cl) must be taken from each to obtain 90 cl of a 30% solution?

LIFE SCIENCE

15.   If two rats in a diet experiment have a combined weight of 800 grams and one weighs 200 grams more than the other, how much does each weigh?

16.   If two monkeys have a combined weight of 10 kg and one weighs 2 kg more than the other, how much does each weigh?

**17.   Animals in an experiment are to be kept on a strict diet. Each animal is to receive, among other things, 20 grams of protein and 6 grams of fat. The laboratory technician is able to purchase two food mixes of the following compositions:

	Protein	Fat
Mix 1	10%	6%
Mix 2	20%	2%

How many grams of each mix should be used to obtain the right diet for a single animal?

**18.   A farmer placed an order with a chemical company for a chemical fertilizer that would contain, among other things, 120 lb of nitrogen and 90 lb of phosphoric acid. The company had two mixtures on hand with the following compositions:

	Nitrogen	Phosphoric acid
Mixture A	20%	10%
Mixture B	6%	6%

How many pounds of each mixture should the chemist mix to fill the order?

PHYSICS-ENGINEERING

19.   Where should the fulcrum be placed on a 12-ft bar if it is to balance with a 14-lb weight on one and a 42-lb weight on the other end?

HINT:

| 14 | x | y | 42 | $14x = 42y$ |

EARTH SCIENCES

20. Where should the fulcrum be placed on a 240-cm bar to balance 30 kg on one end and 50 kg on the other end?

An earthquake produces primary and secondary waves. Near the surface of the earth the primary wave travels at about 8 km/sec and the secondary wave at about 5 km/sec. If a primary wave is recorded at a station 15 sec before the arrival of the secondary wave, how long did each wave travel and how far is the epicenter of the quake from the station? HINT: If x is the time for the secondary wave and y is the time for the primary wave, then $x - y = 15$. The second equation is obtained from the fact that each wave travels the same distance.

**22. Repeat Problem 21 for a time difference of 21 sec.

MUSIC

23. If a guitar string is divided in the ratio of 4:5, a major third chord will result. What will be the length of each part if a 36-in string is used? HINT: Use the proportion $x/y = \frac{4}{5}$ for one of the equations.

24. If a guitar string is divided in the ratio of 5:8, a minor sixth chord will result. How would you divide a 39-in string to produce a minor sixth?

PUZZLES

25. A parking meter takes only nickels and dimes. If it contains 50 coins with a total value of $3.50, how many of each type of coin is in the meter?

26. An all-day parking meter takes only dimes and quarters. If it contains 100 coins worth $14.50, how many of each type of coin is in the meter?

*27. A packing carton contains 144 small packages, some weighing $\frac{1}{4}$ lb each and the others $\frac{1}{2}$ lb each. How many of each type are in the carton if the total contents of the carton weigh 51 lb?

*28. Repeat Problem 27 if the small packages weigh $\frac{1}{3}$ lb and $\frac{3}{4}$ lb, respectively, and the total contents weigh 73 lb?

**29. If 1 flask and 4 mixing dishes balance 16 test tubes and 2 mixing dishes, and if 2 flasks balance 2 test tubes and 6 mixing dishes, how many test tubes will balance 1 flask, and how many test tubes will balance 1 mixing dish?

4.8 CHAPTER REVIEW: IMPORTANT TERMS AND SYMBOLS, REVIEW EXERCISE, PRACTICE TEST

Important terms and symbols

irrational number (*4.1*)   real number (*4.1*)   real number line (*4.1*)   inequality (*4.1*) $<, >$ (*4.1*)   inequality statements (*4.1*)   solution set (*4.1*)   graph (*4.1*) inequality properties (*4.2*)   solving inequality statements (*4.2*) cartesian coordinate system (*4.3*)   coordinate axes (*4.3*) coordinates (*4.3*)   abscissa (*4.3*)   ordinate (*4.3*) quadrants (*4.3*)   rectangular coordinate system (*4.3*) equation of a line (*4.4*)   linear equations (*4.4*) linear systems (*4.6, 4.7*)   equivalent systems (*4.6*) inconsistent systems (*4.6*)   dependent systems (*4.6*) solving by graphing (*4.5*)   solving by elimination (*4.6*)

Exercise 4.8 Review exercise

A 1. Graph on a real number line:
(A) $-3 < x \leq 2$, x an integer
(B) $-3 < x \leq 2$, x a real number

Solve.

2. $\dfrac{x}{-3} > -2$

3. $-4x \leq 12$

4. $3x + 9 \leq -3 - x$

5. $-14 \leq 3x - 2 \leq 7$

Graph in a rectangular coordinate system.

6. $y = 2x - 3$

7. $y = \dfrac{x}{2} + 2$

8. $2x + y = 6$

9. $4x - 3y = 12$

10. Solve graphically: $x - y = 5$
$\qquad\qquad\qquad\qquad x + y = 7$

Solve by elimination method.

11. $2x + 3y = 7$
$\quad\ \ 3x - y = 5$

12. $3x + 2y = 1$
$\quad\ \ 5x + 6y = 7$

13. What numbers satisfy the condition, "5 less than 5 times the number is less than or equal to 10"? Write an inequality and solve.

14. Solve using two equations and two unknowns: if you have 30 nickels and dimes in your pocket worth $2.30, how many of each do you have?

B 15. Indicate true (T) or false (F).
(A) $\sqrt{2}$ is a real number.
(B) -5 is a real number.
(C) 3.47 is a real number.
(D) $-\frac{2}{3}$ is a real number.

175

16. Indicate true (T) or false (F).
(A) $-\frac{3}{4}$ is a rational number.
(B) 5 is an integer.
(C) -4 is a real number.
(D) $\sqrt{3}$ is an irrational number.

Solve and graph on a real number line.

17. $3x - 9 < 7x - 5$

18. $2x - (3x + 2) > 5 - 2(3 - 2x)$

19. $\frac{x}{2} - \frac{x-1}{3} \geq -1$

20. $-9 \leq \frac{2}{3}x - 5 < 7$

Graph in a rectangular coordinate system.

21. $y = \frac{1}{3}x - 2$

22. $4x - 3y = 10$

23. Solve graphically: $2x - 3y = -3$
$3x + y = 12$

Solve by the elimination method.

24. $6u + 4v = -2$
$5u + 3v = -1$

25. $5m - 3n = 4$
$-2m + 4n = -10$

26. A chemical is to be kept between 59 and 86°F (that is, $59 \leq F \leq 86$). What is the temperature range in Celsius degrees ($F = \frac{9}{5}C + 32$)? Set up a double inequality and solve.

27. Part of $6,000 is to be invested at 10 percent and the rest at 6 percent. How much should be invested at each rate if the total annual return from both investments is to be $440? Set up two equations with two unknowns and solve.

28. A chemical storeroom contains a 50% alcohol solution and a 70% solution. How much of each should be used to obtain 100 milliliters (ml) of a 66% solution? Set up two equations with two unknowns and solve.

C **29.** Solve graphically: $2x - 6y = -3$
$-\frac{2}{3}x + 2y = 1$

30. Solve by the elimination method: $x - 4y = 12$
$-\frac{x}{4} + y = 4$

31. To develop a certain roll of photographic film, the temperature of the solution must be kept between 20 and 25°C (that is, $20 \leq C \leq 25$). Find the temperature range in Fahrenheit degrees [$C = \frac{5}{9}(F - 32)$]. Set up a double inequality and solve.

32. Wishing to log some flying time, you have rented an airplane for 2 hr. You decide to fly due east until you have to turn around in order to be back at the airport at the end of the 2 hr. The cruising speed of the plane is 120 mph in still air.

(A) If there is a 30-mph wind blowing from the east, how long should you head east before you turn around, and how long will it take you to get back?
(B) How far from the airport were you when you turned back?
Solve using two-equation–two-unknown method.

**Practice test
Chapter 4**

1. Indicate true (T) or false (F):
(A) -3 is a real number.
(B) $\frac{2}{3}$ is an irrational number.
(C) 4 is a rational number.
(D) 0 is a real number.
(E) $\sqrt{2}$ is an irrational number.

2. Graph on a real number line:
(A) $-4 < x \le -1$, x an integer
(B) $-4 < x \le -1$, x a real number

3. Solve and graph on a real number line:
(A) $\dfrac{x}{-3} < -2$
(B) $-5 < 2x - 3 \le 9$

4. Solve $2(x - 4) - (3x + 2) \ge 2x - 1$.

5. Solve $1 - \dfrac{2x - 1}{2} < \dfrac{x}{3}$.

6. Graph in a rectangular coordinate system: $y = \frac{1}{2}x - 1$

7. Solve graphically: $\begin{aligned} -x + 3y &= 6 \\ 3x - y &= 6 \end{aligned}$

8. Solve by elimination: $\begin{aligned} 3u + 2v &= 3 \\ -4u + 5v &= 19 \end{aligned}$

9. Solve by elimination: $\begin{aligned} 6x - 8y &= -1 \\ -3x + 4y &= \frac{3}{2} \end{aligned}$

10. Solve $15 \le \frac{5}{9}(F - 32) \le 30$.

11. A post office gave you \$1.85 in change consisting of nickels and dimes. If there were 25 coins in all, how many of each type did you receive? Solve using two-equation–two-unknown method.

12. A chemical storeroom has a 40% and an 80% acid solution in stock. How much of each should be used to get 4 liters of a 50% solution? Solve using two-equation–two-unknown method.

5

POLYNOMIALS—
BASIC OPERATIONS

5.1 POLYNOMIALS

Algebraic expressions can be classified in a variety of ways for more efficient study. We will start with the important class of expressions called polynomials.

POLYNOMIALS

$$2x^2 - 5x + 8 \qquad 2x - 1 \qquad x \qquad 5 \qquad 0$$

$$x^2 - \tfrac{1}{3}xy + \sqrt{2}y^2 \qquad 2x^3 - 3x^2y - 4xy^2 + y^3$$

NONPOLYNOMIALS

$$\frac{3x - 1}{2x^2 + 3x - 5} \qquad 2^x \qquad x^3 - \frac{2\sqrt{x}}{y} + 3y^4 \qquad \frac{1}{x}$$

Most of the algebraic expressions we have considered so far are polynomials. What do all of the polynomials above have in common that is not shared by the nonpolynomials? We see that polynomials (in one and two variables) are constructed by adding or subtracting terms of the form ax^n or bx^my^n where a and b are real number coefficients and m and n are natural number exponents. In a polynomial a variable cannot appear in a denominator, as an exponent, or within a radical sign.

The polynomial form, particularly in one and two variables, is encountered with great frequency at all levels in mathematics and science. As a consequence, it is a form that receives a great deal of attention in beginning and intermediate algebra.

It is convenient to identify certain types of polynomials. The concept of degree is used for this purpose. The **degree of a term** in a polynomial is the power of the variable present in the term. If more than one variable is present as a factor, then the sum of the powers of the variables in the term is the degree of the term. The **degree of a polynomial** is the degree of the term with the highest degree in the polynomial.

EXAMPLE 1 **(A)** $3x^5$ is of degree 5.

(B) $7x^2y^3$ is of degree 5.

(C) In $4x^7 - 3x^5 + 2x^3 - 1$, the highest degree term is the first, with degree 7; thus, the degree of the polynomial is 7.

(D) The degree of each of the first three terms in $x^2 - 2xy + y^2 + 2x - 3y + 2$ is 2; the fourth and fifth terms each have degree 1; thus, this is a second-degree polynomial.

PROBLEM 1 **(A)** What is the degree of $5x^3$? Of $2x^3y^4$?

(B) What is the degree of the polynomial $6x^3 - 2x^2 + x - 1$? Of $2x^2 - 3xy + y^2 + x - y + 1$?

Any nonzero real constant is defined to be a polynomial of degree 0. Thus 5 is a polynomial of degree 0. The number 0 is also a polynomial, but is not assigned a degree.

We also call a one-term polynomial a **monomial,** a two-termed polynomial a **binomial,** and a three-termed polynomial a **trinomial.**

$3x^2 - 2x + 1$	$2x - 3y$	$3x^4y^2$	8
trinomial	binomial	monomial	monomial
degree 2	degree 1	degree 6	degree 0

For polynomials to be useful, we must know how to add, subtract, multiply, divide, and factor them. We have already spent some time on these operations with simpler polynomials. We now review and extend these processes to more complex forms.

ANSWERS TO MATCHED PROBLEMS

1. (A) 3, 7; (B) 3, 2

5.2 ADDITION, SUBTRACTION, AND MULTIPLICATION

In the preceding chapters you worked many problems involving addition, subtraction, and multiplication of polynomials. Considerable use was made of the associative and commutative properties of real numbers, as well as the distributive property. The procedures will be reviewed and extended in this section.

Addition and subtraction

The following examples illustrate two methods for adding and subtracting polynomials.

EXAMPLE 2 Add:

$$x^4 - 3x^3 + x^2 \qquad -x^3 - 2x^2 + 3x \qquad \text{and} \qquad 3x^2 - 4x - 5$$

Solution *Method 1* Add horizontally.

$$(x^4 - 3x^3 + x^2) + (-x^3 - 2x^2 + 3x) + (3x^2 - 4x - 5)$$
$$= x^4 - 3x^3 + x^2 - x^3 - 2x^2 + 3x + 3x^2 - 4x - 5$$
$$= x^4 - 4x^3 + 2x^2 - x - 5$$

Method 2 Add vertically.
Line up like terms and add coefficients. This method is generally preferred when you have several polynomials to add.

$$\begin{array}{l} x^4 - 3x^3 + \ x^2 \\ \ - \ x^3 - 2x^2 + 3x \\ \underline{\qquad\quad 3x^2 - 4x - 5} \\ x^4 - 4x^3 + 2x^2 - \ x - 5 \end{array}$$

PROBLEM 2   Add horizontally, then add vertically:

$$3x^4 - 2x^3 - 4x^2 \qquad x^3 - 2x^2 - 5x \qquad \text{and} \qquad x^2 + 7x - 2$$

EXAMPLE 3   Subtract $4x^2 - 3x + 5$ from $x^2 - 8$.

Solution   *Method 1*   Work horizontally.
This method is often preferred for subtraction.

$$(x^2 - 8) - (4x^2 - 3x + 5)$$ Notice which polynomial goes on the right.

$$= x^2 - 8 - 4x^2 + 3x - 5$$ Clear parentheses—be careful of signs.

$$= -3x^2 + 3x - 13$$ Combine like terms.

Method 2   Work vertically.
Notice which polynomial goes on the bottom.

$$
\begin{array}{r}
x^2 \qquad\quad - 8 \\
4x^2 - 3x + 5 \\
\hline
-3x^2 + 3x - 13
\end{array}
$$
Line up like terms.
Change signs (mentally) and add.

PROBLEM 3   Subtract $2x^2 - 5x + 4$ from $5x^2 - 6$.

    Sometimes a horizontal arrangement is more convenient, and sometimes a vertical arrangement is more convenient. You should be able to work either way, letting the situation dictate the choice.

Multiplication

The distributive property is the important principle behind multiplying polynomials, and leads directly to the mechanical rule:

Mechanics of Multiplying Polynomials
<hr>

To multiply two polynomials, multiply each term of the first one by each term of the second one; then add like terms.

EXAMPLE 4   Multiply $(x - 3)(x^2 - 2x + 3)$.

Solution   *Method 1*   Horizontal arrangement.

$$(x - 3)(x^2 - 2x + 3)$$ Use distributive property from right to left.

$$= x(x^2 - 2x + 3) - 3(x^2 - 2x + 3)$$ Use distributive property from left to right.

$$= x^3 - 2x^2 + 3x - 3x^2 + 6x - 9$$ Combine like terms.

$$= x^3 - 5x^2 + 9x - 9$$

Method 2 Vertical arrangement.

This method is probably preferred by most students for this type of problem.

$$x^2 - 2x + 3$$
$$\underline{x \quad - 3}$$
$$x^3 - 2x^2 + 3x$$
$$\underline{\quad - 3x^2 + 6x - 9}$$
$$x^3 - 5x^2 + 9x - 9$$

Notice that we start multiplying from the left first; that is, by x. Then multiply by -3, line up like terms, and add.

NOTE: The product of a first- and second-degree polynomial produces a third-degree polynomial. What do you think the product of two second-degree polynomials would produce? Two first-degree polynomials? (Answer: fourth-degree polynomial; second-degree polynomial.)

PROBLEM 4   Multiply $(3x^2 - 2x + 1)(2x^2 + 3x - 2)$ vertically.

ANSWERS TO MATCHED PROBLEMS

2. $3x^4 - x^3 - 5x^2 + 2x - 2$
3. $3x^2 + 5x - 10$
4. $6x^4 + 5x^3 - 10x^2 + 7x - 2$

EXERCISE 5.2

A *Add:*

1. $3x - 5$ and $2x + 3$
2. $6x + 5$ and $3x - 8$
3. $2x + 3$, $-4x - 2$, and $7x - 4$
4. $7x - 5$, $-x + 3$, and $-8x - 2$
5. $2x^2 - 3x + 1$, $2x - 3$, and $4x^2 + 5$
6. $5x^2 + 2x - 7$, $2x^2 + 3$, and $-3x - 8$

Subtract:

7. $2x + 3$ from $5x + 7$
8. $5x + 2$ from $6x + 4$
9. $4x - 9$ from $2x + 3$
10. $3x - 8$ from $2x - 7$
11. $x^2 - 3x - 5$ from $2x^2 - 6x - 5$
12. $2y^2 - 6y + 1$ from $y^2 - 6y - 1$

Multiply:

13. $(2x - 3)(x + 2)$
14. $(3x - 5)(2x + 1)$

15. $(2x - 1)(x^2 - 3x + 5)$

16. $(3y + 2)(2y^2 + 5y - 3)$

17. $(x - 3y)(x^2 - 3xy + y^2)$

18. $(m + 2n)(m^2 - 4mn - n^2)$

B *Name the degree of each polynomial.*

19. $2x - 3$ **20.** $4x^2 - 2x + 3$

21. $3x^3 - x + 7$ **22.** $2x - y$

23. $x^2 - 3xy + y^2$ **24.** $x^3 - 2x^2y + xy^2 - 3y^3$

25. $2x^6 - 3x^5 + x^2 - x + 1$ **26.** $x^5 - 2x^2 + 5$

Add:

27. $3x^3 - 2x^2 + 5$, $3x^2 - x - 3$, and $2x + 4$

28. $2x^4 - x^2 - 7$, $3x^3 + 7x^2 + 2x$, and $x^2 - 3x - 1$

Subtract:

29. $3x^3 - 2x^2 - 5$ from $2x^3 - 3x + 2$

30. $5x^3 - 3x + 1$ from $2x^3 + x^2 - 1$

31. Subtract the sum of the last two polynomials from the sum of the first two: $2x^2 - 4xy + y^2$, $3xy - y^2$, $x^2 - 2xy - y^2$, and $-x^2 + 3xy - 2y^2$.

32. Subtract the sum of the first two polynomials from the sum of the last two: $3m^3 - 2m + 5$, $4m^2 - m$, $3m^2 - 3m - 2$, and $m^3 + m^2 + 2$.

Multiply:

33. $(a + b)(a^2 - ab + b^2)$

34. $(a - b)(a^2 + ab + b^2)$

35. $(x + 2y)^3$

36. $(2m - n)^3$

37. $(x^2 - 3x + 5)(2x^2 + x - 2)$

38. $(2m^2 + 2m - 1)(3m^2 - 2m + 1)$

39. $(2x^2 - 3xy + y^2)(x^2 + 2xy - y^2)$

40. $(a^2 - 2ab + b^2)(a^2 + 2ab + b^2)$

C *Simplify:*

41. $(3x - 1)(x + 2) - (2x - 3)^2$

42. $(2x + 3)(x - 5) - (3x - 1)^2$

43. $2(x - 2)^3 - (x - 2)^2 - 3(x - 2) - 4$

44. $(2x - 1)^3 - 2(2x - 1)^2 + 3(2x - 1) + 7$

5.3 SPECIAL PRODUCTS OF THE FORM $(ax + b)(cx + d)$

For reasons that will become clear shortly, it is essential that you learn to multiply first-degree factors of the type $(3x + 2)(2x - 1)$ and $(2x - y)(x + 3y)$ mentally. To discover relationships that will make this possible, let us first multiply $(3x + 2)$ and $(2x - 1)$ using a vertical arrangement.

$$\begin{array}{r} 3x + 2 \\ 2x - 1 \\ \hline 6x^2 + 4x \\ -3x - 2 \\ \hline 6x^2 + x - 2 \end{array}$$

Now let us use a horizontal arrangement and try to discover a method that will enable us to carry out the multiplication mentally. We multiply each term in the first binomial times each term in the second binomial to obtain:

F	O	I	L
First product	Outer product	Inner product	Last product

$$(2x - 1)(3x + 2) = 6x^2 \quad + 4x \quad - 3x \quad -2$$

The inner and outer products are like terms, hence combine into one term. Thus,

$$(2x - 1)(3x + 2) = 6x^2 + x - 2$$

To speed up the process we combine the inner and outer products above mentally. The method is called the **FOIL method.** We note again that the product of two first-degree polynomials is a second-degree polynomial.

A simple three-step process for carrying out the FOIL method is illustrated in Example 5.

EXAMPLE 5

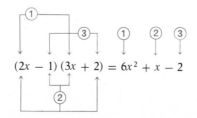

$$(2x - 1)(3x + 2) = 6x^2 + x - 2$$

The like terms are picked up in step 2 and are combined mentally.

(B)

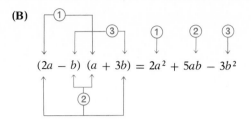

$$(2a - b)(a + 3b) = 2a^2 + 5ab - 3b^2$$

(C)

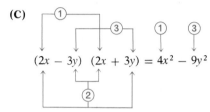

$$(2x - 3y)(2x + 3y) = 4x^2 - 9y^2$$

Notice that the middle term dropped out since its coefficient is 0.

PROBLEM 5   Multiply mentally:

(A) $(2x + 3)(x - 1)$ **(B)** $(a - 2b)(2a + 3b)$ **(C)** $(x - 2y)(x + 2y)$

In Section 5.5 we will consider the reverse problem: given a second-degree polynomial, such as $2x^2 - 5x - 3$ or $3m^2 - 7mn + 2n^2$, find first-degree factors with integer coefficients that will produce these second-degree polynomials as products. To be able to factor second-degree polynomial forms with any degree of efficiency, it is important that you know how to mentally multiply first-degree factors of the types illustrated in this section quickly and accurately.

ANSWERS TO   **5.** (A) $2x^2 + x - 3$;   (B) $2a^2 - ab - 6b^2$;   (C) $x^2 - 4y^2$
MATCHED
PROBLEMS

EXERCISE 5.3   *Multiply mentally:*

A   1. $(x + 1)(x + 2)$ 2. $(y + 3)(y + 1)$
        3. $(y + 3)(y + 4)$ 4. $(x + 3)(x + 2)$
        5. $(x - 5)(x - 4)$ 6. $(m - 2)(m - 3)$
        7. $(n - 4)(n - 3)$ 8. $(u - 5)(u - 3)$
        9. $(s + 7)(s - 2)$ 10. $(t - 6)(t + 4)$
       11. $(m - 12)(m + 5)$ 12. $(a + 8)(a - 4)$
       13. $(u - 3)(u + 3)$ 14. $(t + 4)(t - 4)$
       15. $(x + 8)(x - 8)$ 16. $(m - 7)(m + 7)$
       17. $(y + 7)(y + 9)$ 18. $(x + 8)(x + 11)$
       19. $(c - 9)(c - 6)$ 20. $(u - 8)(u - 7)$

21. $(x - 12)(x + 4)$ 22. $(y - 11)(y + 7)$

23. $(a + b)(a - b)$ 24. $(m - n)(m + n)$

25. $(x + y)(x + 3y)$ 26. $(m + 2n)(m + n)$

B 27. $(x + 2)(3x + 1)$ 28. $(x + 3)(2x + 3)$

29. $(4t - 3)(t - 2)$ 30. $(2x - 1)(x - 4)$

31. $(3y + 7)(y - 3)$ 32. $(t + 4)(2t - 3)$

33. $(2x - 3y)(x + 2y)$ 34. $(3x + 2y)(x - 3y)$

35. $(2x - 1)(3x + 2)$ 36. $(3y - 2)(3y - 1)$

37. $(3y + 2)(3y - 2)$ 38. $(2m - 7)(2m + 7)$

39. $(5s - 1)(s + 7)$ 40. $(a - 6)(5a + 6)$

41. $(3m + 7n)(2m - 5n)$ 42. $(6x - 4y)(5x + 3y)$

43. $(4n - 7)(3n + 2)$ 44. $(5x - 6)(2x + 7)$

45. $(2x - 3y)(3x - 2y)$ 46. $(2s - 3t)(3s - t)$

Since $(a + b)^2 = a^2 + 2ab + b^2$ and $(a - b)^2 = a^2 - 2ab + b^2$, we can formulate a simple **mechanical rule for squaring any binomial:** *The first and last terms in the product are the squares of the first and second terms respectively in the binomial being squared; the middle term in the product is twice the product of the two terms in the binomial. Use this rule to find the following squares.*

47. $(x + 3)^2$ 48. $(x - 4)^2$

49. $(2x - 3)^2$ 50. $(3x + 2)^2$

51. $(2x - 5y)^2$ 52. $(3x + 4y)^2$

53. $(4a + 3b)^2$ 54. $(5m - 3n)^2$

C *Figure out a pattern and use it to mentally multiply the following:*

55. $(x - 1)(x^2 + 2x - 1)$

56. $(x + 2)(x^2 - 3x + 2)$

57. $(2x - 3)(3x^2 - 2x + 4)$

58. $(3x + 2)(2x^2 - 3x - 2)$

59. $(x^2 - 3x + 2)(x^2 + x - 3)$

60. $(x^2 + 2x - 3)(x^2 - 3x - 2)$

5.4 FACTORING OUT COMMON FACTORS

You have already had quite a bit of experience in factoring out common factors in polynomials. The distributive property of real numbers in the form

$$ab + ac = a(b + c)$$

is the important property behind the process.

EXAMPLE 6   Factoring out factors common to all terms:

(A) $6x^2 + 15x = 3x \cdot 2x + 3x \cdot 5 = 3x(2x + 5)$

(B) $2u^3v - 6u^2v^2 + 8uv^3 = 2uv \cdot u^2 - 2uv \cdot 3uv + 2uv \cdot 4v^2$

$$= 2uv(u^2 - 3uv + 4v^2)$$

PROBLEM 6   Factor out factors common to all terms.

(A) $12y^2 - 28y$ (B) $6m^4 - 15m^3n + 9m^2n^2$

Now look closely at the following four examples and try to determine what they have in common.

$$2xy + 3y = y(2x + 3)$$

Because of the commutative property, a common factor may be taken out either on the left or the right.

$$2xA + 3A = A(2x + 3)$$

$$2x(x - 4) + 3(x - 4) = (x - 4)(2x + 3)$$

$$2x(3x + 1) + 3(3x + 1) = (3x + 1)(2x + 3)$$

The factoring involved in each example is essentially the same. The only difference is the nature of the common factors being taken out. In the first two examples the common factors are especially simple. In the last two examples think of $(x - 4)$ and $(3x + 1)$ as single numbers, just as A represents a single number in the second example.

EXAMPLE 7   Removing factors common to all terms.

(A) $5x(x - 1) - (x - 1) = 5x(x - 1) - 1(x - 1) = (x - 1)(5x - 1)$

(B) $3x(2x - y) - 2y(2x - y) = (2x - y)(3x - 2y)$

PROBLEM 7   Remove factors common to all terms.

(A) $3m(m + 2) - (m + 2)$ (B) $2u(u + 3v) - 3v(u + 3v)$

Factoring by grouping

Some polynomials can be factored by grouping terms in such a way that we obtain results that look like Example 7. We can then complete the factoring following the procedures used there. This process will prove useful in Section 5.6 where an efficient method is developed for factoring second-degree polynomials into the product of two first-degree polynomials.

EXAMPLE 8   Factor by grouping.

(A) $2x^2 - 8x + 3x - 12$

(B) $5x^2 - 5x - x + 1$

(C) $6x^2 - 3xy - 4xy + 2y^2$

Solution   (A) $2x^2 - 8x + 3x - 12$ Group the first two and last two terms.

$= (2x^2 - 8x) + (3x - 12)$ Remove common factors from each group.

$= 2x(x - 4) + 3(x - 4)$ The common factor $(x - 4)$ can be taken out.

$= (x - 4)(2x + 3)$ The factoring is complete.

(B) $5x^2 - 5x - x + 1$ Group first two and last two terms. Notice what happens to the signs in the second grouping. (When we clear parentheses we must get back to where we started.)

$= (5x^2 - 5x) - (x - 1)$ Remove common factors from each group.

$= 5x(x - 1) - 1(x - 1)$ The common factor $(x - 1)$ can be taken out.

$= (x - 1)(5x - 1)$ The factoring is complete.

(C) $6x^2 - 3xy - 4xy + 2y^2$ Group the first two and last two terms.

$= (6x^2 - 3xy) - (4xy - 2y^2)$ Signs change inside second parentheses.

$= 3x(2x - y) - 2y(2x - y)$ The common factor $(2x - y)$ can be taken out.

$= (2x - y)(3x - 2y)$ The factoring is complete.

PROBLEM 8   Factor by grouping.

(A) $6x^2 + 2x + 9x + 3$

(B) $3m^2 + 6m - m - 2$

(C) $2u^2 + 6uv - 3uv - 9v^2$

ANSWERS TO
MATCHED
PROBLEMS

6. (A) $4y(3y - 7)$;   (B) $3m^2(2m^2 - 5mn + 3n^2)$
7. (A) $(m + 2)(3m - 1)$;   (B) $(u + 3v)(2u - 3v)$
8. (A) $(3x + 1)(2x + 3)$;   (B) $(m + 2)(3m - 1)$;   (C) $(u + 3v)(2u - 3v)$

EXERCISE 5.4

Write in factored form by removing factors common to all terms.

A

1. $2xA + 3A$	2. $xM - 4M$	3. $10x^2 + 15x$
4. $9y^2 - 6y$	5. $14u^2 - 6u$	6. $20m^2 + 12m$
7. $6u^2 - 10uv$	8. $14x^2 - 21xy$	9. $10m^2n - 15mn^2$
10. $9u^2v + 6uv^2$	11. $2x^3y - 6x^2y^2$	12. $6x^2y^2 - 3xy^3$
13. $3x(x + 2) + 5(x + 2)$	14. $4y(y + 3) + 7(y + 3)$	

15. $3m(m - 4) - 2(m - 4)$

16. $x(x - 1) - 4(x - 1)$

17. $x(x + y) - y(x + y)$

18. $m(m - n) + n(m - n)$

B **19.** $6x^4 - 9x^3 + 3x^2$

20. $6m^4 - 8m^3 - 2m^2$

21. $8x^3y - 6x^2y^2 + 4xy^3$

22. $10u^3v + 20u^2v^2 - 15uv^3$

23. $8x^4 - 12x^3y + 4x^2y^2$

24. $9m^4 - 6m^3n - 6m^2n^2$

25. $3x(2x + 3) - 5(2x + 3)$

26. $2u(3u - 8) - 3(3u - 8)$

27. $x(x + 1) - (x + 1)$

28. $3u(u - 1) - (u - 1)$

29. $4x(2x - 3) - (2x - 3)$

30. $3y(4y - 5) - (4y - 5)$

Replace question marks with algebraic expressions that will make both sides equal.

31. $3x^2 - 3x + 2x - 2 = (3x^2 - 3x) + (\ ?\ )$

32. $2x^2 + 4x + 3x + 6 = (2x^2 + 4x) + (\ ?\ )$

33. $3x^2 - 12x - 2x + 8 = (3x^2 - 12x) - (\ ?\ )$

34. $2y^2 - 10y - 3y + 15 = (2y^2 - 10y) - (\ ?\ )$

35. $8u^2 + 4u - 2u - 1 = (8u^2 + 4u) - (\ ?\ )$

36. $6x^2 + 10x - 3x - 5 = (6x^2 + 10x) - (\ ?\ )$

Factor out common factors from each group, then complete the factoring if possible.

37. $(3x^2 - 3x) + (2x - 2)$

38. $(2x^2 + 4x) + (3x + 6)$

39. $(3x^2 - 12x) - (2x - 8)$

40. $(2y^2 - 10y) - (3y - 15)$

41. $(8u^2 + 4u) - (2u + 1)$

42. $(6x^2 + 10x) - (3x + 5)$

Factor as the product of two first-degree polynomials using grouping. (These problems are related to Problems 37–42.)

43. $3x^2 - 3x + 2x - 2$

44. $2x^2 + 4x + 3x + 6$

45. $3x^2 - 12x - 2x + 8$

46. $2y^2 - 10y - 3y + 15$

47. $8u^2 + 4u - 2u - 1$

48. $6x^2 + 10x - 3x - 5$

Factor as the product of two first-degree factors using grouping.

49. $2m^2 - 8m + 5m - 20$

50. $5x^2 - 10x + 2x - 4$

51. $6x^2 - 9x - 4x + 6$

52. $12x^2 + 8x - 9x - 6$

C **53.** $3u^2 - 12u - u + 4$

54. $6m^2 + 4m - 3m - 2$

55. $6u^2 + 3uv - 4uv - 2v^2$

56. $2x^2 - 4xy - xy + 2y^2$

57. $6x^2 + 3xy - 10xy - 5y^2$

58. $4u^2 - 16uv - 3uv + 12v^2$

5.5 FACTORING SECOND-DEGREE POLYNOMIALS

We now turn our attention to factoring second-degree polynomials such as

$$2x^2 - 5x - 3$$

$$2x^2 + 3xy - 2y^2$$

into the product of two first-degree polynomials with integer coefficients. By now it should be very easy for you to obtain (by mental multiplication) the products

$$(x - 3)(x + 2) = x^2 - x - 6$$

$$(x - 3y)(x + 2y) = x^2 - xy - 6y^2$$

but can you reverse the process? Can you, for example, find integers a, b, c, and d so that

$$2x^2 - 5x - 3 = (ax + b)(cx + d)$$

Representing a second-degree polynomial with integers as coefficients as the product of two first-degree polynomials with integer coefficients is not as easy as multiplying first-degree polynomials. In this section we will develop a method of attack that is relatively easy to understand, but not always easy to apply. In the next section we will develop an approach to the problem that builds on factoring by grouping discussed in the last section. This approach is a little harder to understand, but once the method is understood, it is fairly easy to apply.

Let us start with a very simple polynomial whose factors you will likely guess at once:

$$x^2 + 6x + 8$$

Our problem is to find two first-degree factors with integer coefficients, if they exist. To start we write

$$x^2 + 6x + 8 = (x \qquad)(x \qquad)$$

The coefficients of x on the right are both 1, since the coefficient of x^2 on the left is 1. Since the constant terms in each factor must be factors of $+8$, they are either both positive or both negative. They both must be positive since the middle term on the left is positive. (If the middle term were negative, they would both have to be negative.) Think this through. We always put in the parentheses what we know for certain first. In this case,

The coefficients of x are both 1.

The signs of both constant terms are $+$.

Thus, we are able to write

$$x^2 + 6x + 8 = (x + \qquad)(x + \qquad)$$

Now, what are the constant terms? Since they are positive integer factors of 8, we list the possibilities

$$\frac{8}{\begin{array}{l} 1 \cdot 8 \\ 2 \cdot 4 \end{array}}$$

If we try the first pair (mentally) we obtain the first and last terms, but not the middle term. We next try 2 and 4 which gives us the middle term as well as the first and last terms. Thus,

$$x^2 + 6x + 8 = (x + 2)(x + 4)$$

Because of the commutative property, we could also write

$$x^2 + 6x + 8 = (x + 4)(x + 2)$$

Let us try another polynomial. Find first-degree factors with integer coefficients for

$$2x^2 - 7x + 6$$

Again, we write

$$2x^2 - 7x + 6 = (\ x \qquad)(\ x \qquad)$$

leaving space for the coefficients of x and the constant terms, and then insert what we know for certain:

The coefficient of one x is 2 and the other is 1.

The signs of both constant terms are — (why?).

Thus,

$$2x^2 - 7x + 6 = (2x - \ )(x - \ )$$

The constant terms both must be factors of 6. The possibilities are

$$\frac{6}{\begin{array}{l} 1 \cdot 6 \\ 6 \cdot 1 \\ 2 \cdot 3 \\ 3 \cdot 2 \end{array}}$$ All pairs result in the first and last terms of $2x^2 - 7x + 6$, but will any give the middle term, $-7x$?

Testing each pair (this is why you need to do binomial multiplication mentally), we find that the last pair gives the middle term. Thus,

$$2x^2 - 7x + 6 = (2x - 3)(x - 2)$$

Before you conclude that all second-degree polynomials with integer coefficients have first-degree factors with integer coefficients, consider the following simple polynomial:

$$x^2 + x + 2$$

Proceeding as above, we write

$$x^2 + x + 2 = (x + \ )(x + \ )$$

$$\frac{2}{\begin{matrix}1\cdot 2\\2\cdot 1\end{matrix}}$$

and find that neither combination produces the middle term x. Hence, we conclude that

$$x^2 + x + 2$$

has no first-degree factors with integer coefficients, and we say the polynomial is not factorable (using integers).

EXAMPLE 9 Factor each polynomial, if possible, using integer coefficients.

(A) $2x^2 + 3xy - 2y^2$

(B) $x^2 - 3x + 4$

(C) $6x^2 + 5xy - 4y^2$

Solution (A) $2x^2 + 3xy - 2y^2$

$= (2x \quad y)(x \quad y)$ Signs must be opposite. (We can reverse this
                                  choice if we get $-3xy$ instead of $+3xy$ for the
$= (2x + y)(x - \ y)$ middle term.)

Now what are the factors of 2 (the coefficient of y^2)?

$$\frac{2}{\begin{matrix}1\cdot 2\\2\cdot 1\end{matrix}}$$

The first choice gives us $-3xy$ for the middle term—close, but not there—so we reverse our choice of signs to obtain

$$2x^2 + 3xy - 2y^2 = (2x - y)(x + 2y)$$

(B) $x^2 - 3x + 4 = (x - \ )(x - \ )$

$$\frac{4}{\begin{matrix}2\cdot 2\\1\cdot 4\\4\cdot 1\end{matrix}}$$

No choice produces the middle term; hence,

$$x^2 - 3x + 4$$

is not factorable (using integer coefficients).

(C) $6x^2 + 5xy - 4y^2 = (\ x + \ y)(\ x - \ y)$

The signs must be opposite in the factors, since the third term is negative. (We can reverse our choice of signs later, if necessary.)

We now write all factors of 6 and of 4:

6	4
$2 \cdot 3$	$2 \cdot 2$
$3 \cdot 2$	$1 \cdot 4$
$1 \cdot 6$	$4 \cdot 1$
$6 \cdot 1$	

and try each choice on the left with each on the right—a total of 24 combinations (if we count reversing the signs) that give us the first and last terms in $6x^2 + 5xy - 4y^2$. The question is, does any combination also give us the middle term, $5xy$? After trial and error and, perhaps, some educated guessing among the choices, we find that $3 \cdot 2$ matched with $4 \cdot 1$ gives us the middle term. Thus,

$$6x^2 + 5xy - 4y^2 = (3x + 4y)(2x - y)$$

If none of the 24 combinations (including reversing our sign choice) produced the middle term, then we would conclude that the polynomial is not factorable using integer coefficients.

PROBLEM 9   Factor each polynomial, if possible, using integer coefficients.

(A) $x^2 - 8x + 12$

(B) $x^2 + 2x + 5$

(C) $2x^2 + 7xy - 4y^2$

(D) $4x^2 - 15xy - 4y^2$

ANSWERS TO MATCHED PROBLEMS   **9.** (A) $(x - 2)(x - 6)$; (B) Not factorable using integer coefficients; (C) $(2x - y)(x + 4y)$; (D) $(4x + y)(x - 4y)$

Concluding remarks

It is about here that many students begin to lose interest in factoring, particularly with problems like that found in Example 9**C**. They quickly observe that if the coefficients of the first and last terms get larger and larger with more and more factors, the number of combinations that need to be checked increases very rapidly. And it is quite possible in most practical situations that none of the combinations will work. It is important, however, that you understand the approach presented above, since it will work for most of the simpler factoring problems you will encounter. If you have the time and inclination, read the next (optional) section for a systematic approach to the problem of factoring that will reduce substantially the amount of trial and error and even tell you whether the polynomial can be factored before you proceed too far.

In conclusion, we point out that if a, b, and c are selected at random out of the integers, the probability that

$$ax^2 + bx + c$$

is not factorable in the integers is much greater than the probability that it is. But even being able to factor some second-degree polynomials leads to marked simplification of some algebraic expressions and an easy way to solve some second-degree equations, as will be seen later.

EXERCISE 5.5

Factor in the integers, if possible. If not factorable, say so.

A

1. $x^2 + 5x + 4$
2. $x^2 + 4x + 3$
3. $x^2 + 5x + 6$
4. $x^2 + 7x + 10$
5. $x^2 - 4x + 3$
6. $x^2 - 5x + 4$
7. $x^2 - 7x + 10$
8. $x^2 - 5x + 6$
9. $y^2 + 3y + 3$
10. $y^2 + 2y + 2$
11. $y^2 - 2y + 6$
12. $x^2 - 3x + 5$
13. $x^2 + 8xy + 15y^2$
14. $x^2 + 9xy + 20y^2$
15. $x^2 - 10xy + 21y^2$
16. $x^2 - 10xy + 16y^2$
17. $u^2 + 4uv + v^2$
18. $u^2 + 5uv + 3v^2$
19. $3x^2 + 7x + 2$
20. $2x^2 + 7x + 3$
21. $3x^2 - 7x + 4$
22. $2x^2 - 7x + 6$

B

23. $3x^2 - 14x + 8$
24. $2y^2 - 13y + 15$
25. $3x^2 - 11xy + 6y^2$
26. $2x^2 - 7xy + 6y^2$
27. $n^2 - 2n - 8$
28. $n^2 + 2n - 8$
29. $x^2 - 4x - 6$
30. $x^2 - 3x - 8$
31. $3x^2 - x - 2$
32. $6m^2 + m - 2$
33. $x^2 + 4xy - 12y^2$
34. $2x^2 - 3xy - 2y^2$
35. $3u^2 - 11u - 4$
36. $8u^2 + 2u - 1$
37. $6x^2 + 7x - 5$
38. $2m^2 - 3m - 20$
39. $3s^2 - 5s - 2$
40. $2s^2 + 5s - 3$
41. $3x^2 + 2xy - 3y^2$
42. $2x^2 - 3xy - 4y^2$
43. $5x^2 - 8x - 4$
44. $12x^2 + 16x - 3$
45. $6u^2 - uv - 2v^2$
46. $6x^2 - 7xy - 5y^2$
47. $8x^2 + 6x - 9$
48. $6x^2 - 13x + 6$
49. $3u^2 + 7uv - 6v^2$
50. $4m^2 + 10mn - 6n^2$
51. $4u^2 - 19uv + 12v^2$
52. $12x^2 - xy - 6y^2$

C

53. $12x^2 - 40xy - 7y^2$
54. $15x^2 + 17xy - 4y^2$
55. $12x^2 + 19xy - 10y^2$
56. $24x^2 - 31xy - 15y^2$

5.6 ac TEST AND FACTORING SECOND-DEGREE POLYNOMIALS (OPTIONAL)

We continue our discussion of factoring second-degree polynomials of the type

$$ax^2 + bx + c$$
$$ax^2 + bxy + cy^2$$

(1)

with integer coefficients into the product of two first-degree factors with integer coefficients. In the last section we found that the number of cases that had to be tested tended to increase very rapidly as the coefficients a and c increased in size. And then, in realistic situations, it turns out that it is quite likely that none of the combinations tested will work. It would be useful to know ahead of time if the polynomials in (1) are, in fact, factorable before we start looking for the factors. We now provide a test, called the ac test for factorability, that not only tells us if the polynomials in (1) can be factored using integer coefficients, but, in addition, leads to a direct way of factoring those that are factorable.

---

ac Test for Factorability

If in equations (1) the product ac has two integer factors p and q whose sum is the coefficient of the middle term b, that is, if integers p and q exist so that

$$pq = ac \quad \text{and} \quad p + q = b$$

(2)

then equations (1) have first-degree factors with integer coefficients. If no integers p and q exist that satisfy (2), then equations (1) will not have first-degree factors with integer coefficients.

Once we find integers p and q in the ac test, if they exist, then our work is almost finished, since we can then write equations (1), splitting the middle term, in the forms

$$ax^2 + px + qx + c$$
$$ax^2 + pxy + qxy + cy^2$$

(3)

and the factoring can be completed in a couple of steps using factoring by grouping discussed at the end of Section 5.4.

We now make the discussion concrete through several examples.

EXAMPLE 10   Factor, if possible, using integer coefficients.

(A) $2x^2 + 11x - 6$

(B) $4x^2 - 7x + 4$

(C) $6x^2 + 5xy - 4y^2$

Solution   (A) $2x^2 + 11x - 6$

Step 1 Use the *ac* test

$ac = (2)(-6) = -12$

Try to find two integer factors of -12 whose sum is $b = 11$. We write (or think) all two-integer factors of -12:

$$\frac{pq}{}$$
$(-3)(4)$
$(3)(-4)$
$(2)(-6)$
$(-2)(6)$
$(-1)(12)$
$(1)(-12)$

and check to see if any of these pairs add up to 11, the coefficient of the middle term. We see that the next to the last pair works; that is,

$$\overset{p \quad q}{(-1)(12)} = \overset{ac}{-12} \quad \text{and} \quad \overset{p \quad q}{(-1) + (12)} = \overset{b}{11}$$

We can conclude, because of the *ac* test, that $2x^2 + 11x - 6$ can be factored using integer coefficients.

Step 2 Now split the middle term in the original equation using $p = -1$ and $q = 12$ as coefficients for x. This is possible, since $p + q = b$.

$$2x^2 + \overset{b}{11x} - 6 = 2x^2 \overset{p}{-1x} + \overset{q}{12x} - 6$$

Step 3 Factor the result obtained in Step 2 by grouping (this will always work if we can get to Step 2, and it doesn't matter whether the values for p and q are reversed).

$2x^2 - x + 12x - 6$ Group first two and last two terms.

$= (2x^2 - x) + (12x - 6)$ Factor out common factors.

$= x(2x - 1) + 6(2x - 1)$ Factor out the common factor $(2x - 1)$.

$= (2x - 1)(x + 6)$ The factoring is complete.

Thus,

$2x^2 + 11x - 6 = (2x - 1)(x + 6)$

The process above can be reduced to a few key operational steps when all the commentary is eliminated and some of the process is done mentally. The only trial and error occurs in Step 1, and with a little practice that step will go fairly fast.

(B) $4x^2 - 7x + 4$

Compute ac: $ac = (4)(4) = 16$

Write (or think) all two integer factors of 16, and try to find a pair whose sum is -7, the coefficient of the middle term.

pq
(4)(4)
$(-4)(-4)$
(2)(8)
$(-2)(-8)$
(1)(16)
$(-1)(-16)$

None of these adds up to $-7 = b$; thus, according to the ac test,

$4x^2 - 7x + 4$

is not factorable using integer coefficients.

(C) $6x^2 + 5xy - 4y^2$

Compute ac: $ac = (6)(-4) = -24$

Does -24 have two integer factors whose sum is $5 = b$? A little trial and error (either mentally or by listing) gives us

$$\overset{p\quad q\qquad ac}{(8)(-3) = -24} \quad \text{and} \quad \overset{p\quad q\quad b}{8 + (-3) = 5}$$

Now we split the middle term $5xy$ into $8xy - 3xy$ (using the p and q just found) and write

$$6x^2 + \overset{b}{5xy} - 4y^2 = 6x^2 + \overset{p}{8xy} - \overset{q}{3xy} - 4y^2$$

then complete the factoring by grouping:

$$6x^2 + 8xy - 3xy - 4y^2 = (6x^2 + 8xy) - (3xy + 4y^2)$$
$$= 2x(3x + 4y) - y(3x + 4y)$$
$$= (3x + 4y)(2x - y)$$

Thus,

$$6x^2 + 5xy - 4y^2 = (3x + 4y)(2x - y)$$

PROBLEM 10　Factor, if possible, using integer coefficients.

(A)　$4x^2 + 4x - 3$

(B)　$6x^2 - 3x - 4$

(C)　$6x^2 - 25xy + 4y^2$

ANSWERS TO
MATCHED
PROBLEMS

10.　(A) $(2x - 1)(2x + 3)$;　　(B) Not factorable;　　(C) $(6x - y)(x - 4y)$

EXERCISE 5.6

Factor, if possible, using integer coefficients. Use the ac test and proceed as in Example 10.

A
1. $3x^2 - 7x + 4$
2. $2x^2 - 7x + 6$
3. $x^2 + 4x - 6$
4. $x^2 - 3x - 8$
5. $2x^2 + 5x - 3$
6. $3x^2 - 5x - 2$
7. $3x^2 - 5x + 4$
8. $2x^2 - 11x + 6$

B
9. $3x^2 - 14x + 8$
10. $2y^2 - 13y + 15$
11. $6x^2 + 7x - 5$
12. $5x^2 - 8x - 4$
13. $6x^2 - 4x - 5$
14. $5x^2 - 7x - 4$
15. $2m^2 - 3m - 20$
16. $12x^2 + 16x - 3$
17. $3u^2 - 11u - 4$
18. $8u^2 + 2u - 1$
19. $6u^2 - uv - 2v^2$
20. $6x^2 - 7xy - 5y^2$
21. $3x^2 + 2xy - 3y^2$
22. $2x^2 - 3xy - 4y^2$
23. $8x^2 + 6x - 9$
24. $6x^2 - 5x - 6$
25. $4m^2 + 10mn - 6n^2$
26. $3u^2 + 7uv - 6v^2$
27. $3u^2 - 8uv - 6v^2$
28. $4m^2 - 9mn - 6n^2$
29. $4u^2 - 19uv + 12v^2$
30. $12x^2 - xy - 6y^2$

C
31. $12x^2 - 40xy - 7y^2$
32. $15x^2 + 17xy - 4y^2$
33. $18x^2 - 9xy - 20y^2$
34. $15m^2 + 2mn - 24n^2$
35. Find all integers b such that $x^2 + bx + 12$ can be factored.
36. Find all positive integers c under 15 so that $x^2 - 7x + c$ can be factored.

5.7 MORE FACTORING

In this last section on factoring we will consider a couple of additional factoring forms and combinations of forms we have already considered.

Sum and difference of two squares

Look at the following products and try to determine what they all have in common:

$$(x - 3)(x + 3) = x^2 - 9$$

$$(2x + 4)(2x - 4) = 4x^2 - 16$$

$$(x - 2y)(x + 2y) = x^2 - 4y^2$$

$$(A - B)(A + B) = A^2 - B^2$$

We note that the binomial factors on the left of each are the same except for the signs. Thus, when each pair is multiplied, the middle term drops out. Looking at the expressions on the right we see that each is the difference of two squares. Writing the last result in reverse order, we obtain a factoring formula for the difference of two squares. If we try to factor the sum of two squares, $A^2 + B^2$, we find that it cannot be factored using integer coefficients. (Try it to see why.)

Sum and Difference of Two Squares

$A^2 + B^2$ cannot be factored using integer coefficients (unless A and B have common factors)

$$A^2 - B^2 = (A - B)(A + B)$$

EXAMPLE 11 (A) $x^2 - y^2 = (x - y)(x + y)$

(B) $4x^2 - 9 = (2x - 3)(2x + 3)$

(C) $u^2 + v^2$ is not factorable using integer coefficients

(D) $9m^2 - 25n^2 = (3m - 5n)(3m + 5n)$

PROBLEM 11 Factor, if possible, using integer coefficients.

(A) $x^2 - 4$ (B) $4x^2 - 9y^2$

(C) $4m^2 + n^2$ (D) $16x^2 - 5$

Combined forms

We now consider some examples that involve common factors in combination with other factoring considered earlier.

As a general principle:

Always remove common factors first before proceeding to other methods.

EXAMPLE 12 Factor as far as possible using integer coefficients.

(A) $4x^3 - 14x^2 + 6x$

(B) $18x^3 - 8x$

(C) $8x^3y + 20x^2y^2 - 12xy^3$

(D) $3y^3 + 6y^2 + 6y$

Solution (A) $4x^3 - 14x^2 + 6x$ Remove common factors first.

$= 2x(2x^2 - 7x + 3)$ Factor further, if possible.

$= 2x(2x - 1)(x - 3)$ Factoring is complete.

(B) $18x^3 - 8x$ Remove common factors.

$= 2x(9x^2 - 4)$ Factor the difference of two squares.

$= 2x(3x - 2)(3x + 2)$ Factoring is complete.

(C) $8x^3y + 20x^2y^2 - 12xy^3$

$= 4xy(2x^2 + 5xy - 3y^2)$

$= 4xy(2x - y)(x + 3y)$

(D) $3y^3 + 6y^2 + 6y$

$= 3y(y^2 + 2y + 2)$ Cannot be factored further using integer coefficients.

PROBLEM 12   Factor as far as possible using integer coefficients.

(A) $3x^3 - 15x^2y + 18xy^2$ (B) $3x^3 - 48x$

(C) $3x^3y + 3x^2y - 36xy$ (D) $4x^3 + 12x^2 + 12x$

Factoring by grouping

Occasionally, polynomial forms of a more general nature than we considered in Section 5.4 can be factored by appropriate grouping of terms. The following example illustrates the process.

EXAMPLE 13   (A) $x^2 + xy + 2x + 2y$ Group the first two and last two terms.

$= (x^2 + xy) + (2x + 2y)$ Remove common factors.

$= x(x + y) + 2(x + y)$ Since each term has the common factor $(x + y)$, we can complete the factoring.

$= (x + y)(x + 2)$ Notice the factors are not first-degree polynomials of the same type.

(B) $x^2 - 2x - xy + 2y$

$= (x^2 - 2x) - (xy - 2y)$ Be careful of signs here.

$= x(x - 2) - y(x - 2)$

$= (x - 2)(x - y)$

PROBLEM 13   Factor by grouping terms.

(A) $x^2 - xy + 5x - 5y$ **(B)** $x^2 + 4x - xy - 4y$

ANSWERS TO MATCHED PROBLEMS

11. (A) $(x - 2)(x + 2)$;   (B) $(2x - 3y)(2x + 3y)$;   (C) Not factorable using integers;   (D) Not factorable using integers

12. (A) $3x(x - 2y)(x - 3y)$;   (B) $3x(x - 4)(x + 4)$ (C) $3xy(x - 3)$ $(x + 4)$;   (D) $4x(x^2 + 3x + 3)$

13. (A) $(x - y)(x + 5)$;   (B) $(x + 4)(x - y)$

EXERCISE 5.7

Factor as far as possible using integer coefficients.

A   **1.** $6x^3 + 9x^2$ **2.** $8x^2 + 2x$

3. $u^4 + 6u^3 + 8u^2$ **4.** $m^5 + 8m^4 + 15m^3$

5. $x^3 - 5x^2 + 6x$ **6.** $x^3 - 7x^2 + 12x$

7. $x^2 - 4$ **8.** $x^2 - 1$

9. $4x^2 - 1$ **10.** $9x^2 - 4$

11. $u^2 + v^2$ **12.** $m^2 + 64$

13. $2x^2 - 8$ **14.** $3x^2 - 3$

B   **15.** $9x^2 - 16y^2$ **16.** $25x^2 - 1$

17. $6u^2v^2 - 3uv^3$ **18.** $2x^3y - 6x^2y^3$

19. $4x^3y - xy^3$ **20.** $x^3y - 9xy^3$

21. $3x^4 + 27x^2$ **22.** $2x^3 + 8x$

23. $6x^2 + 36x + 48$ **24.** $4x^2 - 28x + 48$

25. $3x^3 - 6x^2 + 15x$ **26.** $2x^3 - 2x^2 + 8x$

27. $9u^2 + 4v^2$ **28.** $x^2 + 16y^2$

29. $12x^3 + 16x^2y - 16xy^2$ **30.** $9x^2y + 3xy^2 - 30y^3$

31. $x^2 + 3x + xy + 3y$ **32.** $xy + 2x + y^2 + 2y$

33. $x^2 - 3x - xy + 3y$ **34.** $x^2 - 5x + xy - 5y$

35. $2ac + bc - 6ad - 3bd$ **36.** $2ac + 4bc - ad - 2bd$

37. $2mu + 2nu - mv - nv$ **38.** $3wx + 6wy - xz - 2yz$

C   **39.** $4x^3y + 14x^2y^2 + 6xy^3$ **40.** $3x^3y - 15x^2y^2 + 18xy^3$

41. $60x^2y^2 - 200xy^3 - 35y^4$ **42.** $60x^4 + 68x^3y - 16x^2y^2$

By noting that

$$(A - B)(A^2 + AB + B^2) = A^3 - B^3$$

$$(A + B)(A^2 - AB + B^2) = A^3 + B^3$$

we have factoring formulas for the sum and difference of two cubes. Use these formulas for the following problems:

43. $x^3 - 8$ **44.** $x^3 + 1$

45. $x^3 + 27$ **46.** $8y^3 - 1$

47. $x^6 - y^6$

5.8 ALGEBRAIC LONG DIVISION

There are times when it is useful to find quotients of polynomials by a long-division process similar to that used in arithmetic. Several examples will illustrate the process.

EXAMPLE 14 **(A)** Divide: $2x^2 + 5x - 12$ by $x + 4$

Solution

$x + 4\overline{)2x^2 + 5x - 12}$

Both polynomials are arranged in descending powers of the variable if this is not already done.

$$\begin{array}{r} 2x \\ x + 4\overline{)2x^2 + 5x - 12} \end{array}$$

Divide the first term of the divisor into the first term of the dividend, i.e., what must x be multiplied by so that the product is exactly $2x^2$? Answer: $2x$

$$\begin{array}{r} \overset{\frown}{(2x)} \\ x + 4\overline{)2x^2 + 5x - 12} \\ \underline{2x^2 + 8x} \\ -3x - 12 \end{array}$$

Multiply the divisor by $2x$, line up like terms, subtract, and bring down -12 from above.

$$\begin{array}{r} 2x - 3 \\ x + 4\overline{)2x^2 + 5x - 12} \\ \underline{2x^2 + 8x} \\ -3x - 12 \\ \underline{-3x - 12} \\ 0 \end{array}$$

Repeat the process above until the degree of the remainder is less than that of the divisor, or there is no remainder.

CHECK

$(x + 4)(2x - 3) = 2x^2 + 5x - 12$

(B) Divide: $x^3 + 8$ by $x + 2$

Solution

$$\begin{array}{r} x^2 - 2x + 4 \\ x + 2\overline{)x^3 + 0x^2 + 0x + 8} \\ \underline{x^3 + 2x^2} \\ -2x^2 + 0x \\ \underline{-2x^2 - 4x} \\ 4x + 8 \\ \underline{4x + 8} \\ 0 \end{array}$$

Insert, with 0 coefficients, any missing terms of lower degree than 3, and proceed as in part **A**.

The check is left to the reader.

(C) Divide: $3 - 7x + 6x^2$ by $3x + 1$

Solution

$$3x + 1 \overline{) \begin{array}{l} 2x - 3 \\ 6x^2 - 7x + 3 \\ \underline{6x^2 + 2x} \\ -9x + 3 \\ \underline{-9x - 3} \\ 6 = \text{R} \quad \text{(remainder)} \end{array}}$$

Arrange $3 - 7x + 6x^2$ in descending powers of x, then proceed as above until the degree of the remainder is less than the degree of the divisor.

CHECK

Just as in arithmetic, when there is a remainder we check by adding the remainder to the product of the divisor and quotient. Thus

$$(3x + 1)(2x - 3) + 6 \overset{?}{=} 6x^2 - 7x + 3$$

$$6x^2 - 7x - 3 + 6 \overset{?}{=} 6x^2 - 7x + 3$$

$$6x^2 - 7x + 3 \overset{\checkmark}{=} 6x^2 - 7x + 3$$

PROBLEM 14 Divide, using the long-division process, and check:

(A) $(2x^2 + 7x + 3)/(x + 3)$ **(B)** $(x^3 - 8)/(x - 2)$

(C) $(2 - x + 6x^2)/(3x - 2)$

ANSWERS TO MATCHED PROBLEMS

14. **(A)** $2x + 1$;   **(B)** $x^2 + 2x + 4$;   **(C)** $2x + 1$, R $= 4$

EXERCISE 5.8

Divide, using the long-division process. Check the answers.

A
1. $(x^2 + 5x + 6)/(x + 3)$
2. $(x^2 + 6x + 8)/(x + 4)$
3. $(2x^2 + x - 6)/(x + 2)$
4. $(3x^2 - 5x - 2)/(x - 2)$
5. $(2x^2 - 3x - 4)/(x - 3)$
6. $(3x^2 - 11x - 1)/(x - 4)$
7. $(2m^2 + m - 10)/(2m + 5)$
8. $(3y^2 + 5y - 12)/(3y - 4)$
9. $(6x^2 + 5x - 6)/(3x - 2)$
10. $(8x^2 - 14x + 3)/(2x - 3)$
11. $(6x^2 + 11x - 12)/(3x - 2)$
12. $(6x^2 + x - 13)/(2x + 3)$
13. $(3x^2 + 13x - 12)/(3x - 2)$
14. $(2x^2 - 7x - 1)/(2x + 1)$

B
15. $(x^2 - 4)/(x - 2)$
16. $(y^2 - 9)/(y + 3)$
17. $(m^2 - 7)/(m - 3)$
18. $(u^2 - 18)/(u + 4)$
19. $(8c + 4 + 5c^2)/(c + 2)$
20. $(4a^2 - 22 - 7a)/(a - 3)$
21. $(9x^2 - 8)/(3x - 2)$
22. $(8x^2 + 7)/(2x - 3)$
23. $(5y^2 - y + 2y^3 - 6)/(y + 2)$
24. $(x - 5x^2 + 10 + x^3)/(x + 2)$
25. $(x^3 - 1)/(x - 1)$

26. $(x^3 + 27)/(x + 3)$
27. $(x^4 - 16)/(x + 2)$
28. $(x^5 + 32)/(x - 2)$
29. $(3y - y^2 + 2y^3 - 1)/(y + 2)$
30. $(3 + x^3 - x)/(x - 3)$

C 31. $(4x^4 - 10x - 9x^2 - 10)/(2x + 3)$
32. $(9x^4 - 2 - 6x - x^2)/(3x - 1)$
33. $(16x - 5x^3 - 4 + 6x^4 - 8x^2)/(2x - 4 + 3x^2)$
34. $(8x^2 - 7 - 13x + 24x^4)/(3x + 5 + 6x^2)$

5.9 CHAPTER REVIEW: IMPORTANT TERMS AND SYMBOLS, REVIEW EXERCISE, PRACTICE TEST

Important terms and symbols

polynomial (*5.1*)   degree (*5.1*)   monomial (*5.1*)
binomial (*5.1*)   trinomial (*5.1*)   addition and subtraction
of polynomials (*5.2*)   multiplication of polynomials (*5.2*)
FOIL method of multiplying (*5.3*)   Factoring by
grouping (*5.4, 5.7*)   factoring trinomials (*5.5*)   *ac* test (*5.6*)
sum and difference of squares (*5.7*)   algebraic long
division (*5.8*)

**Exercise 5.9
Review exercise**

A 1. Add: $2x^2 + x - 3$, $2x - 3$, and $3x^2 + 2$.
2. Subtract: $3x^2 - x - 2$ from $5x^2 - 2x + 5$.
3. Divide, using algebraic long division: $(6x^2 + 5x - 2)/(2x - 1)$.

Perform the indicated operations and simplify.

4. $(3x - 2)(2x + 5)$ 5. $(2u - 3v)(3u + 4v)$
6. $(2x^2 - 3x + 1) + (3x^3 - x^2 + 5)$ 7. $(2x^2 - 3x + 1) - (3x^3 - x^2 + 5)$

Factor, if possible, using integer coefficients.

8. $x^2 - 9x + 14$ 9. $3x^2 - 10x + 8$
10. $x^2 - 3x - 3$ 11. $4x^2y - 6xy^2$
12. $x^3 - 5x^2 + 6x$ 13. $4u^2 - 9$
14. $x(x - 1) + 3(x - 1)$ 15. $m^2 + 4n^2$

B 16. What is the degree of the polynomial $3x^5 - 2x^3 + 7x^2 - x + 2$? What is the degree of the fourth term? Of the fifth term?
17. Subtract $2x^2 - 5x - 6$ from the product $(2x - 1)(2x + 1)$.

Divide, using algebraic long division.

18. $(2x^2 - 7x - 1)/(2x + 1)$ 19. $(2 - 10x + 9x^3)/(3x - 2)$

Perform the indicated operations and simplify.

20. $(2x - 3)(2x^2 - 3x + 2)$
21. $(9x^2 - 4)(3x^2 + 7x - 6)$
22. $(a + b)(a^2 - ab + b^2)$
23. $[(3x^2 - x + 1) - (x^2 - 4)] - [(2x - 5)(x + 3)]$

Factor, if possible, using integer coefficients.

24. $3u^2 - 12$
25. $2x^2 - xy - 3y^2$
26. $x^2 - xy + y^2$
27. $6y^3 + 3y^2 - 45y$
28. $2x^3 - 4x^2y - 10xy^2$
29. $12x^3y + 27xy^3$
30. $x^2(x - 1) - 9(x - 1)$

Factor (using integer coefficients) by grouping.

31. $x^2 - xy + 4x - 4y$
32. $x^2 + xy - 3x - 3y$
33. $2u^2 - 3u + 6u - 9$
34. $6x^2 + 4x - 3x - 2$

C 35. Simplify: $[-2xy(x^2 - 4y^2)] - [-2xy(x - 2y)(x + 2y)]$.
36. Divide, using algebraic long division: $(20x - 14x^2 + 2x^4 + 4)/(6 + 2x)$.

Factor, if possible, using integer coefficients.

37. $36x^3y + 24x^2y^2 - 45xy^3$
38. $12u^4 - 12u^3v - 20u^3v^2$
39. $6ac + 4bc - 12ad - 8bd$
40. $12ux - 15vx - 4u + 5v$
41. $8x^3 + 1$
42. Find all integers b such that $2x^2 + bx - 4$ can be factored using integer coefficients.

**Practice test
Chapter 5**

1. (A) What is the degree of the third term in $6x^4 - 7x^2 + 3x - 5$? (B) What is the degree of the polynomial?
2. Add $x^2 - x + 1$, $3x + 2$, and $2x^2 - 5$.
3. Subtract $2x^2 - 2x + 1$ from $4x^2 - 3x + 5$.
4. Simplify $(2x - 5)(x - 3) - (3x - 2)^2$.
5. Divide, using algebraic long division: $(1 - 8x + 8x^3)/(2x - 1)$.

Factor, if possible, using integer coefficients.

6. (A) $y^2 - 4$
   (B) $3m^4 + 12m^2$
7. $3x(2x - y) - 2y(2x - y)$

8. $3x^2 + 10xy - 8y^2$

9. $3x^2 + 5x - 4$

10. $18x^4y - 9x^3y^2 - 3x^2y^3$

11. $u^2 - 5u - uv + 5v$

12. Find all integers b such that $x^2 + bx + 6$ can be factored using integer coefficients.

6

ALGEBRAIC FRACTIONS

6.1 RATIONAL EXPRESSIONS

Fractional forms in which the numerator and denominator are polynomials are called **rational expressions.**

$$\frac{1}{x} \qquad \frac{3}{y-5} \qquad \frac{x-2}{2x^2-2x+5} \qquad \frac{x^2-3xy+y^2}{x^2-y^2}$$

are all rational expressions.

In Chapter 3 we worked with simple fractional forms. In this chapter we will use the same basic ideas established there on more complex fractional forms.

The **fundamental principle of real fractions**

$$\frac{ak}{bk} = \frac{a}{b} \qquad b, k \neq 0 \tag{1}$$

will play an important role in our work.

6.2 REDUCING TO LOWEST TERMS

If the numerator and denominator in a quotient of two polynomials (a rational form) contain a common factor, it may be canceled out using the fundamental principle of fractions (1). Several examples should make the process clear. In the examples it is important to keep in mind:

▶ *Common factors cancel.*
▶ *Common terms do not cancel.*

EXAMPLE 1 **(A)**
$$\frac{5(x+3)}{2x(x+3)} = \frac{5(\cancel{x+3})}{2x(\cancel{x+3})} \qquad \text{Cancel common factors.}$$

$$= \frac{5}{2x}$$

(B)
$$\frac{6x^2-3x}{3x} = \frac{\cancel{3x}(2x-1)}{\cancel{3x}} \qquad \text{Factor top; then cancel common factors.}$$

$$= 2x - 1 \qquad \text{NOTE: } \frac{6x^2-\cancel{3x}}{\cancel{3x}} \text{ is wrong (why?).}$$

(C)
$$\frac{x^2y-xy^2}{x^2-xy} = \frac{\cancel{xy}(\cancel{x-y})}{\cancel{x}(\cancel{x-y})} \qquad \text{Factor top and bottom, then cancel common factors.}$$

$$= y$$

(D) $\dfrac{4x^3 + 10x^2 - 6x}{2x^3 - 18x} = \dfrac{\overset{1}{\cancel{2x}}(2x^2 + 5x - 3)}{\underset{1}{\cancel{2x}}(x^2 - 9)}$

$\qquad\qquad = \dfrac{(2x - 1)\overset{1}{\cancel{(x+3)}}}{(x - 3)\underset{1}{\cancel{(x+3)}}}$

$\qquad\qquad = \dfrac{2x - 1}{x - 3}$

PROBLEM 1   Eliminate common *factors* from numerator and denominator:

(A) $\dfrac{3x(x^2 + 2)}{2(x^2 + 2)}$ $\qquad\qquad\qquad$ **(B)** $\dfrac{4m}{8m^2 - 4m}$

(C) $\dfrac{x^2 - 3x}{x^2y - 3xy}$ $\qquad\qquad\qquad$ **(D)** $\dfrac{2x^3 - 8x}{4x^3 - 14x^2 + 12x}$

ANSWERS TO
MATCHED
PROBLEMS

1. **(A)** $\dfrac{3x}{2}$; $\qquad$ **(B)** $\dfrac{1}{2m - 1}$; $\qquad$ **(C)** $\dfrac{1}{y}$; $\qquad$ **(D)** $\dfrac{x + 2}{2x - 3}$

EXERCISE 6.2

Eliminate common factors from numerator and denominator.

A   **1.** $\dfrac{2x^2}{6x}$ $\qquad\qquad\qquad\qquad$ **2.** $\dfrac{9y}{3y^3}$

3. $\dfrac{A}{A^2}$ $\qquad\qquad\qquad\qquad$ **4.** $\dfrac{B^2}{B}$

5. $\dfrac{x + 3}{(x + 3)^2}$ $\qquad\qquad\qquad$ **6.** $\dfrac{(y - 1)^2}{y - 1}$

7. $\dfrac{8(y - 5)^2}{2(y - 5)}$ $\qquad\qquad\qquad$ **8.** $\dfrac{4(x - 1)}{12(x - 1)^3}$

9. $\dfrac{2x^2(x + 7)}{6x(x + 7)^3}$ $\qquad\qquad\qquad$ **10.** $\dfrac{15y^3(x - 9)^3}{5y^4(x - 9)^2}$

11. $\dfrac{x^2 - 2x}{2x - 4}$ $\qquad\qquad\qquad$ **12.** $\dfrac{2x^2 - 10x}{4x - 20}$

13. $\dfrac{9y - 3y^2}{3y}$ $\qquad\qquad\qquad$ **14.** $\dfrac{2x^2 - 4x}{2x}$

15. $\dfrac{m^2 - mn}{m^2n - mn^2}$ $\qquad\qquad\qquad$ **16.** $\dfrac{a^2b + ab^2}{ab + b^2}$

17. $\dfrac{(2x - 1)(2x + 1)}{3x(2x + 1)}$ $\qquad\qquad\qquad$ **18.** $\dfrac{(x + 3)(2x + 5)}{2x^2(2x + 5)}$

19. $\dfrac{(2x + 1)(x - 5)}{(3x - 7)(x - 5)}$

20. $\dfrac{(3x + 2)(x + 9)}{(2x - 5)(x + 9)}$

B 21. $\dfrac{x^2 + 5x + 6}{2x^2 + 6x}$

22. $\dfrac{x^2 + 6x + 8}{3x^2 + 12x}$

23. $\dfrac{x^2 - 9}{x^2 + 6x + 9}$

24. $\dfrac{x^2 - 4}{x^2 + 4x + 4}$

25. $\dfrac{x^2 - 4x + 4}{x^2 - 5x + 6}$

26. $\dfrac{x^2 - 6x + 9}{x^2 - 5x + 6}$

27. $\dfrac{2x^2 + 5x - 3}{4x^2 - 1}$

28. $\dfrac{9x^2 - 4}{3x^2 + 7x - 6}$

29. $\dfrac{9x^2 - 3x + 6}{3}$

30. $\dfrac{2 - 6x - 4x^2}{2}$

31. $\dfrac{10 + 5m - 15m^2}{5m}$

32. $\dfrac{12t^2 + 4t - 8}{4t}$

33. $\dfrac{4m^3n - 2m^2n^2 + 6mn^3}{2mn}$

34. $\dfrac{6x^3y - 12x^2y^2 - 9xy^3}{3xy}$

35. $\dfrac{x^2 - x - 6}{x - 3}$

36. $\dfrac{x^2 + 2x - 8}{x - 2}$

37. $\dfrac{4x^2 - 9y^2}{4x^2y + 6xy^2}$

38. $\dfrac{a^2 - 16b^2}{4ab - 16b^2}$

39. $\dfrac{x^2 - xy + 2x - 2y}{x^2 - y^2}$

40. $\dfrac{u^2 + uv - 2u - 2v}{u^2 + 2uv + v^2}$

41. $\dfrac{x^2y - 8xy + 15y}{xy - 3y}$

42. $\dfrac{m^3 + 7m^2 + 10m}{m^2 + 5m}$

C 43. $\dfrac{6x^3 + 28x^2 - 10x}{12x^3 - 4x^2}$

44. $\dfrac{12x^3 - 78x^2 - 42x}{16x^4 + 8x^3}$

45. $\dfrac{x^3 - 8}{x^2 - 4}$

46. $\dfrac{y^3 + 27}{2y^3 - 6y^2 + 18y}$

6.3 MULTIPLICATION AND DIVISION

The earlier treatment of multiplication and division of rational numbers in Chapter 3 extends naturally to rational forms and real number fractions in general.

Multiplication of Real Fractions

For a, b, c, and d any real numbers, b, $d \neq 0$,

$$\frac{a}{b} \cdot \frac{c}{d} = \frac{a \cdot c}{b \cdot d}$$

This definition coupled with the fundamental principle of real fractions

$$\frac{ak}{bk} = \frac{a}{b} \qquad b, k \neq 0$$

provide the basic tools for multiplying and reducing rational expressions to lowest terms.

EXAMPLE 2 **(A)** $\dfrac{3a^2b}{4c^2d} \cdot \dfrac{8c^2d^3}{9ab^2} = \dfrac{(3a^2b)\cdot(8c^2d^3)}{(4c^2d)\cdot(9ab^2)} = \dfrac{24a^2bc^2d^3}{36ab^2c^2d} = \dfrac{(2ad^2)(\cancel{12abc^2d})}{(3b)(\cancel{12abc^2d})} = \dfrac{2ad^2}{3b}$

This process is easily shortened to the following when it is realized that, in effect, any factor in a numerator may "cancel" any like factor in a denominator. Thus,

$$\frac{\overset{1 \cdot a \cdot 1}{\cancel{3a^2b}}}{\underset{1 \cdot 1 \cdot 1}{\cancel{4c^2d}}} \cdot \frac{\overset{2 \cdot 1 \cdot d^2}{\cancel{8c^2d^3}}}{\underset{3 \cdot 1 \cdot b}{\cancel{9ab^2}}} = \frac{2ad^2}{3b}$$

(B) $(x^2 - 4) \cdot \dfrac{2x-3}{x+2} = \dfrac{\overset{1}{(\cancel{x+2})}(x-2)}{1} \cdot \dfrac{2x-3}{\underset{1}{\cancel{x+2}}}$ Factor numerators and denominators, cancel common factors, then multiply.

$$= (x-2)(2x-3)$$

(C) $\dfrac{4a^2 - 9b^2}{4a^2 + 12ab + 9b^2} \cdot \dfrac{6a^2b}{8a^2b^2 - 12ab^3} = \dfrac{\overset{1}{(\cancel{2a-3b})}\overset{1}{(\cancel{2a+3b})}}{\underset{(2a+3b)}{(\cancel{2a+3b})^2}} \cdot \dfrac{\overset{3a}{\cancel{6a^2b}}}{\underset{2b}{\cancel{4ab^2}(\cancel{2a-3b})}}$

$$= \frac{3a}{2b(2a+3b)}$$

PROBLEM 2   Multiply and reduce to lowest terms:

(A) $\dfrac{4x^2y^3}{9w^2z} \cdot \dfrac{3wz^2}{2xy^4}$ **(B)** $\dfrac{x+5}{x^2-9} \cdot (x+3)$

(C) $\dfrac{x^2 - 9y^2}{x^2 - 6xy + 9y^2} \cdot \dfrac{6x^2y}{2x^2 + 6xy}$

It follows from the definition of division (recall: $A \div B = Q$ if and only if $A = BQ$ and Q is unique) that:

Division of Real Fractions

For a, b, c, and d any real numbers, b, d, $c \neq 0$,

divisor   reciprocal of divisor

$$\frac{a}{b} \div \frac{c}{d} = \frac{a}{b} \cdot \frac{d}{c}$$

That is, to divide one fraction by another, multiply by the reciprocal of the divisor.

EXAMPLE 3   **(A)** $\dfrac{6a^2b^3}{5cd} \div \dfrac{3a^2c}{10bd} = \dfrac{6a^2b^3}{5cd} \cdot \dfrac{10bd}{3a^2c}$ Invert divisor and proceed as in Example 2(A).

$$= \frac{4b^4}{c^2}$$

(B) $(x + 4) \div \dfrac{2x^2 - 32}{6xy} = \dfrac{x + 4}{1} \cdot \dfrac{6xy}{2(x - 4)(x + 4)} = \dfrac{3xy}{x - 4}$

(C) $\dfrac{10x^3y}{3xy + 9y} \div \dfrac{4x^2 - 12x}{x^2 - 9} = \dfrac{10x^3y}{3y(x + 3)} \cdot \dfrac{(x + 3)(x - 3)}{4x(x - 3)} = \dfrac{5x^2}{6}$

PROBLEM 3   Divide and reduce to lowest terms:

(A) $\dfrac{8w^2z^2}{9x^2y} \div \dfrac{4wz}{6xy^2}$ **(B)** $\dfrac{2x^2 - 8}{4x} \div (x + 2)$

(C) $\dfrac{x^2 - 4x + 4}{4x^2y - 8xy} \div \dfrac{x^2 + x - 6}{6x^2 + 18x}$

ANSWERS TO
MATCHED
PROBLEMS

2. **(A)** $\dfrac{2xz}{3wy}$;   **(B)** $\dfrac{x + 5}{x - 3}$;   **(C)** $\dfrac{3xy}{x - 3y}$

3. **(A)** $\dfrac{4wyz}{3x}$;   **(B)** $\dfrac{x - 2}{2x}$;   **(C)** $\dfrac{3}{2y}$

EXERCISE 6.3

Perform the indicated operations and simplify.

A   1. $\dfrac{15}{16} \cdot \dfrac{24}{27}$ 2. $\dfrac{6}{7} \cdot \dfrac{28}{9}$

3. $\dfrac{36}{8} \div \dfrac{9}{4}$ 4. $\dfrac{4}{6} \div \dfrac{24}{8}$

5. $\dfrac{y^4}{3u^5} \cdot \dfrac{2u^3}{3y}$ 6. $\dfrac{6x^3y}{7u} \cdot \dfrac{14u^3}{12xy}$

7. $\dfrac{uvw}{5xyz} \div \dfrac{5vy}{uwxz}$

8. $\dfrac{3c^2d}{a^3b^3} \div \dfrac{3a^3b^3}{cd}$

9. $\dfrac{x+3}{2x^2} \cdot \dfrac{4x}{x+3}$

10. $\dfrac{3x^2y}{x-y} \cdot \dfrac{x-y}{6xy}$

11. $\dfrac{a^2-a}{a-1} \cdot \dfrac{a+1}{a}$

12. $\dfrac{x+3}{x^3+3x^2} \cdot \dfrac{x^3}{x-3}$

13. $\dfrac{4x}{x-4} \div \dfrac{8x^2}{x^2-6x+8}$

14. $\dfrac{x-2}{4y} \div \dfrac{x^2+x-6}{12y^2}$

B 15. $\dfrac{d^5}{3a} \div \left(\dfrac{d^2}{6a^2} \cdot \dfrac{a}{4d^3}\right)$

16. $\left(\dfrac{d^5}{3a} \div \dfrac{d^2}{6a^2}\right) \cdot \dfrac{a}{4d^3}$

17. $\dfrac{2x^2+4x}{12x^2y} \cdot \dfrac{6x}{x^2+6x+8}$

18. $\dfrac{6x^2}{4x^2y-12xy} \cdot \dfrac{x^2+x-12}{3x^2+12x}$

19. $\dfrac{2y^2+7y+3}{4y^2-1} \div (y+3)$

20. $(t^2-t-12) \div \dfrac{t^2-9}{t^2-3t}$

21. $\dfrac{x^2-6x+9}{x^2-x-6} \div \dfrac{x^2+2x-15}{x^2+2x}$

22. $\dfrac{m+n}{m^2-n^2} \div \dfrac{m^2-mn}{m^2-2mn+n^2}$

23. $-(x^2-4) \cdot \dfrac{3}{x+2}$

24. $-(x^2-3x) \cdot \dfrac{x-2}{x-3}$

C 25. $\dfrac{2-m}{2m-m^2} \cdot \dfrac{m^2+4m+4}{m^2-4}$

26. $\dfrac{9-x^2}{x^2+5x+6} \cdot \dfrac{x+2}{x-3}$

27. $\left(\dfrac{x^2-xy}{xy+y^2} \div \dfrac{x^2-y^2}{x^2+2xy+y^2}\right) \div \dfrac{x^2-2xy+y^2}{x^2y+xy^2}$

28. $\dfrac{x^2-xy}{xy+y^2} \div \left(\dfrac{x^2-y^2}{x^2+2xy+y^2} \div \dfrac{x^2-2xy+y^2}{x^2y+xy^2}\right)$

29. $(x^2-1)/(x-1)$ and $x+1$ name the same real number for (*all, all but one, no*) replacements of x by real numbers.

30. $(x^2-x-6)/(x-3) = x+2$, except for what values of x?

31. Can you evaluate the following arithmetic problem in less than 3 min?

$$\dfrac{(108{,}641)^2 - (108{,}643)^2}{(108{,}642)(108{,}646) - (108{,}644)^2}$$
HINT: Let $108{,}641 = x$, $108{,}642 = x+1$, and so on.

6.4 ADDITION AND SUBTRACTION

We again generalize the procedure used for rational numbers in Chapter 3. If you have any difficulty in this section, you should review Section 3.3; then return to this section and it will probably make more sense.

The methods of addition and subtraction of rational expressions follow from the following properties of real fractions:

Addition and Subtraction of Real Fractions

For a, b, c, and k any real numbers, b, $k \neq 0$,

1. $\dfrac{a}{b} + \dfrac{c}{b} = \dfrac{a + c}{b}$

2. $\dfrac{a}{b} - \dfrac{c}{b} = \dfrac{a - c}{b}$

3. $\dfrac{a}{b} = \dfrac{ka}{kb}$

Thus, if the denominator of two rational expressions are the same, we may either add or subtract the expressions by adding or subtracting the numerators and placing the result over the common denominator. If the denominators are not the same, we use property **3** (the fundamental principle of fractions) to change the form of each fraction so that the denominators are the same, then use properties **1** or **2**.

Even though any common denominator will do, we save wasted effort by using the least common denominator (LCD). If the LCD is not obvious (often it is), then we proceed as we did in Chapter 3 by factoring each denominator. Recall, the LCD is the least common multiple of all the denominators.

Finding the Least Common Denominator (LCD)

Step 1. Factor each denominator completely using integer coefficients.

Step 2. The LCD must contain each *different* factor that occurs in all of the denominators to the highest power it occurs in any one denominator.

EXAMPLE 4

$\dfrac{4x - 12}{4x(x - 3)} - \dfrac{2x - 6}{4x(x - 3)}$ — When numerators have more than one term, place the terms in parentheses.

$= \dfrac{(4x - 12)}{4x(x - 3)} - \dfrac{(2x - 6)}{4x(x - 3)}$ — Since the denominators are the same, use property **2** to subtract.

$= \dfrac{(4x - 12) - (2x - 6)}{4x(x - 3)}$ — Simplify numerator. Watch signs!

$= \dfrac{4x - 12 - 2x + 6}{4x(x - 3)}$ — Sign errors are frequently made where arrow points.

$$= \frac{2x - 6}{4x(x - 3)}$$

Factor numerator and reduce to lowest terms.

$$= \frac{2(x - 3)}{4x(x - 3)}$$

NOTE: $\quad \dfrac{2x - 6}{4x(x - 3)} = \dfrac{\overset{1}{2x - 6}}{\underset{2}{4x(x - 3)}}$

$$= \frac{1}{2x}$$

is wrong. (Why?)

PROBLEM 4   Combine into a single fraction, then reduce to lowest terms.

$$\frac{7y - 5}{6y(y - 1)} - \frac{4y - 2}{6y(y - 1)}$$

EXAMPLE 5   Combine into a single fraction, then reduce to lowest terms.

$$\frac{5}{6(x - 1)} + \frac{2}{9(x - 1)^2}$$

Solution   First find the LCD:

$$6(x - 1) = 2 \cdot 3(x - 1)$$ Factor each denominator completely.

$$9(x - 1)^2 = 3^2(x - 1)^2$$

$$\text{LCD} = 2 \cdot 3^2(x - 1)^2$$ LCD must contain each different factor to the highest power it occurs in any one denominator.

$$= 18(x - 1)^2$$

Now use the fundamental principle of fractions, property **3**, to get each denominator to look like the LCD.

$$\frac{5}{6(x - 1)} + \frac{2}{9(x - 1)^2}$$

$$= \frac{3(x - 1) \cdot 5}{3(x - 1) \cdot 6(x - 1)} + \frac{2 \cdot 2}{2 \cdot 9(x - 1)^2}$$

$$= \frac{15(x - 1)}{18(x - 1)^2} + \frac{4}{18(x - 1)^2}$$ Since the denominators are now the same, use property **1** to add.

$$= \frac{15(x - 1) + 4}{18(x - 1)^2}$$ Simplify numerator.

$$= \frac{15x - 15 + 4}{18(x - 1)^2}$$

$$= \frac{15x - 11}{18(x - 1)^2}$$ We are through, since the numerator does not factor.

PROBLEM 5 Combine into a single fraction, then reduce to lowest terms.

$$\frac{3}{8(y-3)^2} - \frac{1}{6(y-3)}$$

EXAMPLE 6 Combine into a single fraction, then reduce to lowest terms.

$$\frac{3}{x^2-4} - \frac{1}{x^2+4x+4}$$

Solution $\dfrac{3}{(x-2)(x+2)} - \dfrac{1}{(x+2)^2}$
Factor each denominator. The LCD is $(x-2)(x+2)^2$.

$$= \frac{(x+2)3}{(x+2)(x-2)(x+2)} - \frac{(x-2)1}{(x-2)(x+2)^2}$$
Make each denominator look like the LCD.

$$= \frac{3(x+2)}{(x-2)(x+2)^2} - \frac{(x-2)}{(x-2)(x+2)^2}$$

$$= \frac{3(x+2)-(x-2)}{(x-2)(x+2)^2}$$

$$= \frac{3x+6-x+2}{(x-2)(x+2)^2} = \frac{2x+8}{(x-2)(x+2)^2} \quad \text{or} \quad \frac{2(x+4)}{(x-2)(x+2)^2}$$

PROBLEM 6 Combine into a single fraction, then reduce to lowest terms.

$$\frac{2}{x^2-6x+9} - \frac{1}{x^2-9}$$

ANSWERS TO MATCHED PROBLEMS   **4.** $\dfrac{1}{2y}$ **5.** $\dfrac{21-4y}{24(y-3)^2}$ **6.** $\dfrac{x+9}{(x-3)^2(x+3)}$

EXERCISE 6.4

Combine into single fractions, then reduce to lowest terms.

A **1.** $\dfrac{3m}{2m^2} - \dfrac{1}{2m^2}$ **2.** $\dfrac{7x}{5x^2} - \dfrac{2}{5x^2}$

3. $\dfrac{5x}{2x} - \dfrac{3x-2}{2x}$ **4.** $\dfrac{9m}{4m} - \dfrac{5m-8}{4m}$

5. $\dfrac{4x}{2x+1} + \dfrac{2}{2x+1}$ **6.** $\dfrac{5a}{a+1} + \dfrac{5}{a+1}$

7. $\dfrac{5x}{x(x-1)} - \dfrac{2x+3}{x(x-1)}$ **8.** $\dfrac{7y-4}{6(y-3)} - \dfrac{4y+5}{6(y-3)}$

9. $\dfrac{2x}{4x^2-9} - \dfrac{3}{4x^2-9}$ **10.** $\dfrac{y}{y^2-9} - \dfrac{3}{y^2-9}$

11. $\dfrac{1}{y+3} - \dfrac{2}{3y}$

12. $\dfrac{2}{x+2} - \dfrac{3}{2x}$

13. $\dfrac{3}{x+1} - \dfrac{2}{x-2}$

14. $\dfrac{1}{x-2} + \dfrac{1}{x+3}$

B 15. $\dfrac{3y+8}{4y^2} - \dfrac{2y-1}{y^3} - \dfrac{5}{8y}$

16. $\dfrac{4t-3}{18t^3} + \dfrac{3}{4t} - \dfrac{2t-1}{6t^2}$

17. $\dfrac{t+1}{t-1} - 1$

18. $2 + \dfrac{x-1}{x-3}$

19. $\dfrac{4}{2x-3} - \dfrac{2x+1}{(2x-3)(x+2)}$

20. $\dfrac{3}{x+3} - \dfrac{3x+1}{(x-1)(x+3)}$

21. $\dfrac{a}{a-1} - \dfrac{2}{a^2-1}$

22. $\dfrac{m}{m+1} - \dfrac{2}{m^2-1}$

23. $\dfrac{1}{x^2+2x+1} - \dfrac{1}{x^2-1}$

24. $\dfrac{1}{y^2+4x+4} - \dfrac{1}{y^2+2y}$

25. $\dfrac{1}{y+2} + 3 - \dfrac{2}{y-2}$

26. $5 + \dfrac{a}{a+1} - \dfrac{a}{a-1}$

27. $\dfrac{2t}{3t^2-48} + \dfrac{t}{4t+t^2}$

28. $\dfrac{s}{s^2-4} + \dfrac{1}{2s^2+4s}$

29. $\dfrac{2}{x+3} - \dfrac{1}{x-3} + \dfrac{2x}{x^2-9}$

30. $\dfrac{2x}{x^2-y^2} + \dfrac{1}{x+y} - \dfrac{1}{x-y}$

C 31. $\dfrac{1}{3x^2+3x} + \dfrac{1}{4x^2} - \dfrac{1}{3x^2+6x+3}$

32. $\dfrac{1}{3m(m-1)} + \dfrac{1}{m^2-2m+1} - \dfrac{1}{5m^2}$

For the next four problems note that $a - b = -(b - a)$, *thus* $1 - x = -(x - 1)$, *and so on.*

33. $\dfrac{3}{x-1} + \dfrac{2}{1-x}$

34. $\dfrac{5}{y-3} - \dfrac{2}{3-y}$

35. $\dfrac{1}{5x-5} - \dfrac{1}{3x-3} + \dfrac{1}{1-x}$

36. $\dfrac{x+7}{ax-bx} + \dfrac{y+9}{by-ay}$

6.5 EQUATIONS INVOLVING FRACTIONAL FORMS

We consider two types of equation, one type with constants in denominators and a second type with some variables in denominators.

Equations with constants in denominators

We have already considered equations involving constants in denominators such as

$$\frac{2}{3} - \frac{x-4}{2} = \frac{5x}{6}$$

(see Section 3.4), and found we could easily convert such an equation into an equivalent equation with integer coefficients by multiplying both sides by the LCM of the denominators—in this case 6.

$$6 \cdot \frac{2}{3} - 6 \cdot \frac{(x-4)}{2} = 6 \cdot \frac{5x}{6}$$ Multiply each side by 6 to clear fractions.

Since 6 is the LCM of the denominator, each denominator will cancel, leaving

$$4 - 3(x - 4) = 5x$$

and we finish the solution in a few simple steps:

$$4 - 3x + 12 = 5x$$ A sign error frequently occurs where the arrow points.

$$-3x + 16 = 5x$$

$$-8x = -16$$

$$x = 2$$

Equations with variables in denominators

If an equation involves a variable in one or more denominators, such as

$$\frac{3}{x} - \frac{1}{2} = \frac{4}{x}$$

we may proceed in essentially the same way as above as long as we are careful to

avoid any value that makes a denominator in the equation zero.

EXAMPLE 7   Solve $\dfrac{3}{x} - \dfrac{1}{2} = \dfrac{4}{x}$.

Solution

$$\frac{3}{x} - \frac{1}{2} = \frac{4}{x} \qquad x \neq 0$$

We first note that $x \neq 0$, then multiply both sides by $2x$, the LCM of the denominators. If 0 turns up later as an apparent solution, it must be rejected.

$$2x \cdot \frac{3}{x} - 2x \cdot \frac{1}{2} = 2x \cdot \frac{4}{x}$$

All denominators cancel.

$$6 - x = 8$$

$$-x = 2$$

$$x = -2$$

PROBLEM 7   Solve $\dfrac{2}{3} - \dfrac{2}{x} = \dfrac{4}{x}$.

EXAMPLE 8   Solve $\dfrac{3x}{x-2} - 4 = \dfrac{14-4x}{x-2}$.

Solution

$$\frac{3x}{x-2} - 4 = \frac{14-4x}{x-2} \qquad x \neq 2$$

$$(x-2)\frac{3x}{(x-2)} - 4(x-2) = (x-2)\frac{(14-4x)}{(x-2)}$$

$$3x - 4(x-2) = 14 - 4x$$

$$3x - 4x + 8 = 14 - 4x$$

$$-x + 8 = 14 - 4x$$

$$3x = 6$$

$$x = 2$$

The original equation has no solution.

> If 2 turns up later as an apparent solution, it must be rejected.
>
> Multiply by $(x-2)$, the LCM of the denominators. Also, it is a good idea to place all binomial numerators and denominators in parentheses.
>
> Not a solution to the original equation, since x cannot equal 2.

PROBLEM 8   Solve $\dfrac{2x}{x-1} - 3 = \dfrac{7-3x}{x-1}$.

ANSWERS TO
MATCHED
PROBLEMS

7. $x = 9$ 8. $x = 2$

EXERCISE 6.5

Solve:

A   1. $\dfrac{2}{x} - \dfrac{1}{3} = \dfrac{5}{x}$

2. $\dfrac{1}{2} - \dfrac{2}{x} = \dfrac{3}{x}$

3. $\dfrac{5}{6} - \dfrac{1}{y} = \dfrac{2}{3y}$

4. $\dfrac{1}{x} + \dfrac{2}{3} = \dfrac{1}{2}$

5. $\dfrac{2}{3x} + \dfrac{1}{2} = \dfrac{4}{x} + \dfrac{4}{3}$

6. $\dfrac{1}{m} - \dfrac{1}{9} = \dfrac{4}{9} - \dfrac{2}{3m}$

7. $\dfrac{1}{2t} + \dfrac{1}{8} = \dfrac{2}{t} - \dfrac{1}{4}$

8. $\dfrac{4}{3k} - 2 = \dfrac{k+4}{6k}$

B   9. $\dfrac{9}{L+1} - 1 = \dfrac{12}{L+1}$

10. $\dfrac{7}{y-2} - \dfrac{1}{2} = 3$

11. $\dfrac{3}{2x-1} + 4 = \dfrac{6x}{2x-1}$

12. $\dfrac{5x}{x+5} = 2 - \dfrac{35}{x+5}$

13. $\dfrac{3N}{N-2} - \dfrac{9}{4N} = 3$

14. $\dfrac{2E}{E-1} = 2 + \dfrac{5}{2E}$

15. $5 + \dfrac{2x}{x-3} = \dfrac{6}{x-3}$

16. $\dfrac{6}{x-2} = 3 + \dfrac{3x}{x-2}$

17. $\dfrac{5}{x-3} = \dfrac{33-x}{x^2-6x+9}$

18. $\dfrac{D^2+2}{D^2-4} = \dfrac{D}{D-2}$

19. $\dfrac{n-5}{6n-6} = \dfrac{1}{9} - \dfrac{n-3}{4n-4}$

20. $\dfrac{1}{3} - \dfrac{s-2}{2s+4} = \dfrac{s+2}{3s+6}$

C **21.** $\dfrac{2}{x-2} = 3 - \dfrac{5}{2-x}$

22. $\dfrac{3x}{x-4} - 2 = \dfrac{3}{4-x}$

23. $\dfrac{5x-22}{x^2-6x+9} - \dfrac{11}{x^2-3x} - \dfrac{5}{x} = 0$

24. $\dfrac{1}{x^2-x-2} - \dfrac{3}{x^2-2x-3} = \dfrac{1}{x^2-5x+6}$

6.6 FORMULAS AND EQUATIONS WITH SEVERAL VARIABLES

One of the immediate applications you will have for algebra in other courses is the changing of formulas or equations to alternate equivalent forms. In the process we will make frequent use of the symmetric property of equality introduced in Section 1.4:

If $a = b$, then $b = a$.   Symmetric property of equality.

Thus, if

$A = P + Prt$

then

$P + Prt = A$

or if

$F = \frac{9}{5}C + 32$

then

$\frac{9}{5}C + 32 = F$

In general,

> An equation can always be reversed without changing any signs.

EXAMPLE 9   Solve the formula $c = wrt/1{,}000$ for t. (The formula gives the cost of using an electrical appliance, where w = power in watts, r = rate per kilowatt-hour, t = time in hours.)

Solution $c = \dfrac{wrt}{1{,}000}$ Start with the given formula.

$\dfrac{wrt}{1{,}000} = c$ Reverse the equation to get t on the left side.

$$\frac{1{,}000}{wr} \cdot \frac{wrt}{1{,}000} = \frac{1{,}000}{wr} \cdot c$$

Multiply both sides by $\dfrac{1{,}000}{wr}$ to isolate t on the left.

$$t = \frac{1{,}000c}{wr}$$

We have solved for t.

PROBLEM 9   Solve the formula in Example 9 for w.

EXAMPLE 10   Solve the formula $A = P + Prt$ for r (simple interest formula).

Solution

$$A = P + Prt$$

$$P + Prt = A$$

Reverse the equation, then perform operations to isolate r on the left side.

$$Prt = A - P$$

$$\frac{Prt}{Pt} = \frac{A - P}{Pt}$$

$$r = \frac{A - P}{Pt}$$

PROBLEM 10   Solve the formula $A = P + Prt$ for t.

EXAMPLE 11   Solve the formula $A = P + Prt$ for P.

Solution

$$A = P + Prt$$

$$P + Prt = A$$ Reverse the equation.

NOTE: If we write $P = A - Prt$ we have not solved for P. To solve for P is to isolate P on the left side with a coefficient of 1. In general, if the variable we are solving for appears on both sides of an equation, we have not solved for it! Since P is a common factor to both terms on the left, we factor P out and complete the problem.

$$P(1 + rt) = A$$

Divide both sides by $(1 + rt)$ to isolate P.

$$\frac{P(1 + rt)}{(1 + rt)} = \frac{A}{(1 + rt)}$$

Note that P appears only on the left side.

$$P = \frac{A}{1 + rt}$$

PROBLEM 11   Solve $A = xy + xz$ for x.

9. $w = \dfrac{1,000c}{rt}$ **10.** $t = \dfrac{A - P}{Pr}$ **11.** $x = \dfrac{A}{y + z}$

EXERCISE 6.6

The following formulas and equations are widely used in science or mathematics.

A
1. Solve $A = P + I$ for I (*Simple interest*)
2. Solve $R = R_1 + R_2$ for R_2 (*Electric circuits—resistance in series*)
3. Solve $d = rt$ for r (*Distance-rate-time*)
4. Solve $d = 1,100t$ for t (*Sound distance in air*)
5. Solve $I = Prt$ for t (*Simple interest*)
6. Solve $C = 2\pi r$ for r (*Circumference of a circle*)
7. Solve $C = \pi D$ for π (*Circumference of a circle*)
8. Solve $e = mc^2$ for m (*Mass-energy equation*)
9. Solve $ax + b = 0$ for x (*First-degree polynomial equation in one variable*)
10. Solve $p = 2a + 2b$ for a (*Perimeter of a rectangle*)
11. Solve $s = 2t - 5$ for t (*Slope-intercept form for a line*)
12. Solve $y = mx + b$ for m (*Slope-intercept form for a line*)

B
13. Solve $3x - 4y - 12 = 0$ for y (*Linear equation in two variables*)
14. Solve $Ax + By + C = 0$ for y (*Linear equation in two variables*)
15. Solve $I = \dfrac{E}{R}$ for E (*Electric circuits—Ohm's law*)
16. Solve $m = \dfrac{b}{a}$ for a (*Optics—magnification*)
17. Solve $C = \dfrac{100B}{L}$ for L (*Anthropology—cephalic index*)
18. Solve $IQ = \dfrac{(100)(MA)}{(CA)}$ for (CA) (*Psychology—intelligence quotient*)
19. Solve $F = G\dfrac{m_1m_2}{d^2}$ for m_1 (*Gravitational force between two masses*)
20. Solve $F = G\dfrac{m_1m_2}{d^2}$ for G (*Gravitational force between two masses*)
21. Solve $A = \dfrac{h}{2}(b_1 + b_2)$ for h (*Area of a trapezoid*)
22. Solve $A = \dfrac{h}{2}(b_1 + b_2)$ for b_2 (*Area of a trapezoid*)
23. Solve $C = \frac{5}{9}(F - 32)$ for F   (*Celsius-Fahrenheit*)
24. Solve $F = \frac{9}{5}C + 32$ for C   (*Fahrenheit-Celsius*)

C **25.** Solve $\dfrac{1}{f} = \dfrac{1}{a} + \dfrac{1}{b}$ for f (*Optics—focal length*)

26. Solve $\dfrac{1}{R} = \dfrac{1}{R_1} + \dfrac{1}{R_2}$ for R_1 (*Electric circuits*)

27. Solve $y = \dfrac{3x + 1}{2x - 1}$ for x in terms of y.

28. Solve $y = \dfrac{x - 3}{3x - 2}$ for x in terms of y.

6.7 COMPLEX FRACTIONS

A fractional form with fractions in its numerator or denominator is called a **complex fraction.** It is often necessary to represent a complex fraction as a **simple fraction,** that is (in all cases we will consider), as the quotient of two polynomials. The process does not involve any new concepts. It is a matter of applying old concepts in the right way. In particular, we will find the fundamental principle of fractions

$$\frac{a}{b} = \frac{ka}{kb} \qquad b, \, k \neq 0 \tag{2}$$

of considerable use. Several examples should clarify the process.

EXAMPLE 12 Express as simple fractions.

(A) $\dfrac{\frac{2}{3}}{\frac{3}{4}}$

(B) $\dfrac{1\frac{1}{2}}{3\frac{2}{3}}$

Solution **(A)** Use fundamental principle of fractions (2) and multiply numerator and denominator by a number divisible by both 3 and 4, that is, 12, the LCD of the internal fractions.

$$\frac{\frac{2}{3}}{\frac{3}{4}} = \frac{12 \cdot \frac{2}{3}}{12 \cdot \frac{3}{4}} \qquad \text{Multiply by 12, then cancel denominators.}$$

$$= \frac{4 \cdot 2}{3 \cdot 3}$$

$$= \frac{8}{9}$$

(B) Recall $1\frac{1}{2}$ and $3\frac{2}{3}$ represent sums and not products; that is, $1\frac{1}{2} = 1 + \frac{1}{2}$ and $3\frac{2}{3} = 3 + \frac{2}{3}$. Thus

$$\frac{1\frac{1}{2}}{3\frac{2}{3}} = \frac{1 + \frac{1}{2}}{3 + \frac{2}{3}} \qquad \text{Write mixed fractions as sums.}$$

$$= \frac{6(1 + \frac{1}{2})}{6(3 + \frac{2}{3})}$$

Multiply top and bottom by 6 the LCD of all fractions within the main fraction.

$$= \frac{6 \cdot 1 + 6 \cdot \frac{1}{2}}{6 \cdot 3 + 6 \cdot \frac{2}{3}}$$

The denominators 2 and 3 cancel.

$$= \frac{6 + 3}{18 + 4} = \frac{9}{22}$$

A simple fraction.

PROBLEM 12   Express as simple fractions.

(A) $\dfrac{\frac{3}{5}}{\frac{1}{4}}$

(B) $\dfrac{2\frac{3}{4}}{4\frac{1}{3}}$

EXAMPLE 13   Express as simple fractions.

(A) $\dfrac{1 - \frac{1}{x^2}}{1 + \frac{1}{x}}$

(B) $\dfrac{\frac{a}{b} - \frac{b}{a}}{\frac{a}{b} + 2 + \frac{b}{a}}$

Solution   (A) $\dfrac{1 - \frac{1}{x^2}}{1 + \frac{1}{x}} = \dfrac{x^2 \left(1 - \frac{1}{x^2}\right)}{x^2 \left(1 + \frac{1}{x}\right)}$

Multiply top and bottom by x^2, the LCD of all internal fractions.

$$= \frac{x^2 \cdot 1 - x^2 \cdot \frac{1}{x^2}}{x^2 \cdot 1 + x^2 \cdot \frac{1}{x}}$$

$$= \frac{x^2 - 1}{x^2 + x}$$

Factor top and bottom to reduce to lowest terms.

$$= \frac{\overset{1}{(x - 1)(\cancel{x + 1})}}{\underset{1}{x(\cancel{x + 1})}}$$

$$= \frac{x - 1}{x}$$

(B) $\dfrac{\frac{a}{b} - \frac{b}{a}}{\frac{a}{b} + 2 + \frac{b}{a}} = \dfrac{ab \left(\frac{a}{b} - \frac{b}{a}\right)}{ab \left(\frac{a}{b} + 2 + \frac{b}{a}\right)}$

LCD of all internal fractions is ab

$$= \frac{ab \cdot \dfrac{a}{b} - ab \cdot \dfrac{b}{a}}{ab \cdot \dfrac{a}{b} + ab \cdot 2 + ab \cdot \dfrac{b}{a}}$$

$$= \frac{a^2 - b^2}{a^2 + 2ab + b^2}$$ Reduce to lowest terms.

$$= \frac{(a - b)(a + b)}{(a + b)^2}$$

$$= \frac{a - b}{a + b}$$ A simple fraction.

PROBLEM 13   Express as simple fractions.

(A) $\dfrac{1 - \dfrac{1}{3x}}{1 - \dfrac{1}{9x^2}}$

(B) $\dfrac{\dfrac{x}{y} + 1 - \dfrac{2y}{x}}{\dfrac{x}{y} - \dfrac{y}{x}}$

ANSWERS TO MATCHED PROBLEMS

12. (A) $\dfrac{12}{5}$;   (B) $\dfrac{33}{52}$

13. (A) $\dfrac{3x}{3x + 1}$;   (B) $\dfrac{x + 2y}{x + y}$

EXERCISE 6.7

Express as simple fractions reduced to lowest terms.

A

1. $\dfrac{\frac{1}{2}}{\frac{2}{3}}$

2. $\dfrac{\frac{1}{4}}{\frac{2}{3}}$

3. $\dfrac{\frac{3}{8}}{\frac{5}{12}}$

4. $\dfrac{\frac{4}{15}}{\frac{5}{6}}$

5. $\dfrac{1\frac{1}{3}}{2\frac{1}{6}}$

6. $\dfrac{3\frac{1}{10}}{2\frac{1}{5}}$

7. $\dfrac{1\frac{2}{9}}{2\frac{5}{6}}$

8. $\dfrac{2\frac{4}{15}}{1\frac{7}{10}}$

B

9. $\dfrac{\dfrac{x}{y}}{\dfrac{1}{y^2}}$

10. $\dfrac{\dfrac{1}{b^2}}{\dfrac{a}{b}}$

11. $\dfrac{\dfrac{y}{2x}}{\dfrac{1}{3x^2}}$

12. $\dfrac{\dfrac{2x}{5y}}{\dfrac{1}{3x}}$

13. $\dfrac{1 + \dfrac{3}{x}}{x - \dfrac{9}{x}}$

14. $\dfrac{1 - \dfrac{2}{x}}{x - \dfrac{4}{x}}$

15. $\dfrac{1 - \dfrac{y^2}{x^2}}{1 - \dfrac{y}{x}}$

16. $\dfrac{\dfrac{a^2}{b^2} - 1}{\dfrac{a}{b} - 1}$

17. $\dfrac{\dfrac{1}{x} + \dfrac{1}{y}}{\dfrac{y}{x} - \dfrac{x}{y}}$

18. $\dfrac{b - \dfrac{a^2}{b}}{\dfrac{1}{a} - \dfrac{1}{b}}$

19. $\dfrac{\dfrac{x}{y} - 2 + \dfrac{y}{x}}{\dfrac{x}{y} - \dfrac{y}{x}}$

20. $\dfrac{1 + \dfrac{2}{x} - \dfrac{15}{x^2}}{1 + \dfrac{4}{x} - \dfrac{5}{x^2}}$

C 21. $\dfrac{\dfrac{a^2}{a - b} - a}{\dfrac{b^2}{a - b} + b}$

22. $\dfrac{n - \dfrac{n^2}{n - m}}{1 + \dfrac{m^2}{n^2 - m^2}}$

23. $\dfrac{\dfrac{m}{m + 2} - \dfrac{m}{m - 2}}{\dfrac{m + 2}{m - 2} - \dfrac{m - 2}{m + 2}}$

24. $\dfrac{\dfrac{y}{x + y} - \dfrac{x}{x - y}}{\dfrac{x}{x + y} + \dfrac{y}{x - y}}$

25. $1 - \dfrac{1}{1 - \dfrac{1}{x}}$

26. $2 - \dfrac{1}{1 - \dfrac{2}{x + 2}}$

APPLICATIONS

27. A formula for the average rate r for a round trip between two points, where the rate going is r_G and the rate returning is r_R, is given by the complex fraction

$$r = \frac{2}{\dfrac{1}{r_G} + \dfrac{1}{r_R}}$$

Express r as a simple fraction.

28. The airspeed indicator on a jet aircraft registers 500 mph. If the plane is traveling with an airstream moving at 100 mph, then the plane's ground speed would be 600 mph—or would it? According to Einstein, velocities must be added according to the following formula:

$$v = \frac{v_1 + v_2}{1 + \dfrac{v_1 v_2}{c^2}}$$

where v is the resultant velocity, c is the speed of light, and v_1 and v_2 are the two velocities to be added. Convert the right side of the equation into a simple fraction.

227

6.8 CHAPTER REVIEW: IMPORTANT TERMS AND SYMBOLS, REVIEW EXERCISE, PRACTICE TEST

Important terms and symbols

rational expression (*6.1*)   fundamental principle of fractions (*6.1*)   multiplication and division (*6.3*) addition and subtraction (*6.4*)   least common denominator, LCD (*6.4*)   complex fraction (*6.7*)

Exercise 6.8 Review exercise

A *Perform the indicated operations and reduce to lowest terms.*

1. $1 + \dfrac{2}{3x}$

2. $\dfrac{2}{x} - \dfrac{1}{6x} + \dfrac{1}{3}$

3. $\dfrac{3x^2(x-3)}{6y} \cdot \dfrac{8y^3}{9(x-3)^2}$

4. $(d-2)^2 \div \dfrac{d^2-4}{d-2}$

5. $\dfrac{2}{3x-1} - \dfrac{1}{2x}$

6. $\dfrac{x}{x-4} - 1$

7. $\dfrac{\frac{1}{4}}{\frac{5}{6}}$

8. $\dfrac{4\frac{2}{3}}{1\frac{1}{2}}$

Solve.

9. $\dfrac{2}{3m} - \dfrac{1}{4m} = \dfrac{1}{12}$

10. $\dfrac{3x}{x-5} - 8 = \dfrac{15}{x-5}$

11. Solve $A = \dfrac{bh}{2}$ for b (*area of a triangle*)

B *Perform the indicated operations and reduce to lowest terms.*

12. $\dfrac{y-2}{y^2-4y+4} \div \dfrac{y^2+2y}{y^2+4y+4}$

13. $\dfrac{6x^4-6x^3}{3xy} \cdot \dfrac{xy+y}{x^4-x^2}$

14. $\dfrac{x+1}{x+2} - \dfrac{x+2}{x+3}$

15. $\dfrac{1}{2x^2} + \dfrac{1}{3x(x-1)}$

16. $\dfrac{m+2}{m-2} - \dfrac{m^2+4}{m^2-4}$

17. $\dfrac{1}{x^2-y^2} - \dfrac{1}{x^2-2xy+y^2}$

18. $\dfrac{x-\dfrac{1}{x}}{1-\dfrac{1}{x^2}}$

19. $\dfrac{\dfrac{x}{y}-\dfrac{y}{x}}{\dfrac{x}{y}+1}$

Solve.

20. $\dfrac{5}{2x+3} - 5 = \dfrac{-5x}{2x+3}$

21. $\dfrac{3}{x} - \dfrac{2}{x+1} = \dfrac{1}{2x}$

22. $s = \dfrac{n(a+L)}{2}$ for L

23. $M = xA + A$ for A

24. If $\frac{1}{2}$ is added to the reciprocal of a number, the sum is 2. Find the number by setting up an equation and solving.

C *Perform the indicated operations and reduce to lowest terms.*

25. $\dfrac{y^2 - y - 6}{y^2 + 4y + 4} \div \dfrac{3 - y}{2 + y}$

26. $\dfrac{2x + 4}{2x - y} + \dfrac{2x - y}{y - 2x}$

27. $\dfrac{\dfrac{3}{x - 1} - 3}{\dfrac{2}{x - 1} + 2}$

28. $2 - \dfrac{2}{2 - \dfrac{2}{x}}$

Solve.

29. $5 - \dfrac{2x}{3 - x} = \dfrac{6}{x - 3}$

30. $y = \dfrac{3x + 1}{2x - 3}$ for x in terms of y

Practice test
Chapter 6

Perform the indicated operations and reduce to lowest terms. Write all complex fractions as simple fractions reduced to lowest terms.

1. $\dfrac{6x^2(x - 1)}{9xy^2} \cdot \dfrac{15y}{(x - 1)^2}$

2. $\dfrac{2x^3 - 50x}{x^2 + 10x + 25} \div \dfrac{2x - 10}{4x^2 + 20x}$

3. $\dfrac{x}{2x - 1} - \dfrac{1}{3}$

4. $\dfrac{3x + 9}{x^2 - 6x + 9} - \dfrac{1}{x - 3}$

5. $\dfrac{1}{2x(x - 3)} - \dfrac{1}{3x^2}$

6. $\dfrac{1 - \dfrac{x^2}{y^2}}{\dfrac{1}{y} - \dfrac{x}{y^2}}$

7. $\dfrac{2 - \dfrac{1}{x - 3}}{3 + \dfrac{1}{x - 3}}$

Solve.

8. $\dfrac{2x}{x - 3} + 2 = \dfrac{6}{x - 3}$

9. $\dfrac{1}{x - 1} - \dfrac{1}{3x} = \dfrac{1}{2x}$

10. $s = 3t + 1$ for t

11. $C = \dfrac{5}{9}(F - 32)$ for F

12. $y = \dfrac{4x + 3}{3x - 1}$ for x in terms of y

7

EXPONENTS
AND RADICALS

**7.1
NATURAL
NUMBER EXPONENTS**

Earlier we defined a number raised to a natural number power. Recall

> ### Natural Number Exponent
>
> For n a natural number
>
> $a^n = a \cdot a \cdots a \qquad n$ factors of a

We then introduced the first law of exponents: If m and n are positive integers and a is a real number, then

$$a^m a^n = a^{m+n}$$

By now you probably use this law almost unconsciously when multiplying polynomial forms.

When more complicated expressions involving exponents are encountered, other exponent laws combined with the first law provide an efficient tool for simplifying and manipulating these expressions. In this section we will introduce and discuss four additional exponent properties.

In each of the following expressions m and n are natural numbers and a and b are real numbers, excluding division by 0, of course.

$$\underset{\substack{3 \\ \text{factors}}}{} \qquad \underset{\substack{4 \\ \text{factors}}}{} \qquad \underset{\substack{3+4 \\ \text{factors}}}{}$$

EXAMPLE 1 $\quad a^3 a^4 = (a \cdot a \cdot a)(a \cdot a \cdot a \cdot a) = (a \cdot a \cdot a \cdot a \cdot a \cdot a \cdot a) = a^{3+4} = a^7$

LAW 1 $\quad a^m a^n = a^{m+n}$

PROBLEM 1 $\quad x^7 x^9 = ?$

$$\overset{\substack{4 \text{ groups of} \\ 3 \text{ factors each}}}{}$$

EXAMPLE 2 $\quad (a^3)^4 = a^3 \cdot a^3 \cdot a^3 \cdot a^3 = (a \cdot a \cdot a)(a \cdot a \cdot a)(a \cdot a \cdot a)(a \cdot a \cdot a)$

$$\underset{\substack{4 \cdot 3 \\ \text{factors}}}{}$$

$$= (a \cdot a \cdot a \cdot a \cdot a \cdot a \cdot a \cdot a \cdot a \cdot a \cdot a \cdot a) = a^{4 \cdot 3} = a^{12}$$

LAW 2 $\quad (a^n)^m = a^{mn}$

PROBLEM 2 $\quad (x^2)^5 = ?$

$$\underset{\substack{4\\ \text{factors of } (ab)}}{} \qquad \underset{\substack{4\\ \text{factors}}}{} \qquad \underset{\substack{4\\ \text{factors}}}{}$$

EXAMPLE 3 $\quad (ab)^4 = (ab)(ab)(ab)(ab) = (a \cdot a \cdot a \cdot a)(b \cdot b \cdot b \cdot b) = a^4 b^4$

LAW 3 $\quad (ab)^m = a^m b^n$

PROBLEM 3 $\quad (xy)^7 = \, ?$

$$\underset{\substack{5\\ \text{factors of } \frac{a}{b}}}{} \qquad\qquad \underset{\substack{5\\ \text{factors of } a}}{}$$

EXAMPLE 4 $\quad \left(\dfrac{a}{b}\right)^5 = \left(\dfrac{a}{b} \cdot \dfrac{a}{b} \cdot \dfrac{a}{b} \cdot \dfrac{a}{b} \cdot \dfrac{a}{b}\right) = \dfrac{a \cdot a \cdot a \cdot a \cdot a}{b \cdot b \cdot b \cdot b \cdot b} = \dfrac{a^5}{b^5}$

$$\underset{\substack{5\\ \text{factors of } b}}{}$$

LAW 4 $\quad \left(\dfrac{a}{b}\right)^m = \dfrac{a^m}{b^m}$

PROBLEM 4 $\quad \left(\dfrac{x}{y}\right)^3 = \, ?$

EXAMPLE 5 **(A)** $\quad \dfrac{a^7}{a^3} = \dfrac{a \cdot a \cdot a \cdot a \cdot a \cdot a \cdot a}{a \cdot a \cdot a} = \dfrac{(a \cdot a \cdot a)(a \cdot a \cdot a \cdot a)}{(a \cdot a \cdot a)} = a^{7-3} = a^4$

(B) $\quad \dfrac{a^3}{a^3} = \dfrac{a \cdot a \cdot a}{a \cdot a \cdot a} = 1$

(C) $\quad \dfrac{a^4}{a^7} = \dfrac{a \cdot a \cdot a \cdot a}{a \cdot a \cdot a \cdot a \cdot a \cdot a \cdot a} = \dfrac{(a \cdot a \cdot a \cdot a)}{(a \cdot a \cdot a \cdot a)(a \cdot a \cdot a)} = \dfrac{1}{a^{7-4}} = \dfrac{1}{a^3}$

LAW 5 $\quad \dfrac{a^m}{a^n} = \begin{cases} a^{m-n} & \textbf{\textit{if } m > n} \\[2mm] 1 & \textbf{\textit{if } m = n} \\[2mm] \dfrac{1}{a^{n-m}} & \textbf{\textit{if } n > m} \end{cases}$

PROBLEM 5 **(A)** $\;\; x^8/x^3 = \,?$ **(B)** $\;\; x^8/x^8 = \,?$ **(C)** $\;\; x^3/x^8 = \,?$

The laws of exponents are theorems, and as such they require proofs. We have only given plausible arguments for each law; formal proofs of these laws require a property of the natural numbers, called the inductive property, which is beyond the scope of this course.

It is very important to observe and remember:

The laws of exponents involve products and quotients, not sums and differences.

Many mistakes are made in algebra by people applying a law of exponents to the wrong algebraic form. For example, $(ab)^3 = a^3b^3$, but $(a + b)^3 \neq a^3 + b^3$. The exponent laws are summarized below for convenient reference.

Laws of Exponents

1. $a^m a^n = a^{m+n}$

2. $(a^n)^m = a^{mn}$

3. $(ab)^m = a^m b^m$

4. $\left(\dfrac{a}{b}\right)^m = \dfrac{a^m}{b^m}$

5. $\dfrac{a^m}{a^n} = \begin{cases} a^{m-n} & \text{if } m > n \\ 1 & \text{if } m = n \\ \dfrac{1}{a^{n-m}} & \text{if } n > m \end{cases}$

EXAMPLE 6 **(A)** $x^{12}x^{13} = x^{12+13} = x^{25}$

(B) $(t^7)^5 = t^{5\cdot 7} = t^{35}$

(C) $(xy)^5 = x^5 y^5$

(D) $\left(\dfrac{u}{v}\right)^3 = \dfrac{u^3}{v^3}$

(E) $\dfrac{x^{12}}{x^4} = x^{12-4} = x^8$

(F) $\dfrac{t^4}{t^9} = \dfrac{1}{t^{9-4}} = \dfrac{1}{t^5}$

PROBLEM 6 Simplify:

(A) $x^8 x^6$ **(B)** $(u^4)^5$ **(C)** $(xy)^9$

(D) $\left(\dfrac{x}{y}\right)^4$ **(E)** $\dfrac{x^{10}}{x^3}$ **(F)** $\dfrac{x^3}{x^{10}}$

EXAMPLE 7 **(A)** $(x^2 y^3)^4 = (x^2)^4 (y^3)^4 = x^8 y^{12}$

(B) $\left(\dfrac{u^3}{v^4}\right)^3 = \dfrac{(u^3)^3}{(v^4)^3} = \dfrac{u^9}{v^{12}}$

(C) $\dfrac{2x^9y^{11}}{4x^{12}y^7} = \dfrac{2}{4} \cdot \dfrac{x^9}{x^{12}} \cdot \dfrac{y^{11}}{y^7} = \dfrac{1}{2} \cdot \dfrac{1}{x^3} \cdot \dfrac{y^4}{1} = \dfrac{y^4}{2x^3}$

NOTE: As before, the "dotted boxes" are used to indicate steps that are usually carried out mentally.

PROBLEM 7 Simplify:

(A) $(u^3v^4)^2$ **(B)** $\left(\dfrac{x^4}{y^3}\right)^4$ **(C)** $\dfrac{9x^7y^2}{3x^5y^3}$

Knowing the rules of the game of chess doesn't make one a good chess player; similarly, memorizing the laws of exponents doesn't necessarily make one good at using them. To acquire skill in their use, one must use these laws in a fairly large variety of problems. Exercise 7.1 should help you acquire this skill.

ANSWERS TO MATCHED PROBLEMS

1. x^{16} 2. x^{10} 3. x^7y^7 4. $\dfrac{x^3}{y^3}$

5. (A) x^5;   (B) 1;   (C) $\dfrac{1}{x^5}$

6. (A) x^{14};   (B) u^{20};   (C) x^9y^9;   (D) $\dfrac{x^4}{y^4}$;   (E) x^7;   (F) $\dfrac{1}{x^7}$

7. (A) u^6v^8;   (B) $\dfrac{x^{16}}{y^{12}}$;   (C) $\dfrac{3x^2}{y}$

EXERCISE 7.1

A *Replace the question marks with appropriate symbols.*

1. $x^7x^5 = x^?$
2. $y^2y^7 = y^?$
3. $x^{10} = x^?x^6$
4. $y^8 = y^3y^?$
5. $(v^2)^3 = ?$
6. $(u^4)^3 = u^?$
7. $y^{12} = (y^6)^?$
8. $x^{10} = (x^?)^5$
9. $(xy)^5 = x^5y^?$
10. $(uv)^7 = ?$
11. $m^3n^3 = (mn)^?$
12. $p^4q^4 = (pq)^?$
13. $\left(\dfrac{x}{y}\right)^4 = \dfrac{x^?}{y^4}$
14. $\left(\dfrac{a}{b}\right)^8 = ?$
15. $\dfrac{x^7}{y^7} = \left(\dfrac{x}{y}\right)^?$
16. $\dfrac{m^3}{n^3} = \left(\dfrac{m}{n}\right)^?$
17. $\dfrac{x^7}{x^3} = x^?$
18. $\dfrac{n^{14}}{n^8} = n^?$
19. $x^3 = \dfrac{x^?}{x^4}$
20. $m^6 = \dfrac{m^8}{m^?}$
21. $\dfrac{a^5}{a^9} = \dfrac{1}{a^?}$
22. $\dfrac{x^4}{x^{11}} = \dfrac{1}{x^?}$

23. $\dfrac{1}{u^2} = \dfrac{u^?}{u^9}$

24. $\dfrac{1}{x^8} = \dfrac{x^4}{x^?}$

Simplify, using appropriate laws of exponents.

25. $(2x^3)(3x^7)$

26. $(5x^2)(2x^9)$

27. $\dfrac{4x^8}{2x^6}$

28. $\dfrac{9x^6}{3x^4}$

29. $\dfrac{4u^3}{2u^7}$

30. $\dfrac{6m^5}{8m^7}$

31. $(cd)^{12}$

32. $(xy)^{10}$

33. $\left(\dfrac{x}{y}\right)^6$

34. $\left(\dfrac{m}{n}\right)^5$

B 35. $(2x^2)(3x^3)(x^4)$

36. $(4y^3)(3y)(y^6)$

37. $(2 \times 10^3)(3 \times 10^{12})$

38. $(5 \times 10^8)(7 \times 10^9)$

39. $(10^4)^5$

40. $(10^7)^2$

41. $(y^4)^5$

42. $(x^3)^2$

43. $(x^2y^3)^4$

44. $(m^2n^5)^3$

45. $\left(\dfrac{a^3}{b^2}\right)^4$

46. $\left(\dfrac{c^2}{d^5}\right)^3$

47. $\dfrac{2x^3y^8}{6x^7y^2}$

48. $\dfrac{9u^8v^6}{3u^4v^8}$

49. $(3a^3b^2)^3$

50. $(2s^2t^4)^4$

51. $2(x^2y)^4$

52. $6(xy^3)^5$

53. $\left(\dfrac{x^2y}{2w^2}\right)^3$

54. $\left(\dfrac{mn^3}{p^2q}\right)^4$

C 55. $\dfrac{(2xy^3)^2}{(4x^2y)^3}$

56. $\dfrac{(4u^3v)^3}{(2uv^2)^6}$

57. $\dfrac{(-2x^2)^3}{(2^2x)^4}$

58. $\dfrac{(9x^3)^2}{(-3x)^2}$

59. $\dfrac{-2^2}{(-2)^2}$

60. $\dfrac{-x^2}{(-x)^2}$

61. $\dfrac{-2^4}{(-2a^2)^4}$

62. $\dfrac{(-x^2)^2}{(-x^3)^3}$

7.2 INTEGER EXPONENTS

How should symbols such as

8^0 and 7^{-3}

be defined? In this section we will extend the meaning of exponent to include 0 and negative integers. Thus, typical scientific expressions such as

The diameter of a red corpuscle is approximately 8×10^{-5} cm.

The amount of water found in the air as vapor is about 9×10^{-6} times that found in seas.

The focal length of a thin lens is given by $f^{-1} = a^{-1} + b^{-1}$.

will then make sense.

In extending the concept of exponent beyond the natural numbers, we will require that any new exponent symbol be defined in such a way that all five laws of exponents for natural numbers continue to hold. Thus, we will need only one set of laws for all types of exponents rather than a new set for each new exponent.

Zero exponents

We will start by defining the 0 exponent. If all the exponent laws must hold even if some of the exponents are 0, then a^0 $(a \neq 0)$ should be defined so that when the first law of exponents is applied,

$$a^0 \cdot a^2 = a^{0+2} = a^2$$

This suggests that a^0 should be defined as 1 for all nonzero real numbers a, since 1 is the only real number that gives a^2 when multiplied by a^2. If we let $a = 0$ and follow the same reasoning, we find that

$$0^0 \cdot 0^2 = 0^{0+2} = 0^2 = 0$$

and 0^0 could be any real number, since $0^2 = 0$; hence 0^0 is not uniquely determined. For this reason and others, we choose not to define 0^0.

Definition of Zero Exponent

For all real numbers $a \neq 0$

$a^0 = 1$

0^0 is not defined

EXAMPLE 8

(A) $5^0 = 1$

(B) $325^0 = 1$

(C) $(\frac{1}{3})^0 = 1$

(D) $t^0 = 1$ $(t \neq 0)$

(E) $(x^2 y^3)^0 = 1$ $(x \neq 0, y \neq 0)$

PROBLEM 8 Simplify:

(A) 12^0 **(B)** 999^0 **(C)** $(\frac{2}{7})^0$

(D) $x^0, x \neq 0$ **(E)** $(m^3 n^3)^0, m, n \neq 0$

Negative integer exponents

To get an idea of how a negative integer exponent should be defined, we can proceed as above. If the first law of exponents is to hold, then a^{-2} $(a \neq 0)$ must be defined so that

$$a^{-2} \cdot a^2 = a^{-2+2} = a^0 = 1$$

Thus a^{-2} must be the reciprocal of a^2; that is

$$a^{-2} = \frac{1}{a^2}$$

This kind of reasoning leads us to the following general definition.

Definition of Negative Integer Exponents

If n is a positive integer and a is a nonzero real number, then

$$a^{-n} = \frac{1}{a^n}$$

Of course, it follows, using equality properties, that

$$a^n = \frac{1}{a^{-n}}$$

EXAMPLE 9 **(A)** $a^{-7} = \dfrac{1}{a^7}$

(B) $\dfrac{1}{x^{-8}} = x^8$

(C) $10^{-3} = \dfrac{1}{10^3}$ or $\dfrac{1}{1,000}$ or 0.001

(D) $\dfrac{x^{-3}}{y^{-5}} = \dfrac{x^{-3}}{1} \cdot \dfrac{1}{y^{-5}} = \dfrac{1}{x^3} \cdot \dfrac{y^5}{1} = \dfrac{y^5}{x^3}$

PROBLEM 9 Write using positive exponents or no exponents:

(A) x^{-5} **(B)** $\dfrac{1}{y^{-4}}$

(C) 10^{-2} **(D)** $\dfrac{m^{-2}}{n^{-3}}$

With the definition of negative exponent and 0 exponent behind us, we can now replace the fifth law of exponents with a simpler form that does not have any restrictions on the relative size of the exponents. Thus

$$\frac{a^m}{a^n} = a^{m-n} = \frac{1}{a^{n-m}}$$

EXAMPLE 10   **(A)** $\dfrac{2^5}{2^8} = 2^{5-8} = 2^{-3}$ or $\dfrac{2^5}{2^8} = \dfrac{1}{2^{8-5}} = \dfrac{1}{2^3}$

        (B) $\dfrac{10^{-3}}{10^6} = 10^{-3-6} = 10^{-9}$ or $\dfrac{10^{-3}}{10^6} = \dfrac{1}{10^{6-(-3)}} = \dfrac{1}{10^{6+3}} = \dfrac{1}{10^9}$

PROBLEM 10   **(A)**   Combine denominator with numerator in $\dfrac{3^4}{3^9}$ and in $\dfrac{x^{-2}}{x^3}$

        (B)   Combine numerator with denominator for each in part **A**.

        Table 1 provides a summary of all of our work on exponents to this point.

TABLE 1
Integer exponents and their laws (summary)

Definition of a^p p an integer and a a real number	Laws of exponents n and m integers, a and b real numbers
1 *If p is a positive integer, then* $a^p = a \cdot a \cdots a$ *p factors of a*	1 $a^m a^n = a^{m+n}$ 2 $(a^n)^m = a^{mn}$
EXAMPLE: $3^5 = 3 \cdot 3 \cdot 3 \cdot 3 \cdot 3$	3 $(ab)^m = a^m b^m$
2 *If $p = 0$, then* $a^p = 1$ $a \neq 0$	4 $\left(\dfrac{a}{b}\right)^m = \dfrac{a^m}{b^m}$
EXAMPLE: $3^0 = 1$	5 $\dfrac{a^m}{a^n} = a^{m-n} = \dfrac{1}{a^{n-m}}$
3 *If p is a negative integer, then* $a^p = \dfrac{1}{a^{-p}}$ $a \neq 0$	
EXAMPLE: 3^{-4} $= \dfrac{1}{3^{-(-4)}}$ $= \dfrac{1}{3^4}$	

EXAMPLE 11   Simplify and express answers using positive exponents only.

        (A) $a^5 a^{-2} = a^{5-2} = a^3$

        (B) $(a^{-3} b^2)^{-2} = (a^{-3})^{-2}(b^2)^{-2} = a^6 b^{-4} = \dfrac{a^6}{b^4}$

        (C) $\left(\dfrac{a^{-5}}{a^{-2}}\right)^{-1} = \dfrac{(a^{-5})^{-1}}{(a^{-2})^{-1}} = \dfrac{a^5}{a^2} = a^3$

        (D) $\dfrac{4x^{-3} y^{-5}}{6x^{-4} y^3} = \dfrac{2x^{-3-(-4)}}{3y^{3-(-5)}} = \dfrac{2x^{-3+4}}{3y^{3+5}} = \dfrac{2x}{3y^8}$

or, changing to positive exponents first,

$$\frac{4x^{-3}y^{-5}}{6x^{-4}y^3} = \frac{2x^4}{3x^3y^3y^5} = \frac{2x}{3y^8}$$

(E) $\dfrac{10^{-4} \cdot 10^2}{10^{-3} \cdot 10^5} = \dfrac{10^{-4+2}}{10^{-3+5}} = \dfrac{10^{-2}}{10^2} = \dfrac{1}{10^4} = \dfrac{1}{10,000} = 0.0001$

(F) $\left(\dfrac{m^{-3}m^3}{n^{-2}}\right)^{-2} = \left(\dfrac{m^{-3+3}}{n^{-2}}\right)^{-2} = \left(\dfrac{m^0}{n^{-2}}\right)^{-2} = \left(\dfrac{1}{n^{-2}}\right)^{-2} = \dfrac{1^{-2}}{(n^{-2})^{-2}} = \dfrac{1}{n^4}$

PROBLEM 11   Simplify and express answers using positive exponents only.

(A) $x^{-2}x^6$

(B) $(x^3y^{-2})^{-2}$

(C) $\left(\dfrac{x^{-6}}{x^{-2}}\right)^{-1}$

(D) $\dfrac{8m^{-2}n^{-4}}{6m^{-5}n^2}$

(E) $\dfrac{10^{-3} \cdot 10^5}{10^{-2} \cdot 10^6}$

(F) $\left(\dfrac{x^{-1}y^2}{y^{-1}}\right)^{-1}$

Common errors

As was stated earlier, laws of exponents involve products and quotients, not sums and differences. Consider:

Correct   *Common error*

$$\dfrac{a^{-2}y}{b} = \dfrac{y}{a^2b} \qquad \dfrac{a^{-2}+y}{b} \neq \dfrac{y}{a^2b}$$

The plus sign in the numerator of the second illustration makes a big difference. Actually, $\dfrac{a^{-2}+y}{b}$ represents a compact way of writing a complex fraction. To simplify, we replace a^{-2} with $\dfrac{1}{a^2}$, then proceed as in Section 6.7.

$$\frac{a^{-2}+y}{b} = \frac{\dfrac{1}{a^2}+y}{b} = \frac{a^2\left(\dfrac{1}{a^2}+y\right)}{a^2 \cdot b}$$

$$= \frac{1+a^2y}{a^2b}$$

Also, consider the following:

Correct   *Common error*

$$(a^{-1}b^{-1})^2 = a^{-2}b^{-2} \qquad (a^{-1}+b^{-1})^2 = a^{-2}+b^{-2}$$

$$= \frac{1}{a^2b^2} \qquad\qquad\qquad = \frac{1}{a^2+b^2}$$

The second illustration contains two errors:

$$(a^{-1} + b^{-1})^2 \neq a^{-2} + b^{-2}$$

and

$$a^{-2} + b^{-2} \neq \frac{1}{a^2 + b^2}$$

The problem is worked correctly here:

EXAMPLE 12   Simplify and express answers using positive exponents only.

(A) $\dfrac{3^{-2} + 2^{-1}}{11} = \dfrac{\frac{1}{3^2} + \frac{1}{2}}{11} = \dfrac{\frac{2}{18} + \frac{9}{18}}{11} = \dfrac{11}{18} \div 11 = \dfrac{11}{18} \cdot \dfrac{1}{11} = \dfrac{1}{18}$

(B) $(a^{-1} - b^{-1})^2 = \left(\dfrac{1}{a} - \dfrac{1}{b}\right)^2 = \left(\dfrac{b-a}{ab}\right)^2 = \dfrac{b^2 - 2ab + a^2}{a^2 b^2}$

PROBLEM 12   Simplify and express answers using positive exponents only.

(A) $\dfrac{2^{-2} + 3^{-1}}{5}$ **(B)** $(x^{-1} + y^{-1})^2$

ANSWERS TO
MATCHED
PROBLEMS

8. All are equal to 1.

9. (A) $\dfrac{1}{x^5}$;   (B) y^4;   (C) $\dfrac{1}{10^2}$ or $\dfrac{1}{100}$ or 0.01;   (D) $\dfrac{n^3}{m^2}$

10. (A) 3^{-5}, x^{-5};   (B) $\dfrac{1}{3^5}$, $\dfrac{1}{x^5}$

11. (A) x^4;   (B) $\dfrac{y^4}{x^6}$;   (C) x^4;   (D) $\dfrac{4m^3}{3n^6}$;

     (E) $\dfrac{1}{10^2}$ or 0.01 or $\dfrac{1}{100}$ (F) $\dfrac{x}{y^3}$

12. (A) $\dfrac{7}{60}$;   (B) $\dfrac{(x+y)^2}{x^2 y^2}$ or $\dfrac{x^2 + 2xy + y^2}{x^2 y^2}$

EXERCISE 7.2

Simplify and write answers using positive exponents only.

A
1. 10^0 **2.** 23^0 **3.** x^0

4. y^0 **5.** 2^{-2} **6.** 3^{-3}

7. x^{-4} **8.** m^{-7} **9.** $\dfrac{1}{3^{-2}}$

10. $\dfrac{1}{4^{-3}}$ **11.** $\dfrac{1}{x^{-3}}$ **12.** $\dfrac{1}{y^{-5}}$

13. $10^{-4} \cdot 10^6$ **14.** $10^7 \cdot 10^{-5}$ **15.** $x^6 x^{-2}$

16. $y^{-3}y^4$ **17.** $m^{-3}m^3$ **18.** u^5u^{-5}

19. $\dfrac{10^8}{10^{-3}}$ **20.** $\dfrac{10^3}{10^{-7}}$ **21.** $\dfrac{a^8}{a^{-4}}$

22. $\dfrac{x^9}{x^{-2}}$ **23.** $\dfrac{b^{-3}}{b^5}$ **24.** $\dfrac{z^{-2}}{z^3}$

25. $\dfrac{10^{-4}}{10^2}$ **26.** $\dfrac{10^{-1}}{10^6}$ **27.** $(2^{-3})^{-2}$

28. $(10^{-4})^{-3}$ **29.** $(x^{-5})^{-2}$ **30.** $(y^{-2})^{-4}$

31. $(x^{-3}y^{-2})^{-1}$ **32.** $(u^{-5}v^{-3})^{-2}$ **33.** $(x^{-2}y^3)^2$

34. $(x^2y^{-3})^2$ **35.** $(x^2y^{-3})^{-1}$ **36.** $(x^{-2}y^3)^{-1}$

B   **37.** $1,231^0$ **38.** $(m^2)^0$ **39.** $\dfrac{10^{-2}}{10^{-4}}$

40. $\dfrac{10^{-3}}{10^{-5}}$ **41.** $\dfrac{x^{-3}}{x^{-2}}$ **42.** $\dfrac{y^{-2}}{y^{-3}}$

43. $\dfrac{10^{23}10^{-11}}{10^{-3}10^{-2}}$ **44.** $\dfrac{10^{-13}10^{-4}}{10^{-21}10^3}$ **45.** $\dfrac{8 \times 10^{-3}}{2 \times 10^{-5}}$

46. $\dfrac{18 \times 10^{12}}{6 \times 10^{-4}}$ **47.** $\left(\dfrac{x^2}{x^{-1}}\right)^2$ **48.** $\left(\dfrac{y}{y^{-2}}\right)^3$

49. $(2cd^2)^{-3}$ **50.** $\dfrac{1}{(3mn)^{-2}}$ **51.** $(3x^3y^{-2})^2$

52. $(2mn^{-3})^3$ **53.** $(x^{-3}y^2)^{-2}$ **54.** $(m^4n^{-5})^{-3}$

55. $(2^{-3}3^2)^{-2}$ **56.** $(2^23^{-3})^{-1}$ **57.** $(10^23^0)^{-2}$

58. $(10^{12}10^{-12})^{-1}$ **59.** $\dfrac{9m^{-4}n^3}{12m^{-1}n^{-1}}$ **60.** $\dfrac{8x^{-3}y^{-1}}{6x^2y^{-4}}$

61. $\dfrac{4x^{-2}y^{-3}}{2x^{-3}y^{-1}}$ **62.** $\dfrac{2a^6b^{-2}}{16a^{-3}b^2}$ **63.** $\left(\dfrac{n^{-3}}{n^{-2}}\right)^{-2}$

64. $\left(\dfrac{x^{-1}}{x^{-8}}\right)^{-1}$ **65.** $\left(\dfrac{x^4y^{-1}}{x^{-2}y^3}\right)^2$ **66.** $\left(\dfrac{m^{-2}n^3}{m^4n^{-1}}\right)^2$

67. $\left(\dfrac{2x^{-3}y^2}{4xy^{-1}}\right)^{-2}$ **68.** $\left(\dfrac{6mn^{-2}}{3m^{-1}n^2}\right)^{-3}$

C   **69.** $(x + y)^{-2}$ **70.** $(a^2 - b^2)^{-1}$ **71.** $\dfrac{2^{-1} + 3^{-1}}{25}$

72. $\dfrac{x^{-1} + y^{-1}}{x + y}$ **73.** $\dfrac{12}{2^{-2} + 3^{-1}}$ **74.** $\dfrac{c - d}{c^{-1} - d^{-1}}$

75. $(2^{-2} + 3^{-2})^{-1}$ **76.** $(x^{-1} + y^{-1})^{-1}$ **77.** $(10^{-2} + 10^{-3})^{-1}$

78. $(x^{-1} - y^{-1})^2$

7.3 SCIENTIFIC NOTATION

Work in science often involves the use of very, very large numbers:

The energy density of a laser beam can go as high as 10,000,000,000,000 watts per square centimeter.

Also involved is the use of very, very small numbers:

The probable mass of a hydrogen atom is
0.000 000 000 000 000 000 000 001 7 gram.

Writing and working with numbers of this type in standard decimal notation is generally awkward. It is often convenient to represent this type of number in **scientific notation; that is,** as the product of a number between 1 and 10 and a power of 10. In fact, any decimal fraction, however large or small, can be represented as the product of a number between 1 and 10 and a power of 10.

EXAMPLE 13   **DECIMAL FRACTIONS AND SCIENTIFIC NOTATION**

$$5 = 5 \times 10^0 \qquad\qquad 0.7 = 7 \times 10^{-1}$$
$$35 = 3.5 \times 10 \qquad\qquad 0.083 = 8.3 \times 10^{-2}$$
$$430 = 4.3 \times 10^2 \qquad\qquad 0.0043 = 4.3 \times 10^{-3}$$
$$5{,}870 = 5.87 \times 10^3 \qquad 0.000687 = 6.87 \times 10^{-4}$$
$$8{,}910{,}000 = 8.91 \times 10^6 \qquad 0.00000036 = 3.6 \times 10^{-7}$$

Can you discover a simple mechanical rule that relates the number of decimal places the decimal is moved with the power of 10 that is used?

PROBLEM 13   Write in scientific notation.

(A) 450   **(B)** 27,000

(C) 0.05   **(D)** 0.0000063

The following three statements are more or less typical of what one is likely to encounter in scientific writing. Knowing the meaning of the power symbol certainly adds to the meaning and interest of these statements. Figures 1 and 2 show how certain power-of-ten forms are related. NOTE: 10^{20} is not just twice 10^{10}.

1. Due to radiation of energy, the sun loses approximately 42×10^5 tons of solar mass per sec.

2. In 1929 a biologist named Vernadsky suggested that all the free oxygen of the earth, about 15×10^{20} grams, is produced by living organisms alone.

3. Edgar Altenburg wrote in his book on genetics, "A human begins life as a single cell, the fertilized egg, which by successive cell divisions forms the 10^{13} cells contained in a grown man. The size of the fertilized egg is approximately the size of the cross section of a human hair. Yet a single fertilized egg, despite its minuteness, contains all of the potentialities of a Shakespeare or a Darwin. . . ."

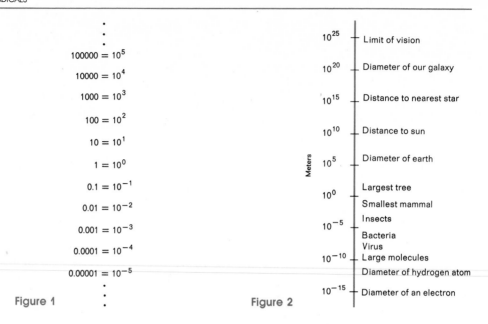

Figure 1   Figure 2

EXAMPLE 14   EVALUATION OF A COMPLICATED ARITHMETIC PROBLEM

$$\frac{(0.26)(720)}{(48,000,000)(0.0013)} = \frac{(2.6 \times 10^{-1})(7.2 \times 10^2)}{(4.8 \times 10^7)(1.3 \times 10^{-3})}$$

$$= \frac{\overset{2}{\cancel{(2.6)}}\overset{6}{\cancel{(7.2)}}}{\underset{4}{\cancel{(4.8)}}\underset{1}{\cancel{(1.3)}}} \cdot \frac{(10^{-1})(10^2)}{(10^7)(10^{-3})}$$

$$= 3 \times 10^{-3} \text{ or } 0.003$$

PROBLEM 14   Convert to power-of-ten form and evaluate: $\dfrac{(42,000)(0.009)}{(600)(0.000021)}$

Astronomy application—looking back in time

We are able to look back in time by looking out into space. Since light travels at a fast but finite rate, we see heavenly bodies not as they exist now, but as they existed sometime in the past. If the distance between the sun and the Earth is approximately 9.3×10^7 miles and if light travels at the rate of approximately 1.86×10^5 miles per sec, we see the sun as it was how many minutes ago?

$$d = rt \qquad t = \frac{d}{r} = \frac{9.3 \times 10^7}{1.86 \times 10^5} = 5 \times 10^2 = 500 \text{ sec}$$

or

$$\frac{500}{60} \approx 8.3 \, \text{min}$$

Hence, we always see the sun as it was 8.3 min ago.

13. (A) 4.5×10^2; (B) 2.7×10^4; (C) 5×10^{-2}; (D) 6.3×10^{-6}

14. 3×10^4 or 30,000

EXERCISE 7.3

A *Write in scientific notation.*

1. 60	**2.** 80	**3.** 600
4. 800	**5.** 600,000	**6.** 80,000
7. 0.06	**8.** 0.008	**9.** 0.00006
10. 0.00000008	**11.** 35	**12.** 52
13. 0.72	**14.** 0.63	**15.** 270
16. 340	**17.** 0.032	**18.** 0.085
19. 5,200	**20.** 6,300	**21.** 0.00072
22. 0.0000068		

Write as a decimal fraction.

23. 5×10^2	**24.** 8×10^2	**25.** 8×10^{-2}
26. 4×10^{-2}	**27.** 6×10^6	**28.** 3×10^5
29. 2×10^{-5}	**30.** 9×10^{-4}	**31.** 7.1×10^3
32. 5.6×10^4	**33.** 8.6×10^{-4}	**34.** 9.7×10^{-3}
35. 8.8×10^6	**36.** 4.3×10^5	**37.** 6.1×10^{-6}
38. 3.8×10^{-7}		

B *Write in scientific notation.*

39. 42,700,000   **40.** 5,460,000,000

41. 0.0000723   **42.** 0.0000000729

43. The distance that light travels in one year is called a light-year. It is approximately 5,870,000,000,000 miles.

44. The energy of a laser beam can go as high as 10,000,000,000,000 watts.

45. The mass of one water molecule is 0.0000000000000000000003 gram.

46. The nucleus of an atom has a diameter of a little more than 1/100,000 that of the whole atom.

Write as a decimal fraction.

47. 3.46×10^9 **48.** 8.35×10^{10}

49. 6.23×10^{-7} **50.** 6.14×10^{-12}

51. The distance from the earth to the sun is approximately 9.3×10^7 miles.

52. The diameter of the sun is approximately 8.65×10^5 miles.

53. The diameter of a red corpuscle is approximately 7.5×10^{-5} cm.

54. The probable mass of a hydrogen atom is 1.7×10^{-24} gram.

Simplify and express answer in scientific notation.

55. $(4 \times 10^5)(2 \times 10^{-3})$ **56.** $(3 \times 10^{-6})(3 \times 10^{10})$

57. $(4 \times 10^{-8})(2 \times 10^5)$ **58.** $(2 \times 10^3)(3 \times 10^{-7})$

59. $\dfrac{9 \times 10^8}{3 \times 10^5}$ **60.** $\dfrac{6 \times 10^{12}}{2 \times 10^7}$

61. $\dfrac{12 \times 10^3}{4 \times 10^{-4}}$ **62.** $\dfrac{15 \times 10^{-2}}{3 \times 10^{-6}}$

Convert each numeral to scientific notation and simplify. Express answer in scientific notation and as a decimal fraction.

63. $\dfrac{(0.0006)(4000)}{0.00012}$ **64.** $\dfrac{(90,000)(0.000002)}{0.006}$

65. $\dfrac{(0.000039)(140)}{(130,000)(0.00021)}$ **66.** $\dfrac{(60,000)(0.000003)}{(0.0004)(1,500,000)}$

C **67.** In 1929 Vernadsky, a biologist, estimated that all the free oxygen of the earth is 1.5×10^{21} grams and that it is produced by life alone. If one gram is approximately 2.2×10^{-3} lb, what is the amount of free oxygen in pounds?

68. If the mass of the earth is 6×10^{27} grams and each gram is 1.1×10^{-6} ton, find the mass of the earth in tons.

69. Some of the designers of high-speed computers are currently thinking of single-addition times of 10^{-7} sec (100 nanosec). How many additions would such a computer be able to perform in 1 sec? In 1 min?

70. If electricity travels in a computer circuit at the speed of light (1.86×10^5 miles per sec), how far will it travel in the time it takes the computer in the preceding problem to complete a single addition? (Size of circuits is becoming a critical problem in computer design.) Give the answer in miles and in feet.

7.4 SQUARE ROOTS AND RADICALS

Going from exponents to radicals in the same chapter may seem unnatural; however, exponents and radicals have a lot more in common than one might first expect. We will comment briefly on this relationship before the end of the chapter. In more advanced courses the relationship is developed in detail.

Definition of square root

In this and the next two sections we will take a careful look at the square root radical

$$\sqrt{}$$

and some of its properties. To start we define a square root of a number:

Definition of Square Root

x is a square root of y if $x^2 = y$

EXAMPLE 15 **(A)** 2 is a square root of 4 since $2^2 = 4$.

(B) -2 is a square root of 4 since $(-2)^2 = 4$.

PROBLEM 15 Find two square roots of 9.

How many square roots of a real number are there? The following theorem, which we state without proof, answers this question.

THEOREM 1 **(A)** Every positive real number has exactly two real square roots, each the opposite of the other.

(B) Negative real numbers have no real number square roots (since no real number squared can be negative—think about this).

(C) The square root of 0 is 0.

  Square Root Notation

For a a positive number,

$\sqrt{a}$ is the positive square root of a

$-\sqrt{a}$ is the negative square root of a

NOTE: $\sqrt{-a}$ is not a real number.

EXAMPLE 16 **(A)** $\sqrt{4} = 2$

(B) $-\sqrt{4} = -2$

(C) $\sqrt{-4}$ is not a real number

(D) $\sqrt{0} = 0$

PROBLEM 16   Evaluate, if possible:

(A) $\sqrt{9}$ (B) $-\sqrt{9}$ (C) $\sqrt{-9}$ (D) $\sqrt{0}$

Irrational numbers

It can be shown that if a is a positive integer that is not the square of an integer, then

$$-\sqrt{a} \quad \text{and} \quad \sqrt{a}$$

are irrational numbers. Thus,

$$-\sqrt{7} \quad \text{and} \quad \sqrt{7}$$

name irrational numbers that are, respectively, the negative and positive square roots of 7.

Square root properties

Note that $\sqrt{4}\sqrt{36} = 2 \cdot 6 = 12$ and $\sqrt{4 \cdot 36} = \sqrt{144} = 12$; therefore,

$$\sqrt{4}\sqrt{36} = \sqrt{4 \cdot 36}$$

Also note that

$$\frac{\sqrt{36}}{\sqrt{4}} = \frac{6}{2} = 3 \quad \text{and} \quad \sqrt{\frac{36}{4}} = \sqrt{9} = 3$$

therefore,

$$\frac{\sqrt{36}}{\sqrt{4}} = \sqrt{\frac{36}{4}}$$

These examples suggest the following general properties:

Properties of Radicals

For a and b nonnegative real numbers,

1. $\sqrt{a^2} = a$

2. $\sqrt{a}\sqrt{b} = \sqrt{ab}$

3. $\dfrac{\sqrt{a}}{\sqrt{b}} = \sqrt{\dfrac{a}{b}}$

To see that the second property holds, let $N = \sqrt{a}$ and $M = \sqrt{b}$, then $N^2 = a$ and $M^2 = b$. Hence,

$$\sqrt{a}\sqrt{b} = NM = \sqrt{(NM)^2} = \sqrt{N^2M^2} = \sqrt{ab}$$

Note how properties of exponents are used. The proof of the quotient part is left as an exercise.

EXAMPLE 17 (A) $\sqrt{5}\sqrt{10} = \sqrt{5 \cdot 10} = \sqrt{50} = \sqrt{25 \cdot 2} = \sqrt{25}\sqrt{2} = 5\sqrt{2}$

(B) $\dfrac{\sqrt{32}}{\sqrt{8}} = \sqrt{\dfrac{32}{8}} = \sqrt{4} = 2$

(C) $\sqrt{\dfrac{7}{4}} = \dfrac{\sqrt{7}}{\sqrt{4}} = \dfrac{\sqrt{7}}{2}$ or $\dfrac{1}{2}\sqrt{7}$

PROBLEM 17 Simplify as in Example 17:

(A) $\sqrt{3}\sqrt{6}$ (B) $\dfrac{\sqrt{18}}{\sqrt{2}}$ (C) $\sqrt{\dfrac{11}{9}}$

Simplest radical form

The foregoing definitions and theorems allow us to change algebraic expressions containing radicals to a variety of equivalent forms. One form that is often useful is called the **simplest radical form.**

Definition of the Simplest Radical Form

An algebraic expression that contains square root radicals is in *simplest radical form* if all three of the following conditions are satisfied:

1. No radicand (the expression within the radical sign) when expressed in completely factored form contains a factor raised to a power greater than 1. ($\sqrt{x^3}$ violates this condition.)

2. No radical appears in a denominator. $\left(\dfrac{3}{\sqrt{5}}\text{ violates this condition.}\right)$

3. No fraction appears within a radical. $\left(\sqrt{\dfrac{2}{3}}\text{ violates this condition.}\right)$

It should be understood that forms other than the simplest radical form may be more useful on occasion. The situation dictates the choice.

EXAMPLE 18 Change to simplest radical form—all variables represent positive real numbers.

(A) $\sqrt{72} = \sqrt{6^2 \cdot 2}$ Violates condition **1**, since 6 is raised to a power greater than 1.

$= \sqrt{6^2}\sqrt{2}$ $\sqrt{ab} = \sqrt{a}\sqrt{b}$

$= 6\sqrt{2}$ $\sqrt{a^2} = a,\ a \geq 0$

(B) $\sqrt{8x^3} = \sqrt{(2^2x^2)(2x)}$

Violates condition **1**. Separate $8x^3$ into a perfect square part, (2^2x^2), and what is left over, $(2x)$, then use multiplication property 2.

$$= \sqrt{2^2x^2}\sqrt{2x} \qquad \sqrt{ab} = \sqrt{a}\sqrt{b}$$

$$= 2x\sqrt{2x} \qquad \sqrt{a^2} = a, a \geq 0$$

PROBLEM 18 Repeat Example 18 for

(A) $\sqrt{32}$ **(B)** $\sqrt{18y^3}$

EXAMPLE 19 Change to simplest radical form—all variables represent positive real numbers.

(A) $\dfrac{3x}{\sqrt{3}}$ **(B)** $\sqrt{\dfrac{x}{2}}$

Solution **(A)** $\dfrac{3x}{\sqrt{3}}$ has a radical in the denominator; hence, it violates condition **2**. To remove the radical from the denominator we multiply top and bottom by $\sqrt{3}$ to obtain $\sqrt{3^2}$ in the denominator:

$$\frac{3x}{\sqrt{3}} = \frac{3x}{\sqrt{3}} \cdot \frac{\sqrt{3}}{\sqrt{3}}$$

$$= \frac{3x\sqrt{3}}{\sqrt{3^2}}$$

$$= \frac{3x\sqrt{3}}{3} = x\sqrt{3}$$

(B) $\sqrt{\dfrac{x}{2}}$ has a fraction within the radical; hence, it violates condition **3**. To remove the fraction from the radical, we multiply the top and bottom of $\dfrac{x}{2}$ inside the radical by 2 to make the denominator a perfect square:

$$\sqrt{\frac{x}{2}} = \sqrt{\frac{2 \cdot x}{2 \cdot 2}}$$

$$= \sqrt{\frac{2x}{2^2}}$$

$$= \frac{\sqrt{2x}}{\sqrt{2^2}} = \frac{\sqrt{2x}}{2}$$

PROBLEM 19 Repeat Example 19 for

(A) $\dfrac{2x}{\sqrt{2}}$ **(B)** $\sqrt{\dfrac{y}{3}}$

Further comments

In the discussion above we restricted variables to nonnegative quantities. If we lift this restriction, then

$$\sqrt{a^2} = a$$

is only correct for certain values of a, and is not true for others. Which values? If a is positive or zero, then it is true; if a is negative, then it is false. For example, let us test $\sqrt{a^2} = a$ for $a = 2$ and for $a = -2$:

$a = 2$	$a = -2$
$\sqrt{2^2} \overset{?}{=} 2$	$\sqrt{(-2)^2} \overset{?}{=} -2$
$\sqrt{4} \overset{?}{=} 2$	$\sqrt{4} \overset{?}{=} -2$
$2 \overset{\checkmark}{=} 2$	$2 \neq -2$

Problems 61 and 62 in Exercise 7.4 suggest how $\sqrt{a^2}$ should be interpreted if we allow a to take on any real value. All the other problems in Exercise 7.4 restrict variables to positive real numbers, thus $\sqrt{a^2} = a$ is correct for this restriction.

ANSWERS TO
MATCHED
PROBLEMS

15. $-3, 3$

16. (A) 3;   (B) -3;   (C) Not a real number;   (D) 0

17. (A) $3\sqrt{2}$;   (B) 3;   (C) $\dfrac{\sqrt{11}}{3}$ or $\tfrac{1}{3}\sqrt{11}$

18. (A) $4\sqrt{2}$;   (B) $3y\sqrt{2y}$

19. (A) $x\sqrt{2}$;   (B) $\dfrac{\sqrt{3y}}{3}$ or $\tfrac{1}{3}\sqrt{3y}$

EXERCISE 7.4

Simplify and express each answer in simplest radical form. All variables represent positive real numbers unless stated to the contrary.

A

1. $\sqrt{16}$ **2.** $\sqrt{25}$ **3.** $-\sqrt{81}$

4. $-\sqrt{49}$ **5.** $\sqrt{x^2}$ **6.** $\sqrt{y^2}$

7. $\sqrt{9m^2}$ **8.** $\sqrt{4u^2}$ **9.** $\sqrt{8}$

10. $\sqrt{18}$ **11.** $\sqrt{x^3}$ **12.** $\sqrt{m^3}$

13. $\sqrt{18y^3}$ **14.** $\sqrt{8x^3}$ **15.** $\sqrt{\dfrac{1}{4}}$

16. $\sqrt{\dfrac{1}{9}}$ **17.** $-\sqrt{\dfrac{4}{9}}$ **18.** $-\sqrt{\dfrac{9}{16}}$

19. $\dfrac{1}{\sqrt{x^2}}$ **20.** $\dfrac{1}{\sqrt{y^2}}$ **21.** $\dfrac{1}{\sqrt{3}}$

22. $\dfrac{1}{\sqrt{5}}$ **23.** $\sqrt{\dfrac{1}{3}}$ **24.** $\sqrt{\dfrac{1}{5}}$

25. $\dfrac{1}{\sqrt{x}}$ **26.** $\dfrac{1}{\sqrt{y}}$ **27.** $\sqrt{\dfrac{1}{x}}$

28. $\sqrt{\dfrac{1}{y}}$ **29.** $\sqrt{25x^2y^4}$ **30.** $\sqrt{49x^4y^2}$

B **31.** $\sqrt{4x^5y^3}$ **32.** $\sqrt{9x^3y^5}$ **33.** $\sqrt{8x^7y^6}$

34. $\sqrt{18x^8y^5}$ **35.** $\dfrac{1}{\sqrt{3y}}$ **36.** $\dfrac{1}{\sqrt{2x}}$

37. $\dfrac{4xy}{\sqrt{2y}}$ **38.** $\dfrac{6x^2}{\sqrt{3x}}$ **39.** $\dfrac{2x^2y}{\sqrt{3xy}}$

40. $\dfrac{3a}{\sqrt{2ab}}$ **41.** $\sqrt{\dfrac{2}{3}}$ **42.** $\sqrt{\dfrac{3}{5}}$

43. $\sqrt{\dfrac{3m}{2n}}$ **44.** $\sqrt{\dfrac{6x}{7y}}$ **45.** $\sqrt{\dfrac{4a^3}{3b}}$

46. $\sqrt{\dfrac{9m^5}{2n}}$

Approximate with decimal fractions using the square root table in the Appendix.

EXAMPLE

(A) $\sqrt{35}\sqrt{40} = \sqrt{35\cdot40} = \sqrt{(2^2\cdot5^2)(2\cdot7)} = 10\sqrt{14} = (10)(3.742) = 37.42$

(B) $\sqrt{\dfrac{7}{5}} = \dfrac{\sqrt{7}}{\sqrt{5}} = \dfrac{\sqrt{7}\sqrt{5}}{\sqrt{5}\sqrt{5}} = \dfrac{\sqrt{35}}{5} = \dfrac{5.916}{5} = 1.183$

47. $\sqrt{6}\sqrt{3}$ **48.** $\sqrt{2}\sqrt{6}$ **49.** $\sqrt{\dfrac{1}{5}}$

50. $\sqrt{\dfrac{1}{3}}$ **51.** $\dfrac{\sqrt{33}}{\sqrt{2}}$ **52.** $\dfrac{\sqrt{23}}{\sqrt{5}}$

C *Express in simplest radical form.*

53. $\dfrac{\sqrt{2x}\sqrt{5}}{\sqrt{20x}}$ **54.** $\dfrac{\sqrt{6}\sqrt{8x}}{\sqrt{3x}}$ **55.** $\sqrt{a^2+b^2}$

56. $\sqrt{m^2+n^2}$ **57.** $\sqrt{x^4-2x^2}$ **58.** $\sqrt{m^3+4m^2}$

59. Explain why the square root of a negative number cannot be a real number.

60. Prove property **3** for radicals.

61. Is $\sqrt{x^2}=x$ true for $x=4$? For $x=-4$?

62. Is $\sqrt{x^2}=|x|$ true for $x=4$? For $x=-4$?

63. If we define the symbol $5^{\frac{1}{2}}$ in such a way that the laws of exponents continue to hold [in particular $(5^{\frac{1}{2}})^2 = 5^{2(\frac{1}{2})} = 5$ or $5^{\frac{1}{2}}\cdot5^{\frac{1}{2}} = 5^{\frac{1}{2}+\frac{1}{2}} = 5$], how should it be defined?

64. If $x^2=y^2$, does it necessarily follow that $x=y$? HINT: Can you find a pair of numbers that make the first equation true, but the second equation false?

65. Find the fallacy in the following "proof" that all real numbers are equal: If m and n are any real numbers, then

$$(m - n)^2 = (n - m)^2$$
$$m - n = n - m$$
$$2m = 2n$$
$$m = n$$

7.5 SUMS AND DIFFERENCES OF RADICALS

Algebraic expressions can often be simplified by combining terms that contain the exact same radical forms. We proceed in essentially the same way that we do when we combine like terms. You will recall that the distributive law played a central role in this process. Remember, we wrote

$$3x + 5x = (3 + 5)x = 8x$$

and we concluded that we could combine like terms by adding their numerical coefficients. We have a similar mechanical rule for radicals:

Mechanical Rule for Adding Radicals

Two terms involving identical radicals can be combined into a single term by adding numerical coefficients.

EXAMPLE 20

(A) $3\sqrt{2} + 5\sqrt{2} = (3 + 5)\sqrt{2} = 8\sqrt{2}$

(B) $2\sqrt{m} - 7\sqrt{m} = (2 - 7)\sqrt{m} = -5\sqrt{m}$

(C) $3\sqrt{x} - 2\sqrt{5} + 4\sqrt{x} - 7\sqrt{5} = 3\sqrt{x} + 4\sqrt{x} - 2\sqrt{5} - 7\sqrt{5}$

$$= 7\sqrt{x} - 9\sqrt{5}$$

PROBLEM 20 Simplify:

(A) $2\sqrt{3} + 4\sqrt{3}$　　　　　　　　**(B)** $3\sqrt{x} - 5\sqrt{x}$

(C) $2\sqrt{y} - 3\sqrt{7} + 4\sqrt{y} - 2\sqrt{7}$

　　Occasionally terms containing radicals can be combined after they have been expressed in simplest radical form.

EXAMPLE 21 **(A)** $4\sqrt{8} - 2\sqrt{18} = 4 \cdot \sqrt{4} \cdot \sqrt{2} - 2 \cdot \sqrt{9} \cdot \sqrt{2}$

$$= 4 \cdot 2 \cdot \sqrt{2} - 2 \cdot 3 \cdot \sqrt{2}$$

$$= 8\sqrt{2} - 6\sqrt{2}$$

$$= 2\sqrt{2}$$

(B) $\quad 2\sqrt{12} - \sqrt{\dfrac{1}{3}} = 2 \cdot \sqrt{4} \cdot \sqrt{3} - \sqrt{\dfrac{1}{3} \cdot \dfrac{3}{3}}$

$$= 4\sqrt{3} - \frac{\sqrt{3}}{3}$$

$$= \left(4 - \frac{1}{3}\right)\sqrt{3}$$

$$= \frac{11}{3}\sqrt{3} \quad \text{or} \quad \frac{11\sqrt{3}}{3}$$

PROBLEM 21   Express in simplest radical form and simplify:

(A) $\;5\sqrt{3} - 2\sqrt{12}$ **(B)** $\;3\sqrt{8} - \sqrt{\tfrac{1}{2}}$

ANSWERS TO MATCHED PROBLEMS

20. (A) $6\sqrt{3}$; (B) $-2\sqrt{x}$; (C) $6\sqrt{y} - 5\sqrt{7}$

21. (A) $\sqrt{3}$; (B) $\dfrac{11\sqrt{2}}{2}$ or $\dfrac{11}{2}\sqrt{2}$

EXERCISE 7.5

Simplify by combining as many terms as possible. All variables represent positive real numbers. Use exact radical forms only.

A 1. $5\sqrt{2} + 3\sqrt{2}$ 2. $7\sqrt{3} + 2\sqrt{3}$
3. $6\sqrt{x} - 3\sqrt{x}$ 4. $12\sqrt{m} - 3\sqrt{m}$
5. $4\sqrt{7} - 3\sqrt{5}$ 6. $2\sqrt{3} + 5\sqrt{2}$
7. $\sqrt{y} - 4\sqrt{y}$ 8. $2\sqrt{a} - 7\sqrt{a}$
9. $3\sqrt{5} - \sqrt{5} + 2\sqrt{5}$ 10. $4\sqrt{7} - 6\sqrt{7} + \sqrt{7}$
11. $2\sqrt{x} - \sqrt{x} + 3\sqrt{x}$ 12. $\sqrt{n} - 4\sqrt{n} - 2\sqrt{n}$
13. $3\sqrt{2} - 2\sqrt{3} - \sqrt{2}$ 14. $\sqrt{5} - 2\sqrt{3} + 3\sqrt{5}$
15. $2\sqrt{x} - \sqrt{y} + 3\sqrt{y}$ 16. $\sqrt{m} - \sqrt{n} - 2\sqrt{n}$

B 17. $\sqrt{8} - \sqrt{2}$ 18. $\sqrt{18} + \sqrt{2}$
19. $\sqrt{27} - 3\sqrt{12}$ 20. $\sqrt{8} - 2\sqrt{32}$
21. $\sqrt{8} + 2\sqrt{27}$ 22. $2\sqrt{12} + 3\sqrt{18}$
23. $\sqrt{4x} - \sqrt{9x}$ 24. $\sqrt{8mn} + 2\sqrt{18mn}$
25. $\sqrt{24} - \sqrt{12} + 3\sqrt{3}$ 26. $\sqrt{8} - \sqrt{20} + 4\sqrt{2}$

C 27. $\sqrt{\tfrac{2}{3}} - \sqrt{\tfrac{3}{2}}$ 28. $\sqrt{\tfrac{1}{8}} + \sqrt{8}$
29. $\sqrt{\dfrac{xy}{2}} + \sqrt{8xy}$ 30. $\sqrt{\dfrac{3uv}{2}} - \sqrt{24uv}$
31. $\sqrt{12} - \sqrt{\tfrac{1}{3}}$ 32. $\sqrt{\tfrac{3}{5}} + 2\sqrt{20}$
33. $\sqrt{\dfrac{1}{2}} + \dfrac{\sqrt{2}}{2} + \sqrt{8}$ 34. $\dfrac{\sqrt{3}}{3} + 2\sqrt{\dfrac{1}{3}} + \sqrt{12}$

7.6 PRODUCTS AND QUOTIENTS INVOLVING RADICALS

We will conclude this chapter by considering several special types of products and quotients that involve radicals. The distributive law plays a central role in our approach to these problems. In the examples that follow all variables represent positive real number.

Special products

The following examples illustrate several types of special products.

EXAMPLE 22   Multiply and simplify:

(A) $\sqrt{2}(\sqrt{2} - 3) = \sqrt{2}\sqrt{2} - 3\sqrt{2} = 2 - 3\sqrt{2}$

(B) $\sqrt{x}(\sqrt{x} - 3) = \sqrt{x}\sqrt{x} - 3\sqrt{x} = x - 3\sqrt{x}$

(C) $(\sqrt{2} - 3)(\sqrt{2} + 5) = \sqrt{2}\sqrt{2} - 3\sqrt{2} + 5\sqrt{2} - 15$

$$= 2 + 2\sqrt{2} - 15$$
$$= 2\sqrt{2} - 13$$

(D) $(\sqrt{x} - 3)(\sqrt{x} + 5) = \sqrt{x}\sqrt{x} - 3\sqrt{x} + 5\sqrt{x} - 15$

$$= x + 2\sqrt{x} - 15$$

(E) $(\sqrt{a} + \sqrt{b})^2 = (\sqrt{a})^2 + 2\sqrt{a}\sqrt{b} + (\sqrt{b})^2$

$$= a + 2\sqrt{ab} + b$$

NOTE: $(\sqrt{a} + \sqrt{b})^2 \neq a + b$

PROBLEM 22   Multiply and simplify:

(A) $\sqrt{3}(2 - \sqrt{3})$ (B) $\sqrt{y}(2 + \sqrt{y})$

(C) $(\sqrt{3} - 1)(\sqrt{3} + 4)$ (D) $(\sqrt{y} + 2)(\sqrt{y} - 5)$

(E) $(\sqrt{x} - \sqrt{y})^2$

EXAMPLE 23   Show that $(2 - \sqrt{3})$ is a solution of the equation $x^2 - 4x + 1 = 0$.

Solution
$$x^2 - 4x + 1 = 0$$
$$(2 - \sqrt{3})^2 - 4(2 - \sqrt{3}) + 1 \overset{?}{=} 0$$
$$4 - 4\sqrt{3} + 3 - 8 + 4\sqrt{3} + 1 \overset{?}{=} 0$$
$$0 \overset{\checkmark}{=} 0$$

PROBLEM 23   Show that $(2 + \sqrt{3})$ is a solution of $x^2 - 4x + 1 = 0$.

Special quotients—rationalizing denominators

Recall that to express $\sqrt{2}/\sqrt{3}$ in simplest radical form, we multiplied the numerator and denominator by $\sqrt{3}$ to clear the denominator of the radical:

$$\frac{\sqrt{2}}{\sqrt{3}} = \frac{\sqrt{2} \cdot \sqrt{3}}{\sqrt{3} \cdot \sqrt{3}} = \frac{\sqrt{6}}{3}$$

The denominator is thus converted to a rational number. The process of converting irrational denominators to rational forms is called **rationalizing the denominator.**

How can we rationalize the binomial denominator in

$$\frac{1}{\sqrt{3} - \sqrt{2}}$$

Multiplying the numerator and denominator by $\sqrt{3}$ or $\sqrt{2}$ does not help. Try it! Recalling the special product

$$(a - b)(a + b) = a^2 - b^2$$

this suggests that we multiply the numerator and denominator by the denominator, only with the middle sign changed. Thus,

$$\frac{1}{\sqrt{3} - \sqrt{2}} = \frac{1(\sqrt{3} + \sqrt{2})}{(\sqrt{3} - \sqrt{2})(\sqrt{3} + \sqrt{2})} = \frac{\sqrt{3} + \sqrt{2}}{3 - 2} = \sqrt{3} + \sqrt{2}$$

EXAMPLE 24 Rationalize denominators and simplify.

(A) $\dfrac{\sqrt{2}}{\sqrt{6} - 2} = \dfrac{\sqrt{2}(\sqrt{6} + 2)}{(\sqrt{6} - 2)(\sqrt{6} + 2)} = \dfrac{\sqrt{12} + 2\sqrt{2}}{6 - 4}$

$$= \frac{2\sqrt{3} + 2\sqrt{2}}{2} = \frac{\cancel{2}(\sqrt{3} + \sqrt{2})}{\cancel{2}} = \sqrt{3} + \sqrt{2}$$

(B) $\dfrac{\sqrt{x} - \sqrt{y}}{\sqrt{x} + \sqrt{y}} = \dfrac{(\sqrt{x} - \sqrt{y})(\sqrt{x} - \sqrt{y})}{(\sqrt{x} + \sqrt{y})(\sqrt{x} - \sqrt{y})} = \dfrac{x - 2\sqrt{xy} + y}{x - y}$

PROBLEM 24 Rationalize denominators and simplify.

(A) $\dfrac{\sqrt{2}}{\sqrt{2} + 3}$

(B) $\dfrac{\sqrt{x} + \sqrt{y}}{\sqrt{x} - \sqrt{y}}$

ANSWERS TO MATCHED PROBLEMS

22. (A) $2\sqrt{3} - 3$; (B) $2\sqrt{y} + y$; (C) $3\sqrt{3} - 1$;
(D) $y - 3\sqrt{y} - 10$ (E) $x - 2\sqrt{xy} + y$
23. $(2 + \sqrt{3})^2 - 4(2 + \sqrt{3}) + 1 = 4 + 4\sqrt{3} + 3 - 8 - 4\sqrt{3} + 1 = 0$
24. (A) $\dfrac{2 - 3\sqrt{2}}{-7}$ or $\dfrac{-2 + 3\sqrt{2}}{7}$; (B) $\dfrac{x + 2\sqrt{xy} + y}{x - y}$

EXERCISE 7.6

Multiply and simplify where possible.

A 1. $4(\sqrt{5} + 2)$
2. $3(\sqrt{3} - 4)$
3. $2(5 - \sqrt{2})$
4. $5(3 - \sqrt{5})$

5. $\sqrt{2}(\sqrt{2} + 3)$ 6. $\sqrt{3}(\sqrt{3} + 2)$

7. $\sqrt{5}(\sqrt{5} - 4)$ 8. $\sqrt{7}(\sqrt{7} - 2)$

9. $\sqrt{3}(2 - \sqrt{3})$ 10. $\sqrt{2}(3 - \sqrt{2})$

11. $\sqrt{x}(\sqrt{x} - 3)$ 12. $\sqrt{y}(\sqrt{y} - 8)$

13. $\sqrt{m}(3 - \sqrt{m})$ 14. $\sqrt{n}(4 - \sqrt{n})$

B   15. $\sqrt{6}(\sqrt{2} - 1)$ 16. $\sqrt{3}(5 + \sqrt{6})$

17. $\sqrt{5}(\sqrt{10} + \sqrt{5})$ 18. $\sqrt{20}(\sqrt{5} - 1)$

19. $(\sqrt{2} - 1)(\sqrt{2} + 3)$ 20. $(2 - \sqrt{3})(3 + \sqrt{3})$

21. $(\sqrt{x} + 2)(\sqrt{x} - 3)$ 22. $(\sqrt{m} - 3)(\sqrt{m} - 4)$

23. $(\sqrt{5} + 2)^2$ 24. $(\sqrt{3} - 3)^2$

25. $(2\sqrt{2} - 5)(3\sqrt{2} + 2)$ 26. $(4\sqrt{3} - 1)(3\sqrt{3} - 2)$

27. $(3\sqrt{x} - 2)(2\sqrt{x} - 3)$ 28. $(4\sqrt{y} - 2)(3\sqrt{y} + 1)$

29. Show that $2 + \sqrt{3}$ is a solution to $x^2 - 4x + 1 = 0$.

30. Show that $2 - \sqrt{3}$ is a solution to $x^2 - 4x + 1 = 0$.

Reduce by removing common factors from numerator and denominator.

31. $\dfrac{8 + 4\sqrt{2}}{12}$ 32. $\dfrac{6 - 2\sqrt{3}}{6}$

33. $\dfrac{-3 - 6\sqrt{5}}{9}$ 34. $\dfrac{-4 + 2\sqrt{7}}{4}$

35. $\dfrac{6 - \sqrt{18}}{3}$ 36. $\dfrac{10 + \sqrt{8}}{2}$

C   *Rationalize denominators and simplify.*

37. $\dfrac{1}{\sqrt{11} + 3}$ 38. $\dfrac{1}{\sqrt{5} + 2}$

39. $\dfrac{2}{\sqrt{5} + 1}$ 40. $\dfrac{4}{\sqrt{6} - 2}$

41. $\dfrac{\sqrt{y}}{\sqrt{y} + 3}$ 42. $\dfrac{\sqrt{x}}{\sqrt{x} - 2}$

43. $\dfrac{\sqrt{3} + 2}{\sqrt{3} - 2}$ 44. $\dfrac{\sqrt{2} - 1}{\sqrt{2} + 2}$

45. $\dfrac{\sqrt{x} + 2}{\sqrt{x} - 3}$ 46. $\dfrac{\sqrt{a} - 3}{\sqrt{a} + 2}$

7.7 CHAPTER REVIEW: IMPORTANT TERMS AND SYMBOLS, REVIEW EXERCISE, PRACTICE TEST

Important terms and symbols

natural number exponent (*7.1*)   laws of exponents (*7.1, 7.2*) zero exponent (*7.2*)   integer exponent (*7.2*)   scientific notation (*7.3*)   square root (*7.4*)   radical (*7.4*)

$\sqrt{}$ (7.4)   properties of radicals (7.4)   simplest radical form (7.4)   sum and difference of radicals (7.5)   products and quotients of radicals (7.6)   rationalizing denominators (7.6)

Exercise 7.7
Review exercise

A *All variables represent positive real numbers.*

1. Evaluate: (A) 2^4; (B) 3^{-2}

2. Evaluate: (A) $\left(\frac{1}{3}\right)^0$; (B) $\frac{1}{3^{-2}}$

Simplify and write answers using positive exponents only.

3. $\left(\frac{2x^2}{3y^3}\right)^2$

4. $(x^2y^{-3})^{-1}$

5. Multiply $(3 \times 10^4)(2 \times 10^{-6})$ and express answer in (A) power-of-ten form and (B) nonpower-of-ten form.

Simplify and express in simplest radical form.

6. $-\sqrt{25}$

7. $\sqrt{4x^2y^4}$

8. $\sqrt{\frac{25}{y^2}}$

9. $4\sqrt{x} - 7\sqrt{x}$

10. $\sqrt{5}(\sqrt{5} + 2)$

B 11. Evaluate: (A) $(3^{-2})^{-1}$; (B) $10^{-21}10^{19}$

12. Evaluate: (A) $\frac{3^{-2}}{3}$; (B) $(25^{-5})(25^5)$

Simplify and write answers using positive exponents only.

13. $\frac{1}{(2x^2y^{-3})^{-2}}$

14. $\frac{3m^4n^{-7}}{6m^2n^{-2}}$

15. Change $\frac{(480,000)(0.005)}{1,200,000}$ to power-of-ten notation and evaluate. Express answer in (A) power-of-ten form (B) nonpower-of-ten form.

Simplify and express in simplest radical form.

16. $\sqrt{36x^4y^7}$

17. $\frac{1}{\sqrt{2y}}$

18. $\sqrt{\frac{3x}{2y}}$

19. $\sqrt{\frac{2}{3}} + \sqrt{\frac{3}{2}}$

20. $(\sqrt{3} - 1)(\sqrt{3} + 2)$

C *Simplify and write answers using positive exponents only.*

21. $\left(\frac{9m^3n^{-3}}{3m^{-2}n^2}\right)^{-2}$

22. $(x^{-1} + y^{-1})^{-1}$

23. The volume of mercury increases linearly with temperature over a fairly wide temperature range (this is why mercury is often used in thermometers). If 1 cc of mercury at $0°C$ is heated to a temperature of $T°C$, its volume is given by the formula

$$V = 1 + (1.8 \times 10^{-4})T$$

Find the volume of the sample at $(2 \times 10^2)°C$ as a decimal fraction.

Simplify and express in simplest radical form.

24. $\dfrac{\sqrt{8m^3n^4}}{\sqrt{12m^2}}$

25. $\sqrt{4x^4 + 16x^2}$

26. $\dfrac{\sqrt{x} - 2}{\sqrt{x} + 2}$

27. If a is a square root of b, then does $a^2 = b$ or does $b^2 = a$?

28. Describe the set $\{x \mid \sqrt{x^2} = |x|, x$ a real number$\}$.

Practice test
Chapter 7

Simplify and write answers using positive exponents only.

1. $\dfrac{2^3 \cdot 2^{-3}}{2^{-2}}$

2. $\dfrac{8x^{-2}y^3}{2x^{-3}y^5}$

3. $\dfrac{1}{(2a^3b^{-2})^{-2}}$

4. $\left(\dfrac{3a^{-2}b}{6a^2b^{-3}}\right)^{-2}$

5. $(u^{-1} - v^{-1})^{-1}$

6. Change each number to scientific notation and evaluate:

$$\dfrac{(0.028)(300,000,000)}{0.000000000014}$$

Simplify and express answer in simplest radical form. (All variables represent positive real numbers.)

7. $\sqrt{12x^4y^5}$

8. $\dfrac{6x}{\sqrt{3x}}$

9. $3\sqrt{2} - 5\sqrt{x} - \sqrt{2}$

10. $\sqrt{\dfrac{5}{3}} - \sqrt{\dfrac{3}{5}}$

11. $(\sqrt{2} + 5)(\sqrt{2} - 3)$

12. $\dfrac{\sqrt{y} + 3}{\sqrt{y} - 3}$

8

QUADRATIC EQUATIONS

8.1 INTRODUCTORY REMARKS

The equation

$$\tfrac{1}{2}x - \tfrac{1}{3}(x + 3) = 2 - x$$

is a first-degree equation in one variable since it can be transformed into the equivalent equation

$$7x - 18 = 0$$

which is a special case of

$$ax + b = 0 \qquad a \neq 0$$

We have solved many equations of this type and found that they always have a single solution. From a mathematical point of view this pretty well takes care of first-degree equations in one variable.

In this chapter we will consider the next class of polynomial equations called quadratic equations. A **quadratic equation** in one variable is any equation that can be written in the form

$$ax^2 + bx + c = 0 \qquad a \neq 0$$

where x is a variable and a, b, and c are constants. We will refer to this form as the **standard form** for the quadratic equation. The equations

$$2x^2 - 3x + 5 = 0$$

$$15 = 180t - 16t^2$$

are both quadratic equations since they are either in the standard form or can be transformed into this form.

Problems that give rise to quadratic equations are many and varied. For example, to find the dimensions of a rectangle with an area of $78\,\text{in}^2$ and a length twice its width, we are led to the equation

$$(2x)(x) = 78$$

or

$$2x^2 - 78 = 0$$

If an arrow is shot vertically in the air (from the ground) with an initial velocity of 176 fps (feet per second) its distance y above the ground t sec after it is released (neglecting air resistance) is given by $y = 176t - 16t^2$. To find the times when y is 0, we are led to the equation

$$176t - 16t^2 = 0$$

or

$$16t^2 - 176t = 0$$

To find the times when the arrow is 16 ft off the ground, we are led to the equation

$$176t - 16t^2 = 16$$

or

$$16t^2 - 176t + 16 = 0$$

We actually have at hand, particularly since the last chapter on exponents and radicals, all of the tools we need to solve these equations—it is a matter of putting this material together in the right way. Putting this material together in the right way is the subject matter for this chapter.

8.2 SOLUTION BY FACTORING

If the coefficients a, b, and c in the quadratic equation

$$ax^2 + bx + c = 0$$

are such that $ax^2 + bx + c$ can be written as the product of two first-degree factors with integer coefficients, then the quadratic equation can be quickly and easily solved. The method of solution by factoring rests on the following property of real numbers:

Zero Property

$ab = 0$ if and only if $a = 0$ or $b = 0$

EXAMPLE 1 Solve $x^2 + 2x - 15 = 0$ by factoring.

Solution $x^2 + 2x - 15 = 0$

$(x - 3)(x + 5) = 0$ $(x - 3)(x + 5) = 0$ if and only if $(x - 3) = 0$ or $(x + 5) = 0$

$x - 3 = 0$ or $x + 5 = 0$

$x = 3$ or $x = -5$

CHECK

$x = 3$: $3^2 + 2(3) - 15 = 9 + 6 - 15 = 0$

$x = -5$: $(-5)^2 + 2(-5) - 15 = 25 - 10 - 15 = 0$

PROBLEM 1 Solve $x^2 - 2x - 8 = 0$ by factoring.

EXAMPLE 2 Solve $2x^2 = 3x$.

Solution $2x^2 = 3x$ Why shouldn't both sides be divided by x?

$2x^2 - 3x = 0$ Write in standard form and factor.

$$x(2x - 3) = 0$$

$x(2x - 3) = 0$ if and only if $x = 0$ or $2x - 3 = 0$

$$x = 0 \quad \text{or} \quad 2x - 3 = 0$$

$$x = 0 \quad \text{or} \quad x = \tfrac{3}{2}$$

CHECK

$$x = 0: \ 2(0)^2 \overset{?}{=} 3(0) \qquad x = \tfrac{3}{2}: \ 2(\tfrac{3}{2})^2 \overset{?}{=} 3(\tfrac{3}{2})$$

$$0 \overset{\checkmark}{=} 0 \qquad\qquad \tfrac{9}{2} \overset{\checkmark}{=} \tfrac{9}{2}$$

PROBLEM 2 Solve $3t^2 = 2t$.

EXAMPLE 3 Solve $2x^2 - 8x + 3 = 0$ by factoring, if possible, using integer coefficients.

Solution $2x^2 - 8x + 3$ cannot be factored using integer coefficients; hence, another method, which we will consider later, must be used.

PROBLEM 3 Solve $x^2 - 3x - 3 = 0$ by factoring, if possible, using integer coefficients.

EXAMPLE 4 Solve $x + \dfrac{7}{2} = \dfrac{2}{x}$.

Solution
$$x + \frac{7}{2} = \frac{2}{x}$$

Multiply both sides by $2x$, the LCM of the denominators. Note: $x \neq 0$.

$$2x \cdot x + 2x \cdot \frac{7}{2} = 2x \cdot \frac{2}{x}$$

Cancel denominators.

$$2x^2 + 7x = 4$$

Write in standard form.

$$2x^2 + 7x - 4 = 0$$

Factor, if possible.

$$(2x - 1)(x + 4) = 0$$

Solve as in Example 1.

$$2x - 1 = 0 \quad \text{or} \quad x + 4 = 0$$

$$2x = 1 \qquad\qquad x = -4$$

$$x = \tfrac{1}{2}$$

Check is left to the reader.

PROBLEM 4 Solve $x = \dfrac{1}{2} + \dfrac{3}{x}$.

1. $x = -2, 4$ **2.** $t = 0, \frac{2}{3}$

3. Cannot be solved by factoring using integer coefficients.

4. $x = -\frac{3}{2}, 2$

EXERCISE 8.2

A *Solve.*

1. $(x - 3)(x - 4) = 0$ **2.** $(x - 9)(x - 4) = 0$

3. $(x + 6)(x - 5) = 0$ **4.** $(x - 9)(x + 3) = 0$

5. $(x + 4)(3x - 2) = 0$ **6.** $(2x - 1)(x + 2) = 0$

7. $(4t + 3)(5t - 2) = 0$ **8.** $(2m + 3)(3m - 2) = 0$

9. $u(4u - 1) = 0$ **10.** $z(3z + 5) = 0$

Solve by factoring.

11. $x^2 - 6x + 5 = 0$ **12.** $x^2 - 5x + 6 = 0$

13. $x^2 - 4x + 3 = 0$ **14.** $x^2 - 8x + 15 = 0$

15. $x^2 - 4x - 12 = 0$ **16.** $x^2 + 4x - 5 = 0$

17. $x^2 - 3x = 0$ **18.** $x^2 + 5x = 0$

19. $4t^2 - 8t = 0$ **20.** $3m^2 + 12m = 0$

21. $x^2 - 25 = 0$ **22.** $x^2 - 36 = 0$

B *Solve each equation by factoring. If an equation cannot be solved by factoring, state this as your answer.*

NOTE: *(A) Clear the equation of fractions (if they are present) by multiplying through by the least common multiple of all of the denominators. (B) Write the equation in standard quadratic form. (C) If all numerical coefficients contain a common factor, divide it out. (D) Test for factorability. (E) If factorable, solve.*

23. $2x^2 = 3 - 5x$ **24.** $3x^2 = x + 2$

25. $3x(x - 2) = 2(x - 2)$ **26.** $2x(x - 1) = 3(x + 1)$

27. $4n^2 = 16n + 128$ **28.** $3m^2 + 12m = 36$

29. $3z^2 - 10z = 8$ **30.** $2y^2 + 15y = 8$

31. $3 = t^2 + 7t$ **32.** $y^2 = 5y - 2$

33. $\frac{u}{4}(u + 1) = 3$ **34.** $\frac{x^2}{2} = x + 4$

35. $y = \frac{9}{y}$ **36.** $\frac{t}{2} = \frac{2}{t}$

37. The width of a rectangle is 8 in less than its length. If its area is 33 in², find its dimensions.

38. Find the base and height of a triangle with area 2 ft² if its base is 3 ft longer than its height $(A = \frac{1}{2}bh)$.

C *Solve.*

39. $y = \dfrac{15}{y-2}$

40. $2x - 3 = \dfrac{2}{x}$

41. $2 + \dfrac{2}{x^2} = \dfrac{5}{x}$

42. $1 - \dfrac{3}{x} = \dfrac{10}{x^2}$

43. The sum of a number and its reciprocal is $\frac{13}{6}$. Find the number(s).

44. The difference between a number and its reciprocal is $\frac{7}{12}$. Find the number(s).

45. A flag has a cross of uniform width centered on a red background (Figure 1). Find the width of the cross so that it takes up exactly half of the total area of a 4- by 3-ft flag.

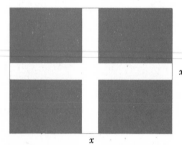

Figure 1

8.3 SOLUTION BY SQUARE ROOT

The method of square root is very fast when it applies, and it leads to a general method for solving all quadratic equations, as will be seen in the next section. A few examples should make the process clear.

EXAMPLE 5 $x^2 - 9 = 0$ Notice that the first-degree term is missing.

$\qquad\qquad x^2 = 9$ What number squared is 9?

$\qquad\qquad x = \pm\sqrt{9}$ Short for $\sqrt{9}$ or $-\sqrt{9}$.

$\qquad\qquad x = \pm 3$

PROBLEM 5   Solve $x^2 = 16$ by the square root method.

EXAMPLE 6 $x^2 - 7 = 0$

$\qquad\qquad x^2 = 7$ What number squared is 7?

$\qquad\qquad x = \pm\sqrt{7}$

PROBLEM 6   Solve $x^2 - 8 = 0$.

EXAMPLE 7 $2x^2 - 3 = 0$

$$2x^2 = 3$$
$$x^2 = \tfrac{3}{2}$$
$$x = \pm\sqrt{\tfrac{3}{2}} \quad \text{or} \quad \pm\frac{\sqrt{6}}{2}$$

PROBLEM 7 Solve $3x^2 - 2 = 0$.

EXAMPLE 8 $(x - 2)^2 = 16$ Solve for $(x - 2)$ first, then solve for x.

$$x - 2 = \pm 4$$
$$x = 2 \pm 4$$
$$x = 6, -2$$

PROBLEM 8 Solve $(x + 3)^2 = 25$.

EXAMPLE 9 $(x + \tfrac{1}{2})^2 = \tfrac{5}{4}$ Solve for $(x + \tfrac{1}{2})$ first, then solve for x.

$$x + \tfrac{1}{2} = \pm\sqrt{\tfrac{5}{4}}$$
$$x + \tfrac{1}{2} = \pm\frac{\sqrt{5}}{2}$$
$$x = -\frac{1}{2} \pm \frac{\sqrt{5}}{2}$$
$$x = \frac{-1 \pm \sqrt{5}}{2}$$

PROBLEM 9 Solve $(x - \tfrac{1}{3})^2 = \tfrac{7}{9}$.

EXAMPLE 10 $x^2 = -9$

No solution in the real numbers, since no real number squared is negative.

PROBLEM 10 Solve $x^2 + 4 = 0$.

ANSWERS TO MATCHED PROBLEMS

5. $x = \pm 4$

6. $x = \pm\sqrt{8}$ or $\pm 2\sqrt{2}$

7. $x = \pm\sqrt{\dfrac{2}{3}}$ or $\pm\dfrac{\sqrt{6}}{3}$

8. $x = -8, 2$

9. $x = \dfrac{1 \pm \sqrt{7}}{3}$

10. No real solutions

EXERCISE 8.3

Solve for all real solutions by the square root method.

A 1. $x^2 = 16$ 2. $x^2 = 49$

   3. $m^2 - 64 = 0$ 4. $n^2 - 25 = 0$

   5. $x^2 = 3$ 6. $y^2 = 2$

   7. $u^2 - 5 = 0$ 8. $x^2 - 11 = 0$

   9. $a^2 = 18$ 10. $y^2 = 8$

   11. $x^2 - 12 = 0$ 12. $n^2 - 27 = 0$

   13. $x^2 = \frac{4}{9}$ 14. $y^2 = \frac{9}{16}$

   15. $9x^2 = 4$ 16. $16y^2 = 9$

   17. $9x^2 - 4 = 0$ 18. $16y^2 - 9 = 0$

B 19. $25x^2 - 4 = 0$ 20. $9x^2 - 1 = 0$

   21. $4t^2 - 3 = 0$ 22. $9x^2 - 7 = 0$

   23. $2x^2 - 5 = 0$ 24. $3m^2 - 7 = 0$

   25. $3m^2 - 1 = 0$ 26. $5n^2 - 1 = 0$

   27. $(y - 2)^2 = 9$ 28. $(x - 3)^2 = 4$

   29. $(x + 2)^2 = 25$ 30. $(y + 3)^2 = 16$

   31. $(y - 2)^2 = 3$ 32. $(y - 3)^2 = 5$

   33. $(x - \frac{1}{2})^2 = \frac{9}{4}$ 34. $(x - \frac{1}{3})^2 = \frac{4}{9}$

   35. $(x - 3)^2 = -4$ 36. $(t + 1)^2 = -9$

C 37. $(x - \frac{3}{2})^2 = \frac{3}{2}$ 38. $(y + \frac{5}{2})^2 = \frac{5}{2}$

   39. Solve for b: $a^2 + b^2 = c^2$ 40. Solve for v: $k = \frac{1}{2}mv^2$

APPLICATIONS

41. The pressure p in lb per sq ft from a wind blowing at v mph is given by $p = 0.003v^2$. If a pressure gauge on a bridge registers a wind pressure of 14.7 lb per sq ft, what is the velocity of the wind?

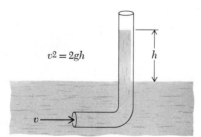

Figure 2

42. One method of measuring the velocity of water in a stream or river is to use an L-shaped tube as indicated in Figure 2. Torricelli's law in physics tells us that the height (in feet) that the water is pushed up into the tube above the surface is related to the water's velocity (in feet per second) by the formula $v^2 = 2gh$, where g is approximately 32 ft per sec per sec. (NOTE: The device can also be used as a simple speedometer for a boat.) How fast is a stream flowing if $h = 0.5$ ft? Find the answer to two decimal places.

8.4 SOLUTION BY COMPLETING THE SQUARE

The factoring and square root methods discussed in the last two sections are fast and easy to use when they apply. Unfortunately many quadratic equations will not yield to either method as stated. For example, the very simple-looking polynomial in

$$x^2 + 6x - 2 = 0$$

cannot be factored in the integers, and at least for now, the square root method does not seem applicable either. The equation requires a new method if it can be solved at all.

In this section we will discuss a method, called "solution by completing the square," that will work for all quadratic equations. In the next section we will use this method to develop a general formula that will be used in the future whenever the methods of the two preceding sections fail.

The method of completing the square is based on the process of transforming the standard quadratic equation,

$$ax^2 + bx + c = 0$$

into the form

$$(x + A)^2 = B$$

where A and B are constants. This last equation can easily be solved (assuming $B \geq 0$) by the square root method discussed in the last section. That is,

$$(x + A)^2 = B$$
$$x + A = \pm \sqrt{B}$$
$$x = -A \pm \sqrt{B}$$

Before considering how the first part is accomplished, let's pause for a moment and consider a related problem: What number must be added to $x^2 + 6x$ so that the result is the square of a linear expression? There is an easy mechanical rule for finding this number based on the squares of the following binomials:

$$(x + m)^2 = x^2 + 2mx + m^2$$

$$(x - m)^2 = x^2 - 2mx + m^2$$

In either case, we see that the third term on the right is the square of one-half of the coefficient of x in the second term on the right. This observation leads directly to the rule:

> To **complete the square** of a quadratic of the form
>
> $$x^2 + bx$$
>
> add the square of one-half of the coefficient of x, that is
>
> $$\left(\frac{b}{2}\right)^2$$

EXAMPLE 11   **(A)**   To complete the square of $x^2 + 6x$, add $(\frac{6}{2})^2$, that is, 9; thus

$$x^2 + 6x + 9 = (x + 3)^2$$

(B)   To complete the square of $x^2 - 3x$, add $(-\frac{3}{2})^2$; that is $\frac{9}{4}$; thus

$$x^2 - 3x + \tfrac{9}{4} = (x - \tfrac{3}{2})^2$$

PROBLEM 11   **(A)**   Complete the square of $x^2 + 10x$ and factor.
(B)   Complete the square of $x^2 + 5x$ and factor.

It is important to note that the rule stated above applies only to quadratic forms where the coefficient of the second-degree term is 1.

Solution of quadratic equations by completing the square

Solving quadratic equations by the method of completing the square is best illustrated by examples. In this course we are only going to be interested in real solutions.

EXAMPLE 12   Solve $x^2 + 6x - 2 = 0$ by the method of completing the square.

Solution $x^2 + 6x - 2 = 0$ Add 2 to both sides of the equation to remove -2 from the left side.

$x^2 + 6x \qquad = 2$ To complete the square of the left side, add the square of one-half of the coefficient of x to each side of the equation.

$x^2 + 6x + 9 = 2 + 9$ Factor the left side.

$(x + 3)^2 = 11$ Proceed as in the last section.

$x + 3 = \pm\sqrt{11}$

$x = -3 \pm \sqrt{11}$

PROBLEM 12   Solve $x^2 - 8x + 10 = 0$ by the method of completing the square.

EXAMPLE 13 Solve $2x^2 - 4x - 3 = 0$ by the method of completing the square.

$$2x^2 - 4x - 3 = 0$$
$$x^2 - 2x - \tfrac{3}{2} = 0$$
$$x^2 - 2x \quad\ = \tfrac{3}{2}$$
$$x^2 - 2x + 1 = \tfrac{3}{2} + 1$$
$$(x - 1)^2 = \tfrac{5}{2}$$
$$x - 1 = \pm\sqrt{\tfrac{5}{2}}$$
$$x - 1 = \pm\frac{\sqrt{10}}{2}$$
$$x = 1 \pm \frac{\sqrt{10}}{2}$$
$$x = \frac{2 \pm \sqrt{10}}{2}$$

Note that the coefficient of x^2 is not 1. Divide through by the leading coefficient and proceed as in the last example.

PROBLEM 13 Solve $2x^2 + 8x + 3 = 0$ by the method of completing the square.

ANSWERS TO MATCHED PROBLEMS

11. (A) $x^2 + 10x + 25 = (x + 5)^2$;   (B) $x^2 + 5x + \tfrac{25}{4} = (x + \tfrac{5}{2})^2$
12. $x = 4 \pm \sqrt{6}$
13. $x = -2 \pm \sqrt{\dfrac{5}{2}}$ or $\dfrac{-4 \pm \sqrt{10}}{2}$

EXERCISE 8.4

A *Complete the square and factor.*

1. $x^2 + 4x$
2. $x^2 + 8x$
3. $x^2 - 6x$
4. $x^2 - 10x$
5. $x^2 + 12x$
6. $x^2 + 2x$

Solve by the method of completing the square.

7. $x^2 + 4x + 2 = 0$
8. $x^2 + 8x + 3 = 0$
9. $x^2 - 6x - 3 = 0$
10. $x^2 - 10x - 3 = 0$

B *Complete the square and factor.*

11. $x^2 + 3x$
12. $x^2 + x$
13. $u^2 - 5u$
14. $m^2 - 7m$

Solve by the method of completing the square.

15. $x^2 + x - 1 = 0$
16. $x^2 + 3x - 1 = 0$

17. $u^2 - 5u + 2 = 0$

18. $n^2 - 3n - 1 = 0$

19. $m^2 - 4m + 8 = 0$

20. $x^2 - 2x + 3 = 0$

21. $2y^2 - 4y + 1 = 0$

22. $2x^2 - 6x + 3 = 0$

23. $2u^2 + 3u - 1 = 0$

24. $3x^2 + x - 1 = 0$

C   25. Solve for x: $x^2 + mx + n = 0$

26. Solve for x: $ax^2 + bx + c = 0$, $a \neq 0$

8.5 THE QUADRATIC FORMULA

The method of completing the square can be used to solve any quadratic equation, but the process is often tedious. If you had a very large number of quadratic equations to solve by completing the square, before you finished you would probably ask yourself if the process could not be made more efficient. Why not take the general equation

$$ax^2 + bx + c = 0 \qquad a \neq 0$$

and solve it once and for all for x in terms of the coefficients a, b, and c by the method of completing the square, and thus, obtain a formula that could be memorized and used whenever a, b, and c are known?

We start by making the leading coefficient 1. How? Multiply both sides of the equation by $1/a$. Thus

$$x^2 + \frac{b}{a}x + \frac{c}{a} = 0$$

Adding $-c/a$ to each side, we get

$$x^2 + \frac{b}{a}x = -\frac{c}{a}$$

Add the square of one-half of the coefficient of x, which is $(b/2a)^2$, to each side to complete the square of the left side. Thus,

$$x^2 + \frac{b}{a}x + \frac{b^2}{4a^2} = \frac{b^2}{4a^2} - \frac{c}{a}$$

Factor the left side and combine the right side into a single term, leaving

$$\left(x + \frac{b}{2a}\right)^2 = \frac{b^2 - 4ac}{4a^2}$$

If $b^2 - 4ac \geq 0$, then by the definition of square root of nonnegative real numbers,

$$x + \frac{b}{2a} = \pm\sqrt{\frac{b^2 - 4ac}{4a^2}}$$

$$x = -\frac{b}{2a} \pm \frac{\sqrt{b^2 - 4ac}}{2a}$$

$$\boxed{x = \frac{-b \pm \sqrt{b^2 - 4ac}}{2a} \qquad a \neq 0}$$

This last equation is called the **quadratic formula.** It should be memorized and used to solve quadratic equations when simpler methods fail. The following examples illustrate the use of the formula. Note that if $b^2 - 4ac$ is negative there are no real solutions.

EXAMPLE 14   Solve $2x + \frac{3}{2} = x^2$ by use of the quadratic formula.

Solution

$2x + \frac{3}{2} = x^2$ Clear the equation of fractions.

$4x + 3 = 2x^2$ Write in standard form.

$2x^2 - 4x - 3 = 0$

$x = \frac{-b \pm \sqrt{b^2 - 4ac}}{2a}$ $\begin{aligned} a &= 2 \\ b &= -4 \\ c &= -3 \end{aligned}$ Write down the quadratic formula, and identify a, b, and c.

$x = \frac{-(-4) \pm \sqrt{(-4)^2 - 4(2)(-3)}}{2(2)}$ Substitute into formula and simplify. Sign errors are easily made at this stage. Be careful!

$x = \frac{4 \pm \sqrt{40}}{4} = \frac{4 \pm 2\sqrt{10}}{4}$

$x = \frac{2(2 \pm \sqrt{10})}{4}$

$x = \frac{2 \pm \sqrt{10}}{2}$

PROBLEM 14   Solve $3x^2 = 2x + 2$ by use of the quadratic formula.

ANSWERS TO MATCHED PROBLEMS   14. $x = \frac{1 \pm \sqrt{7}}{3}$

EXERCISE 8.5   **A** *Specify the constants a, b, and c for each quadratic equation when written in the standard form $ax^2 + bx + c = 0$.*

1. $x^2 + 4x + 2 = 0$ 2. $x^2 + 8x + 3 = 0$
3. $x^2 - 3x - 2 = 0$ 4. $x^2 - 6x - 8 = 0$

5. $3x^2 - 2x + 1 = 0$ 6. $2x^2 - 5x + 3 = 0$

7. $2u^2 = 1 - 3u$ 8. $m = 1 - 3m^2$

9. $2x^2 - 5x = 0$ 10. $3y^2 - 5 = 0$

Solve by use of the quadratic formula.

11. $x^2 + 4x + 2 = 0$ 12. $x^2 + 8x + 3 = 0$

13. $y^2 - 6y - 3 = 0$ 14. $y^2 - 10y - 3 = 0$

B   15. $3t + t^2 = 1$ 16. $x^2 = 1 - x$

17. $2x^2 - 6x + 3 = 0$ 18. $2x^2 - 4x + 1 = 0$

19. $3m^2 = 1 - m$ 20. $3u + 2u^2 = 1$

21. $x^2 = 2x - 3$ 22. $x^2 + 8 = 4x$

23. $2x = 3 + \dfrac{3}{x}$ 24. $x + \dfrac{2}{x} = 6$

C   25. $m^2 = \dfrac{8m - 1}{5}$ 26. $x^2 = 3x + \dfrac{1}{2}$

27. $3u^2 = \sqrt{3}u + 2$ 28. $t^2 - \sqrt{5}t - 11 = 0$

True (T) or false (F)?

29. If $b_2 - 4ac < 0$, the quadratic equation has no real solution.

30. If $b^2 - 4ac > 0$, the quadratic equation has two real solutions.

31. If $b^2 - 4ac = 0$, the quadratic equation has one real solution.

32. If $b^2 - 4ac = 0$, the quadratic equation has two real solutions.

8.6 WHICH METHOD?

In normal practice the quadratic formula is used whenever the square root method or the factoring method do not produce results. These latter methods are generally faster when they apply, and should be used.

Note that any equation of the form

$ax^2 + c = 0$ Note that the bx term is missing.

can always be solved (if solutions exist in the real numbers) by the square root method. And any equation of the form

$ax^2 + bx = 0$ Note that the c term is missing.

can always be solved by factoring since $ax^2 + bx = x(ax + b)$.

It is important to realize, however, that the quadratic formula can always be used and will produce the same results as any of the other methods. For example, let us solve

$2x^2 + 7x - 15 = 0$

Suppose you observe that the polynomial factors. Thus,

$$(2x - 3)(x + 5) = 0$$

$$2x - 3 = 0 \quad \text{or} \quad x + 5 = 0$$

$$x = \tfrac{3}{2} \quad \text{or} \quad x = -5$$

Suppose you had used the quadratic formula instead?

$$x = \frac{-b \pm \sqrt{b^2 - 4ac}}{2a}$$

$$x = \frac{-(7) \pm \sqrt{7^2 - 4(2)(-15)}}{2(2)} \qquad \begin{array}{l} a = 2 \\ b = 7 \\ c = -15 \end{array}$$

$$x = \frac{-7 \pm \sqrt{169}}{4} = \frac{-7 \pm 13}{4}$$

$$x = \tfrac{3}{2}, \ -5$$

The quadratic formula produces the same result as the factoring method (as it should), but with a little more work.

In the exercises for this section the problems will be mixed up, and it will be up to you to use the most efficient method—formula, factoring, or square root—for each particular problem.

EXERCISE 8.6

Use the most efficient method to find all real solutions to each equation.

A
1. $x^2 - x - 6 = 0$
2. $x^2 + 2x - 8 = 0$
3. $x^2 + 7x = 0$
4. $x^2 = 3x$
5. $2x^2 = 32$
6. $3x^2 - 27 = 0$
7. $x^2 + 2x - 2 = 0$
8. $m^2 - 3m - 1 = 0$
9. $2x^2 = 4x$
10. $2y^2 + 3y = 0$
11. $x^2 - 2x = 1$
12. $x^2 - 2 = 2x$

B
13. $u^2 = 3u - \tfrac{3}{2}$
14. $t^2 = \tfrac{3}{2}(t + 1)$
15. $M = M^2$
16. $t(t - 3) = 0$
17. $6y = \dfrac{1 - y}{y}$
18. $2x + 1 = \dfrac{6}{x}$
19. $I^2 - 50 = 0$
20. $72 = u^2$
21. $(B - 2)^2 = 3$
22. $(u + 3)^2 = 5$

C
23. $x^2 + 4 = 0$
24. $(x - 2)^2 = -9$
25. $\dfrac{24}{n} = 12n - 28$
26. $3x = \dfrac{84 - 9x}{x}$
27. $\dfrac{24}{10 + x} + 1 = \dfrac{24}{10 - x}$
28. $\dfrac{1.2}{x - 1} + \dfrac{1.2}{x} = 1$

Solve for the indicated letter in terms of the other letters.

29. $d = \tfrac{1}{2}gt^2$ for t (positive)

30. $a^2 + b^2 = c^2$ for a (positive)

31. $A = P(1 + r)^2$ for r (positive)

32. $P = EI - RI^2$ for I

8.7 ADDITIONAL APPLICATIONS

Many real world problems lead directly to quadratic equations for their solutions. We conclude this chapter with a couple of examples and an exercise that includes a variety of applications from a number of different fields. Since quadratic equations usually have two solutions, it is important to check both of the solutions in the original problem to see if one or both must be rejected. It is often the case that only one of the solutions will make sense in the context of the original application.

To get you started, our first example is a relatively easy word problem involving numbers. The second example is a geometric problem that is slightly more involved. Remember, draw figures, make diagrams, write down related formulas, and so on. Use scratch paper to try out ideas.

EXAMPLE 15 If the reciprocal of a number is subtracted from the original number, the difference is $\frac{8}{3}$. Find the number.

Solution Let x = the number.

$$x - \frac{1}{x} = \frac{8}{3}$$ Write an equation.

$$3x \cdot x - 3x \cdot \frac{1}{x} = 3x \cdot \frac{8}{3}$$ Clear fractions.

$$3x^2 - 3 = 8x$$ Convert to standard form.

$$3x^2 - 8x - 3 = 0$$ Solve by one of the methods discussed in earlier sections.

$$(3x + 1)(x - 3) = 0$$ Factoring works.

$$3x + 1 = 0 \quad \text{or} \quad x - 3 = 0$$

$$3x = -1 \qquad\qquad x = 3$$ Both answers are good, as the reader can easily check.

$$x = -\tfrac{1}{3}$$

PROBLEM 15 The sum of a number and its reciprocal is $\frac{5}{2}$. Find the number.

EXAMPLE 16 A painting measuring 6 by 8 in has a frame of uniform width with a total area equal to the area of the painting. How wide is the frame? Give the answer in simplest radical form and as a decimal fraction to two decimal places.

Solution

Total area of picture and frame $=$ Twice the area of the picture

$$(6 + 2x)(8 + 2x) = 2(6 \cdot 8)$$
$$48 + 28x + 4x^2 = 96$$
$$x^2 + 7x - 12 = 0$$

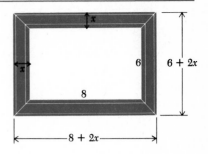

$$x = \frac{-b \pm \sqrt{b^2 - 4ac}}{2a}$$

$a = 1$
$b = 7$
$c = 12$

$$x = \frac{-7 \pm \sqrt{7^2 - 4(1)(-12)}}{2(1)}$$

$$x = \frac{-7 \pm \sqrt{97}}{2}$$ Check to see if one of the answers must be rejected.

The negative answer must be rejected since it has no meaning relative to the original problem; hence

$$x = \frac{-7 + \sqrt{97}}{2} \approx 1.42 \quad \text{in}$$

PROBLEM 16 If the length and width of a 4- by 2-in rectangle are each increased by the same amount, the area of the new rectangle will be twice the old. What are the dimensions to two decimal places of the new rectangle?

ANSWERS TO MATCHED PROBLEMS

15. $\frac{1}{2}$ or   2
16. 5.12 by 3.12 in

EXERCISE 8.7

These problems are not grouped from easy (A) to difficult or theoretical (C). They are grouped somewhat according to type. The most difficult problems are double-starred (**), those of moderate difficulty single-starred (*), and the easier ones are not marked. For each problem set up an appropriate equation and solve.

NUMBER PROBLEMS

1. Find a positive number that is 56 less than its square.
2. Find two consecutive positive even integers whose product is 168.

3. Find all numbers with the property that when the number is added to itself, the sum is the same as when the number is multiplied by itself.

4. Find two numbers such that their sum is 21 and their product is 104.

5. The sum of a number and its reciprocal is $\frac{10}{3}$. Find the number(s).

6. Find all numbers such that 6 times the reciprocal of the number is 1 less than the original number.

BUSINESS AND ECONOMICS

7. If P dollars is invested at r percent compounded annually, at the end of 2 years it will grow to $A = P(1 + r)^2$. At what interest rate will $100 grow to $144 in two years? NOTE: $A = 144$ and $P = 100$.

8. Repeat Problem 7 for $1,000 growing to $1,210 in 2 years.

* 9. Cost equations for manufacturing companies are often quadratic in nature. (At very high or very low outputs the costs are more per unit because of inefficiency of plant operation at these extremes.) If the cost equation for manufacturing transistor radios is $C = x^2 - 10x + 31$, where C is the cost of manufacturing x units per week (both x and C are in thousands), find the output x for a $15,000 weekly cost.

*10. Repeat Problem 9 for a weekly cost of $6,000.

*11. The manufacturing company in Problem 9 sells its transistor radios for $3 each. Thus its revenue equation is $R = 3x$, where R is revenue and x is the number of units sold per week (both in thousands). Find the break-even points for the company, that is, the output x at which revenue equals cost.

*12. Repeat Problem 11 for the company selling its radios for $6 each.

COMMUNICATIONS

*13. The number of telephone connections c possible through a switchboard to which n telephones are connected is given by the formula $c = n(n - 1)/2$. How many telephones n could be handled by a switchboard that had the capacity of 190 connections? HINT: Find n when $c = 190$.

*14. Repeat Problem 13 for a switchboard with a capacity of 435 connections.

GEOMETRY

The following theorem may be used where needed:

PYTHAGOREAN THEOREM. *A triangle is a right triangle if and only if the square of the longest side is equal to the sum of the squares of the two shorter sides.*

$c^2 = a^2 + b^2$

15. Find the length of each side of a right triangle if the second longest side is 1 meter longer than the shortest side and the longest side is 2 meters longer than the shortest side.

16. Find the length of each side of a right triangle if the two shorter sides are 2 and 4 cm shorter than the longest side.

*17. Find r in Figure 3. Express the answer in simplest radical form. (Radius of smaller circle is 1 in.)

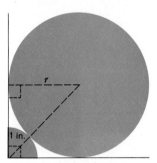

Figure 3

*18. Approximately how far would a person be able to see from the top of a mountain 2 miles high (Figure 4)? Use a calculator or a square root table to estimate the answer to the nearest mile.

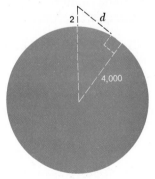

Figure 4

POLICE SCIENCE

19. Skid marks are often used to estimate the speed of a car in an accident. It is common practice for an officer to drive the car in question (if it is still running) at a speed of 20 to 30 mph and skid it to a stop near the original skid marks. It is known (from physics) that the speed of the car and the length of the skid marks are related by the formula

$$\frac{d_a}{v_a^2} = \frac{d_t}{v_t^2}$$

where d_a = length of accident car's skid marks
d_t = length of test car's skid marks
v_a = speed of accident car (to be found)
v_t = speed of test car

Estimate the speed of an accident vehicle if its skid marks are 100 ft and the test car driven at 30 mph produces skid marks of 36 ft.

20. Repeat Problem 19 for an accident vehicle that left skid marks 196 ft long.

RATE-TIME PROBLEMS

21. Two boats travel at right angles to each other after leaving the same dock at the same time; 1 hr later they are 13 miles apart. If one travels 7 mph faster than the other, what is the rate of each? (HINT: See theorem preceding Problem 15.)

**22. Repeat Problem 21 with one boat traveling 1 mph faster than the other and after 1 hour they are 5 miles apart.

**23. A motorboat takes 1 hr longer to go 24 miles up a river than to return. If the boat cruises at 10 mph in still water, what is the rate of the current?

**24. A speedboat takes 1 hr longer to go 60 miles up a river than to return. If the boat can cruise at 25 mph on still water, what is the rate of the current?

8.9 CHAPTER REVIEW: IMPORTANT TERMS AND SYMBOLS, REVIEW EXERCISE, PRACTICE TEST

Important terms and symbols

quadratic equation (8.1)   solution by factoring (8.2)
solution by square root (8.3)   solution by completing the square (8.4)   quadratic formula (8.5)

Exercise 8.8
Review exercise

A *Find all real solutions by factoring or square root methods.*

1. $x^2 = 25$ **2.** $x^2 - 3x = 0$

3. $(2x - 1)(x + 3) = 0$ **4.** $x^2 - 5x + 6 = 0$

5. $x^2 - 2x - 15 = 0$

6. Write $4x = 2 - 3x^2$ in standard form $ax^2 + bx + c = 0$ and identify a, b, and c.

7. Write down the quadratic formula associated with $ax^2 + bx + c = 0$.

8. Use the quadratic formula to solve $x^2 + 3x + 1 = 0$.

9. Solve $x^2 + 3x - 10 = 0$ by any method.

10. Find two positive numbers whose product is 27 if one is 6 more than the other.

B *Find all real solutions by factoring or square root methods.*

11. $3x^2 = 36$ 12. $10x^2 = 20x$

13. $(x - 2)^2 = 16$ 14. $3t^2 - 8t - 3 = 0$

15. $2x = \dfrac{3}{x} - 5$

16. Solve $x^2 - 6x - 3 = 0$ by the completing-the-square method.

17. Solve $3x^2 = 2(x + 1)$ using the quadratic formula.

18. Solve $2x^2 - 2x = 40$ by any method.

19. Divide 18 into two parts so that their product is 72.

20. The perimeter of a rectangle is 22 in. If its area is 30 in², find the length of each side.

C *Find all real solutions by factoring or square root methods.*

21. $2x^2 + 27 = 0$ 22. $(t - \frac{3}{2})^2 = \frac{3}{2}$

23. $\dfrac{8m^2 + 15}{2m} = 13$

24. Solve $2x^2 - 2x - 3 = 0$ by the completing-the-square method.

25. Solve $3x - 1 = \dfrac{2(x + 1)}{x + 2}$ using the quadratic formula.

26. If $b^2 - 4ac > 0$, then the quadratic equation has two real solutions. [True (T) or false (F)?]

Practice test
Chapter 8

In Problems 1–3 find all real solutions by factoring or the square root method.

1. $2y^2 = 16$ 2. $4x^2 = 8x$

3. $2x^2 + 3x - 2 = 0$

4. Write the quadratic formula associated with $ax^2 + bx + c = 0$.

Solve Problems 5 and 6 using the quadratic formula.

5. $x^2 + 3x - 3 = 0$ 6. $4u^2 - 1 = 2u$

7. Solve $x^2 - 4x - 4 = 0$ by completing the square.

Solve using any method discussed in Chapter 8.

8. $3x = \dfrac{2}{x} + 1$ 9. $2x^2 - 10x = 28$

10. $\left(m + \dfrac{1}{2}\right)^2 = \dfrac{5}{4}$

11. If the length and width of a 4-cm by 6-cm rectangle are each increased by the same amount, the area of the new rectangle will be twice the original. What are the dimensions of the new rectangle?

12. A **golden rectangle** is one that has the property that when a square with side equal to the short side of the rectangle is removed from one end, the ratio of the sides of the remaining rectangle is the same as the ratio of the sides of the original rectangle. If the shorter side of the original rectangle is 1, find the shorter side of the remaining rectangle (see Figure 5). This number is called the golden ratio, and it turns up frequently in the history of mathematics.

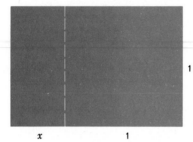

Figure 5

APPENDIXES

APPENDIX A

ARITHMETIC REVIEW

The following three sections provide a very brief review of fractions, decimals, and percent (with practice problems) for those who need it.

A.1 FRACTIONS
Fractional forms

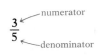

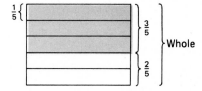

Multiplication

To multiply two fractions, multiply their numerators and place the product over the product of the denominators.

$$\frac{a}{b} \cdot \frac{c}{d} = \frac{a \cdot c}{b \cdot d}$$

EXAMPLE 1 $\dfrac{3}{5} \cdot \dfrac{2}{7} = \dfrac{3 \cdot 2}{5 \cdot 7} = \dfrac{6}{35}$

PROBLEM 1   Multiply $\frac{7}{4} \cdot \frac{5}{6}$.

Reducing to lowest terms

EXAMPLE 2 $\dfrac{9}{12} = \dfrac{3 \cdot 3}{3 \cdot 4} = \dfrac{3}{3} \cdot \dfrac{3}{4} = 1 \cdot \dfrac{3}{4} = \dfrac{3}{4}$

or

$\dfrac{9}{12} = \dfrac{9 \div 3}{12 \div 3} = \dfrac{3}{4}$

Removing a common factor from a numerator and denominator is called **canceling.**

A common factor can be canceled from a numerator and denominator.

$$\frac{ka}{kb} = \frac{\overset{1}{\cancel{k}}a}{\underset{1}{\cancel{k}}b} = \frac{a}{b}$$

Canceling *all* common factors from a numerator and denominator is called **reducing a fraction to lowest terms.**

PROBLEM 2 Reduce to lowest terms: $\frac{12}{15}$.

EXAMPLE 3 Multiply and reduce to lowest terms: $\frac{8}{9} \cdot \frac{3}{4}$.

Solution $\frac{8}{9} \cdot \frac{3}{4} = \frac{\overset{2}{\cancel{8}}}{\underset{3}{\cancel{9}}} \cdot \frac{\overset{1}{\cancel{3}}}{\underset{1}{\cancel{4}}}$

Any factor in a numerator can cancel a like factor in a denominator.

$= \frac{2}{3}$

PROBLEM 3 Multiply and reduce to lowest terms: $\frac{2}{15} \cdot \frac{3}{8}$.

Division

To divide two fractions, invert the divisor and multiply.

$$\frac{a}{b} \div \frac{c}{d} = \frac{a}{b} \cdot \frac{d}{c}$$

invert divisor

EXAMPLE 4 $\frac{5}{8} \div \frac{3}{4} = \frac{5}{8} \cdot \frac{4}{3}$

Invert divisor and multiply. Do not cancel before inverting divisor.

$= \frac{5}{\underset{2}{\cancel{8}}} \cdot \frac{\overset{1}{\cancel{4}}}{3}$

$= \frac{5}{6}$

PROBLEM 4 Divide and reduce to lowest terms: $\frac{7}{16} \div \frac{14}{10}$.

Addition and subtraction

If two fractions have the same denominator, we add or subtract them by adding or subtracting their numerators and placing the result over the common denominator.

$$\frac{a}{b} + \frac{c}{b} = \frac{a + c}{b} \qquad \frac{a}{b} - \frac{c}{b} = \frac{a - c}{b}$$

EXAMPLE 5 **(A)** $\frac{5}{6} + \frac{4}{6} = \frac{5 + 4}{6} = \frac{9}{6} = \frac{3}{2}$

(B) $\dfrac{5}{8} - \dfrac{3}{8} = \dfrac{5-3}{8} = \dfrac{2}{8} = \dfrac{1}{4}$

PROBLEM 5 Perform the indicated operation and reduce to lowest terms:

(A) $\dfrac{11}{24} + \dfrac{4}{24}$ **(B)** $\dfrac{17}{12} - \dfrac{9}{12}$

If two fractions do not have a common denominator, then we must change them so that they do before we can add or subtract. The following important property of fractions is behind this process.

Fundamental Principle of Fractions

We can multiply the numerator and denominator by the same nonzero number.

$$\frac{a}{b} = \frac{ka}{kb} \qquad k \neq 0$$

The most convenient common denominator to use is the **least common multiple (LCM)** of the denominators. The LCM is the smallest number exactly divisible by each denominator. The least common multiple of the denominators is also called the **least common denominator (LCD).**

EXAMPLE 6 **(A)** $\dfrac{3}{4} + \dfrac{2}{3}$ LCD is 12 (the smallest number divisible by 4 and 3). Use the fundamental principle of fractions to make each denominator 12.

$$= \frac{3 \cdot 3}{3 \cdot 4} + \frac{4 \cdot 2}{4 \cdot 3}$$

$$= \frac{9}{12} + \frac{8}{12}$$

$$= \frac{9 + 8}{12}$$

$$= \frac{17}{12}$$

(B) $\dfrac{5}{6} - \dfrac{11}{15}$ LCD $= 30$

$$= \frac{5 \cdot 5}{5 \cdot 6} - \frac{2 \cdot 11}{2 \cdot 15}$$

$$= \frac{25}{30} - \frac{22}{30}$$

$$= \frac{25 - 22}{30}$$

$$= \frac{3}{30} = \frac{1}{10}$$

PROBLEM 6 Perform the indicated operations and reduce to lowest terms.

(A) $\frac{5}{6} + \frac{4}{9}$

(B) $\frac{7}{10} - \frac{3}{25}$

ANSWERS TO MATCHED PROBLEMS

1. $\frac{35}{24}$ 2. $\frac{4}{5}$ 3. $\frac{1}{20}$ 4. $\frac{5}{16}$

5. (A) $\frac{5}{8}$; (B) $\frac{2}{3}$ 6. (A) $\frac{23}{18}$; (B) $\frac{29}{50}$

EXERCISE A.1

Perform the indicated operations and reduce to lowest terms.

1. $\frac{2}{3} \cdot \frac{4}{5}$

2. $\frac{3}{4} \cdot \frac{2}{7}$

3. $\frac{1}{2} \div \frac{2}{3}$

4. $\frac{3}{4} \div \frac{4}{3}$

5. $\frac{4}{9} \cdot \frac{3}{12}$

6. $\frac{5}{12} \cdot \frac{9}{10}$

7. $\frac{10}{12} \div \frac{6}{18}$

8. $\frac{18}{24} \div \frac{12}{9}$

9. $\frac{5}{12} + \frac{3}{12}$

10. $\frac{3}{8} + \frac{7}{8}$

11. $\frac{11}{9} - \frac{5}{9}$

12. $\frac{17}{14} - \frac{9}{14}$

13. $\frac{1}{4} + \frac{2}{3}$

14. $\frac{3}{5} + \frac{1}{2}$

15. $\frac{5}{6} - \frac{3}{4}$

16. $\frac{3}{4} - \frac{1}{3}$

17. $\frac{5}{12} + \frac{3}{8}$

18. $\frac{11}{18} + \frac{5}{12}$

19. $\frac{7}{9} - \frac{5}{12}$

20. $\frac{7}{20} - \frac{9}{30}$

21. $\frac{2}{3} \cdot \left(\frac{3}{4} \div \frac{9}{12} \right)$

22. $\frac{4}{5} \div \left(\frac{8}{10} \div \frac{3}{4} \right)$

23. $\frac{8}{9} \cdot \left(\frac{3}{4} - \frac{2}{3} \right)$

24. $\frac{7}{5} \div \left(\frac{5}{6} - \frac{1}{4} \right)$

A.2 DECIMALS
Decimal fractions

A decimal fraction is a way of representing numbers in decimal form. The base 10 is central to the process. Recall:

hundreds tens units tenths hundredths thousandths

3 4 6 . 2 3 5

$$300 + 40 + 6 + \frac{2}{10} + \frac{3}{100} + \frac{5}{1,000}$$

Thus,

$$0.2 = \frac{2}{10}$$

$$0.03 = \frac{3}{100}$$

$$0.005 = \frac{5}{1,000}$$

$$0.235 = \frac{235}{1,000}$$

To convert a fraction into a decimal fraction, we divide the denominator into the numerator. Consider the following example.

EXAMPLE 7 Convert $\frac{12}{23}$ to a decimal fraction rounded to three decimal places.

Solution

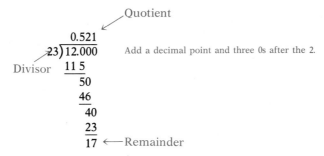

Quotient

Add a decimal point and three 0s after the 2.

Divisor

Remainder

In order to round to three decimal places, we do the following: If the remainder after carrying out the division to three decimal places is greater than or equal to one-half the divisor, we add 1 to the last decimal place in the quotient. If the remainder is less than one-half the divisor, we leave the last decimal place alone. In this case the remainder 17 is more than one-half the divisor 23, so we add 1 to the third decimal place to obtain

$$\frac{12}{23} \approx 0.522 \qquad \approx \text{means approximately equal to}$$

We write 0.522 instead of .522 because the latter might be mistaken for the whole number 522. Placing the 0 to the left of the decimal point keeps the decimal point from getting lost.

PROBLEM 7 Convert $\frac{26}{35}$ to a decimal fraction rounded to two decimal places.

Addition

> To add two or more decimal fractions, line up decimal points and add as in whole number arithmetic. The decimal point is carried straight down to the sum.

EXAMPLE 8   Add: 325.2, 62.25, 3.012

Solution

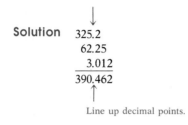

$$
\begin{array}{r}
325.2 \\
62.25 \\
3.012 \\
\hline
390.462
\end{array}
$$

Line up decimal points.

PROBLEM 8   Add: 22.06, 204.135, 3.4

Subtraction

> To subtract one decimal fraction from another, line up decimals and subtract as in whole number arithmetic. The decimal point is carried straight down to the difference.

EXAMPLE 9   Subtract 23.427 from 125.8.

Solution

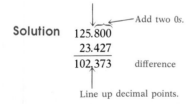

Add two 0s.

$$
\begin{array}{r}
125.800 \\
23.427 \\
\hline
102.373
\end{array}
$$
difference

Line up decimal points.

PROBLEM 9   Subtract 325.63 from 407.5.

Multiplication

> To multiply two decimal fractions, multiply as in whole number arithmetic. The product has as many decimal places as the sum of the number of decimal places used in the two original decimal fractions.

EXAMPLE 10   Multiply 36.24 and 13.6.

Solution

$$3\,6.2\,4$$
$$\underline{1\,3.6}$$
$$2\,1\,7\,4\,4$$
$$1\,0\,8\,7\,2$$
$$\underline{3\,6\,2\,4}$$
$$4\,9\,2.8\,6\,4$$

2 decimal places
1 decimal place } add
3 decimal places

PROBLEM 10   Multiply 103.2 and 26.72.

Division

> To divide one decimal fraction by another, divide as in whole number arithmetic. To locate the decimal in the quotient, move the decimal point in the dividend and divisor as many places to the right as there are decimal places in the divisor. Then move the decimal point straight up from the dividend to the quotient.

EXAMPLE 11   Divide 425.3 by 2.43, and round answer to two decimal places.

Solution

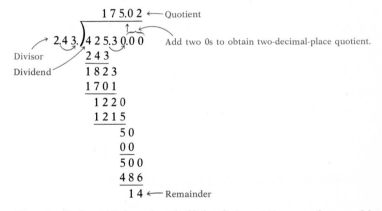

Quotient

Add two 0s to obtain two-decimal-place quotient.

Divisor

Dividend

Remainder

The remainder 14 is less than half the divisor 243, so we do not add 1 to the second decimal place. The quotient rounded to two decimal places is

175.02

PROBLEM 11   Divide 3.74 by 2.4, and round answer to two decimal places.

ANSWERS TO
MATCHED PROBLEMS

7. 0.74   **8.** 229.595   **9.** 81.87
10. 2,757.504   **11.** 1.56

EXERCISE A.2

Add:

1. 3.1, 2.5, 0.2   **2.** 6.4, 0.3, 5.6
3. 23.2, 2.45, 6.012   **4.** 405.03, 21.105, 5.2

Subtract:

5. 25.32 from 43.05	**6.** 6.09 from 13.12
7. 23.56 from 103.2	**8.** 5.69 from 41.2

Multiply:

9. 2.5 by 13	**10.** 24 by 1.6
11. 4.26 by 0.002	**12.** 3.04 by 0.006

Convert to decimal fractions rounded to two decimal places.

13. $\dfrac{5}{6}$ **14.** $\dfrac{7}{9}$

15. $\dfrac{34}{46}$ **16.** $\dfrac{16}{24}$

Divide and round answers to one decimal place.

17. 84 by 2.2	**18.** 68 by 4.5
19. 36.2 by 4.6	**20.** 4.02 by 6.4

A.3 PERCENT
Percent

If 1 is divided into 100 equal parts, then each part is called **1 percent. The** symbol % is read "percent." Thus,

Percent		Fraction		Decimal
23%	=	$\dfrac{23}{100}$	=	0.23
4%	=	$\dfrac{4}{100}$	=	0.04
162%	=	$\dfrac{162}{100}$	=	1.62
0.3%	=	$\dfrac{0.3}{100}$	=	0.003

Percents to decimals

> To convert a percent to a decimal, remove the percent symbol and move the decimal point two places to the left.

EXAMPLE 12

(A) $43\% = 43.\% = 0.43$

(B) $7\% = 07.\% = 0.07$

(C) $234\% = 234.\% = 2.34$

(D) $0.3\% = 00.3\% = 0.003$

PROBLEM 12 Convert each percent to a decimal.

(A) 59%   (B) 2%   (C) 105%   (D) 0.7%

Decimals or whole numbers to percent

> To convert a decimal or whole number to percent, shift the decimal point two places to the right and use the percent symbol.

EXAMPLE 13 (A) $0.23 = 0.23 = 23\%$

(B) $0.06 = 0.06 = 6\%$

(C) $4.37 = 4.37 = 437\%$

(D) $0.064 = 0.064 = 6.4\%$

PROBLEM 13 Convert each decimal to a percent.

(A) 0.71   (B) 0.05   (C) 1.09   (D) 0.003

Percents of various quantities

> To find a percent of a given quantity, we convert the percent to a decimal and multiply.

EXAMPLE 14 (A) Find 23 percent of 45.

Solution 23% of 45 = 0.23 × 45 = 10.35

(B) Find 6.2 percent of 28.

Solution 6.2% of 28 = 0.062 × 28 = 1.736

PROBLEM 14 Find each:

(A) 4% of 64   (B) 108% of 22

ANSWERS TO   **12.** (A) 0.59;   (B) 0.02;   (C) 1.05;   (D) 0.007
MATCHED PROBLEMS **13.** (A) 71%;   (B) 5%;   (C) 109%;   (D) 0.3%
                  14. (A) 2.56;   (B) 23.76

EXERCISE A.3

Change percents to decimals.

1. 67%	**2.** 14%	**3.** 9%
4. 1%	**5.** 216%	**6.** 308%

7.	0.6%	**8.**	0.1%	**9.**	7.4%
10.	2.8%	**11.**	23.1%	**12.**	64.5%

Change decimals to percents.

13.	0.12	**14.**	0.21	**15.**	0.08
16.	0.02	**17.**	3.25	**18.**	6.04
19.	0.007	**20.**	0.004	**21.**	0.072
22.	0.069	**23.**	0.405	**24.**	0.236

Find each.

25.	12% of 403	**26.**	18% of 40	**27.**	6% of 4,000
28.	8% of 2,000	**29.**	125% of 200	**30.**	150% of 44
31.	6.5% of 24	**32.**	4.8% of 36	**33.**	0.4% of 20
34.	0.7% of 80				

APPENDIX B

SQUARES AND SQUARE ROOTS (0 to 199)

n	n^2	$\sqrt{n}$	n	n^2	$\sqrt{n}$	n	n^2	$\sqrt{n}$	n	n^2	$\sqrt{n}$
0	0	0.000	40	1,600	6.325	80	6,400	8.944	120	14,400	10.954
1	1	1.000	41	1,681	6.403	81	6,561	9.000	121	14,641	11.000
2	4	1.414	42	1,764	6.481	82	6,724	9.055	122	14,884	11.045
3	9	1.732	43	1,849	6.557	83	6,889	9.110	123	15,129	11.091
4	16	2.000	44	1,936	6.633	84	7,056	9.165	124	15,376	11.136
5	25	2.236	45	2,025	6.708	85	7,225	9.220	125	15,625	11.180
6	36	2.449	46	2,116	6.782	86	7,396	9.274	126	15,876	11.225
7	49	2.646	47	2,209	6.856	87	7,569	9.327	127	16,129	11.269
8	64	2.828	48	2,304	6.928	88	7,744	9.381	128	16,384	11.314
9	81	3.000	49	2,401	7.000	89	7,921	9.434	129	16,641	11.358
10	100	3.162	50	2,500	7.071	90	8,100	9.487	130	16,900	11.402
11	121	3.317	51	2,601	7.141	91	8,281	9.539	131	17,161	11.446
12	144	3.464	52	2,704	7.211	92	8,464	9.592	132	17,424	11.489
13	169	3.606	53	2,809	7.280	93	8,649	9.644	133	17,689	11.533
14	196	3.742	54	2,916	7.348	94	8,836	9.695	134	17,956	11.576
15	225	3.873	55	3,025	7.416	95	9,025	9.747	135	18,225	11.619
16	256	4.000	56	3,136	7.483	96	9,216	9.798	136	18,496	11.662
17	289	4.123	57	3,249	7.550	97	9,409	9.849	137	18,769	11.705
18	324	4.243	58	3,364	7.616	98	9,604	9.899	138	19,044	11.747
19	361	4.359	59	3,481	7.681	99	9,801	9.950	139	19,321	11.790
20	400	4.472	60	3,600	7.746	100	10,000	10.000	140	19,600	11.832
21	441	4.583	61	3,721	7.810	101	10,201	10.050	141	19,881	11.874
22	484	4.690	62	3,844	7.874	102	10,404	10.100	142	20,164	11.916
23	529	4.796	63	3,969	7.937	103	10,609	10.149	143	20,449	11.958
24	576	4.899	64	4,096	8.000	104	10,816	10.198	144	20,736	12.000
25	625	5.000	65	4,225	8.062	105	11,025	10.247	145	21,025	12.042
26	676	5.099	66	4,356	8.124	106	11,236	10.296	146	21,316	12.083
27	729	5.196	67	4,489	8.185	107	11,449	10.344	147	21,609	12.124
28	784	5.292	68	4,624	8.246	108	11,664	10.392	148	21,904	12.166
29	841	5.385	69	4,761	8.307	109	11,881	10.440	149	22,201	12.207
30	900	5.477	70	4,900	8.367	110	12,100	10.488	150	22,500	12.247
31	961	5.568	71	5,041	8.426	111	12,321	10.536	151	22,801	12.288
32	1,024	5.657	72	5,184	8.485	112	12,544	10.583	152	23,104	12.329
33	1,089	5.745	73	5,329	8.544	113	12,769	10.630	153	23,409	12.369
34	1,156	5.831	74	5,476	8.602	114	12,996	10.677	154	23,716	12.410
35	1,225	5.916	75	5,625	8.660	115	13,225	10.724	155	24,025	12.450
36	1,296	6.000	76	5,776	8.718	116	13,456	10.770	156	24,336	12.490
37	1,369	6.083	77	5,929	8.775	117	13,689	10.817	157	24,649	12.530
38	1,444	6.164	78	6,084	8.832	118	13,924	10.863	158	24,964	12.570
39	1,521	6.245	79	6,241	8.888	119	14,161	10.909	159	25,281	12.610
n	n^2	$\sqrt{n}$	n	n^2	$\sqrt{n}$	n	n^2	$\sqrt{n}$	n	n^2	$\sqrt{n}$

Squares and Square Roots (0 to 199) continued

n	n^2	$\sqrt{n}$	n	n^2	$\sqrt{n}$	n	n^2	$\sqrt{n}$	n	n^2	$\sqrt{n}$
160	25,600	12.649	170	28,900	13.038	180	32,400	13.416	190	36,100	13.784
161	25,921	12.689	171	29,241	13.077	181	32,761	13.454	191	36,481	13.820
162	26,244	12.728	172	29,584	13.115	182	33,124	13.491	192	36,864	13.856
163	26,569	12.767	173	29,929	13.153	183	33,489	13.528	193	37,249	13.892
164	26,896	12.806	174	30,276	13.191	184	33,856	13.565	194	37,636	13.928
165	27,225	12.845	175	30,625	13.229	185	34,225	13.601	195	38,025	13.964
166	27,556	12.884	176	30,976	13.266	186	34,596	13.638	196	38,416	14.000
167	27,889	12.923	177	31,329	13.304	187	34,969	13.675	197	38,809	14.036
168	28,224	12.961	178	31,684	13.342	188	35,344	13.711	198	39,204	14.071
169	28,561	13.000	179	32,041	13.379	189	35,721	13.748	199	39,601	14.107
n	n^2	$\sqrt{n}$	n	n^2	$\sqrt{n}$	n	n^2	$\sqrt{n}$	n	n^2	$\sqrt{n}$

APPENDIX C

ANSWERS TO SELECTED PROBLEMS

EXERCISE 1.1

1. T **3.** T **5.** T **7.** T **9.** $5 \in P$
11. $6 \notin R$ **13.** $P = R$ **15.** $P \neq Q$ **17.** $\{6, 7, 8, 9\}$
19. $\emptyset$ **21.** $\{Su, M, T, W, Th, F, S\}$ **23.** $\{a, b, l\}$
25. $\emptyset$ **27.** $\{2, 4\}$ **29.** $\emptyset$ **31.** $\{1, 2, 3, 4, 5, 6, 7, 8\}$

EXERCISE 1.2

1. 6, 13 **3.** 67, 402 **5.** Even: 14, 28; odd: 9, 33 **7.** Even:
426; odd: 23, 105, 77 **9.** Composite: 6, 9; prime: 2, 11
11. Composite: 12, 27; prime: 17, 23 **13.** 20, 22, 24, 26, 28, 30
15. 21, 23, 25, 27, 29 **17.** 20, 21, 22, 24, 25, 26, 27, 28, 30
19. 23, 29 **21.** $2 \cdot 5$ **23.** $2 \cdot 3 \cdot 5$ **25.** $2 \cdot 2 \cdot 3 \cdot 7$
27. $2 \cdot 2 \cdot 3 \cdot 5$ **29.** $2 \cdot 2 \cdot 3 \cdot 3 \cdot 3$ **31.** $2 \cdot 3 \cdot 5 \cdot 7$
33. 36 **35.** 48 **37.** 24 **39.** 60 **41.** 90
43. 120 **45.** No, no, no, yes **47.** Finite **49.** Infinite
51. Finite

EXERCISE 1.3

1. 13 **3.** 2 **5.** 17 **7.** 2 **9.** 24 **11.** 80
13. 1 **15.** 21 **17.** 8 **19.** 10 **21.** 5 **23.** 2
25. 18 **27.** 12 **29.** 4 **31.** $5x$ **33.** $x + 5$
35. $x - 5$ **37.** $5 - x$ **39.** Constants: $\frac{1}{2}$; variables: A, b, h
41. Constants: none, variables: d, r, t **43.** Constants: none;
variables: I, p, r, t **45.** Constants: 2, 3; variables: x, y
47. Constants: 2 and 3; variables: u and v **49.** $A = 18 \, \text{cm}^2$,
$P = 18 \, \text{cm}$ **51.** $A = 80 \, \text{km}^2$, $P = 36 \, \text{km}$ **53.** 20 **55.** 33
57. 14 **59.** 8 **61.** 5 **63.** 3 **65.** 22
67. 684 km **69.** 600 words **71.** $2x + 3$ **73.** $12x - 3$
75. $3(x - 8)$ **77.** 105 **79.** 16 **81.** 16 **83.** $3(t + 2)$
85. $t + (t + 2) + (t + 4)$ **87.** (A) $s = 16t^2$; (B) Constants: 16,
2; variables: s, t; (C) 1,024 ft

EXERCISE 1.4

1. T **3.** F **5.** T **7.** T **9.** T **11.** $5 = x + 3$
13. $8 = x - 3$ **15.** $18 = 3x$ **17.** $49 = 2x + 7$
19. $52 = 5x - 8$ **21.** $4x = 3x + 3$ **23.** $x + 5 = 3(x - 4)$
25. $x + (x + 1) + (x + 2) = 90$ **27.** $x + (x + 2) = 54$
29. Incorrect use of equal sign. We cannot write John = human
(why?). **31.** $x + 2x = 54$ **33.** $50 = x(x + 10)$

EXERCISE 1.5

1. $x + 10$ **3.** $28ab$ **5.** $16 + a + b$ **7.** xxx
9. $2xxxyy$ **11.** $3wwxyyy$ **13.** x^3 **15.** $2x^3y^2$

17. $3xy^2z^3$ **19.** u^{14} **21.** a^6 **23.** w^{19} **25.** y^{16}
27. 3^{30} (not 9^{30})   **29.** 9^{11} (not 81^{11})   **31.** $30abc$
33. $60uvw$ **35.** $x + y + z + 14$ **37.** $u + v + w + 19$
39. x^7 **41.** y^{10} **43.** $24x^9$ **45.** a^3b^3 **47.** $12x^2y^2$
49. $6x^4y^2$ **51.** (C) is false since $9 - 7 \neq 7 - 9$;   (D) is false since
$14 \div 7 \neq 7 \div 14$ **53.** Commutative $+$ **55.** Associative $\times$
57. Commutative $\times$ **59.** Commutative $+$
61. Commutative $+$ **63.** Commutative $+$

EXERCISE 1.6

1. Both 12   **3.** Both 45   **5.** $4x + 4y$ **7.** $7m + 7n$
9. $6x + 12$ **11.** $10 + 5m$ **13.** $3(x + y)$ **15.** $5(m + n)$
17. $a(x + y)$ **19.** $2(x + 2)$ **21.** $2x + 2y + 2z$
23. $3x + 3y + 3z$ **25.** $7(x + y + z)$ **27.** $2(m + n + 3)$
29. $x + x^2$ **31.** $y + y^3$ **33.** $6x^2 + 15x$ **35.** $2m^4 + 6m^3$
37. $6x^3 + 9x^2 + 3x$ **39.** $10x^3 + 15x^2 + 5x + 10$
41. $6x^5 + 9x^4 + 3x^3 + 6x^2$ **43.** $10x$ **45.** $11u$ **47.** $5xy$
49. $10x^2y$ **51.** $(7 + 2 + 5)x = 14x$ **53.** $x(x + 2)$
55. $u(u + 1)$ **57.** $2x(x^2 + 2)$ **59.** $x(x + y + z)$
61. $3m(m^2 + 2m + 3)$ **63.** $uv(u + v)$ **65.** $8m^5n^4 + 4m^3n^5$
67. $6x^3y^4 + 12x^3y + 3x^2y^3$ **69.** $12x^4yz^4 + 4x^2y^2z^4$
71. $uc + vc + ud + vd$ **73.** $x^2 + 5x + 6$ **75.** $abc(a + b + c)$
77. $4xyz(4x^2z + xy + 3yz^2)$

EXERCISE 1.7

1. 4   **3.** 8   **5.** 1   **7.** 1   **9.** 3   **11.** 2
13. 1   **15.** $3x, 4x$; $2y, 5y$ **17.** $6x^2, 3x^2, x^2$; $x^3, 4x^3$
19. $2u^2v, u^2v$; $3uv^2, 5uv^2$ **21.** $9x$ **23.** $4u$ **25.** $9x^2$
27. $10x$ **29.** $7x + 4y$ **31.** $3x + 5y + 6$ **33.** $m^2n, 5m^2n$;
$4mn^2, mn^2$; $2mn, 3mn$ **35.** $6t^2$ **37.** $4x + 7y + 7z$
39. $11x^3 + 4x^2 + 4x$ **41.** $4x^2 + 3xy + 2y^2$ **43.** $6x + 9$
45. $4t^2 + 8t + 10$ **47.** $3x^3 + 3x^2y + 4xy^2 + 2y^3$ **49.** $8x + 31$
51. $3x^2 + 4x$ **53.** $11t^2 + 13t + 17$ **55.** $3y^3 + 3y^2 + 4y$
57. $6x^2 + 5xy + 6y^2$ **59.** $4x^4 + 3x^2y^2 + 3y^4$
61. $4m^6 + 9m^5 + 4m^4$ **63.** $x + (x + 1) + (x + 2) + (x + 3)$; $4x + 6$
65. $9x^2y^2 + 5x^3y^3$ **67.** $8u^3v^3 + 7u^4v^2$ **69.** $6x^2 + 13x + 6$
71. $2x^2 + 5xy + 2y^2$ **73.** $x^3 + 5x^2 + 11x + 15$
75. $y(y + 2)$; $y^2 + 2y$ **77.** $x + x(x + 2) = 180$; $x^2 + 3x = 180$

EXERCISE 1.8
REVIEW EXERCISE

*The italicized decimal number(s) in parentheses following each answer
indicates the section(s) in which that type of problem is discussed.*

1. (A) $\{11, 13, 15\}$;   (B) $\{11, 13\}$ *(1.1)*   **2.** 2 *(1.3)*   **3.** 11 *(1.3)*
4. 1 *(1.3)*   **5.** 20 *(1.3)*   **6.** x^{25} *(1.5)*   **7.** $6x^8$ *(1.5)*   **8.** 2^{25}
(not 4^{25}) *(1.5)*   **9.** $x^2 + x$ *(1.6)*   **10.** $10x + 15y + 5z$ *(1.6)*
11. $6u^3 + 3u^2$ *(1.6)*   **12.** $9y$ *(1.7)*   **13.** $5m + 5n$ *(1.7)*
14. $7x^2 + 3x$ *(1.7)*   **15.** $8x^2y + 2xy^2$ *(1.7)*   **16.** $3(m + n)$ *(1.7)*

17. $8(u + v + w)$ (1.7) **18.** $x(y + w)$ (1.7) **19.** $4(x + 2w)$ (1.7) **20.** $12x$ (1.3) **21.** $3x + 3$ (1.3) **22.** $2x - 5$ (1.3)
23. (A) 23, 29, 31;   (B) Both $(1.1, 1.2)$ **24.** (A) 3;   (B) 1; (C) 1 $(1.5, 1.7)$ **25.** $2 \cdot 2 \cdot 2 \cdot 3 \cdot 5$ (1.2) **26.** 36 (1.2)
27. 90 (1.2) **28.** 90 (1.2) **29.** 180 (1.2) **30.** 6 (1.3)
31. 6 (1.3) **32.** 12 (1.3) **33.** 20 (1.3) **34.** 24 (1.3)
35. 36 (1.3) **36.** $18x^8$ (1.5) **37.** $12x^3y^5z^4$ (1.5)
38. $6y^5 + 3y^4 + 15y^3$ (1.6) **39.** $19u^2 + 7u + 17$ $(1.6, 1.7)$
40. $8x^2 + 22x$ $(1.6, 1.7)$ **41.** $3x^2 + 7xy + 2y^2$ $(1.6, 1.7)$
42. $u(u^2 + u + 1)$ (1.6) **43.** $3xy(2x + y)$ (1.6)
44. $3m^2(m^3 + 2m^2 + 5)$ (1.6) **45.** $24 = 2x - 6$ (1.4)
46. $3x = x + 12$ (1.4) **47.** $x + (x + 1) + (x + 2) + (x + 3) = 138$ (1.4)
48. $x + (x + 2) + (x + 4) = 78$ (1.4) **49.** $\{27, 51, 61\}$ (1.2)
50. $\{61\}$ (1.2) **51.** 48 (1.3) **52.** 39 (1.3)
53. $10u^5v^4 + 5u^4v^3 + 10u^3v^2$ $(1.5, 1.6)$ **54.** $7x^5 + 11x^3 + 6x^2$ $(1.6, 1.7)$ **55.** $8x^2 + 10x + 3$ $(1.6, 1.7)$ **56.** $3x^2yz(4xz + 3)$ (1.6)
57. $5x^2y^2(4x + y + 3)$ (1.6) **58.** Commutative $\times$ (1.5)
59. Associative $+$ (1.5) **60.** Commutative $+$ (1.5)
61. Associative $+$ (1.5) **62.** $4x = (x + 2) + (x + 4)$ (1.4)

PRACTICE TEST

1. (A) T;   (B) F   **2.** (A) $2 \cdot 2 \cdot 3 \cdot 3 \cdot 5$;   (B) $2 \cdot 3 \cdot 3 \cdot 5 = 90$
3. (A) 12;   (B) 42   **4.** (A) 1;   (B) 4   **5.** (A) $5x^2 + 4x + 6$;
(B) $7x^3y^2 + 2x^2y^3$ **6.** (A) $15x^4y^4$;   (B) $6x^4 + 2x^3 + 8x^2$
7. (A) $6(a + b + c)$;   (B) $4x^2(2x^2 + 3x + 1)$ **8.** (A) $5u^2 + 11u$;
(B) $6x^2 + 11xy + 3y^2$ **9.** (A) $2x - 5$;   (B) $3(x - 2) + 6$ or
$6 + 3(x - 2)$ **10.** (A) $x + 7 = 2x - 5$;
(B) $x + (x + 2) + (x + 4) = 72$

EXERCISE 2.1

1. $-8, -2, +3, +9$ **3.**

5.

7. $+4$ **9.** -10 **11.** -3 **13.** T   **15.** F   **17.** T
19. F   **21.** T   **23.** T   **25.** $+20,270$ **27.** -280
29. -5 **31.** $+27$ **33.** -3 **35.** $+25$ **37.** -10
39. -9 **41.** $+1$ **43.** $+17$ **45.** -3

EXERCISE 2.2

1. -9 **3.** $+2$ **5.** $+4$ **7.** $+6$ **9.** 0
11. Sometimes   **13.** Never   **15.** -11 **17.** -5
19. $+13$ **21.** $+2$ or -2 **23.** No solution   **25.** $+6$
27. $+5$ **29.** -5 **31.** $+5$ **33.** -8 **35.** -7
37. $+5$ **39.** -7 **41.** -5 **43.** $+5$ **45.** $+2$
47. $\{+5\}$ **49.** $\{+3\}$ **51.** $\{-6, +6\}$ **53.** $\emptyset$
55. $\{0\}$ **57.** Set of all nonpositive integers   **59.** Set of all
integers I **61.** Set of all nonnegative integers   **63.** $\{0\}$

EXERCISE 2.3

1. +11 **3.** −3 **5.** −2 **7.** −8 **9.** +3
11. +9 **13.** −6 **15.** −9 **17.** −9 **19.** −2
21. −4 **23.** −5 **25.** −4 **27.** −12 **29.** −622
31. −38 **33.** −668 **35.** −36 **37.** −4 **39.** −4
41. −5 **43.** +5 **45.** −77 **47.** +14 **49.** −6
51. −2 **53.** +3 **55.** \$23 **57.** −1493 ft **59.** 0
61. −m **63.** Commutative property, Associative property,
Theorem 2, Definition of addition

EXERCISE 2.4

1. +5 **3.** +13 **5.** −5 **7.** −5 **9.** +5
11. −6 **13.** −8 **15.** −3 **17.** +14 **19.** −4
21. −6 **23.** −5 **25.** +15 **27.** +87
29. −315 **31.** −245 **33.** +1 **35.** −3 **37.** 0
39. +7 **41.** +4 **43.** +2 **45.** +3
47. $(+29{,}141) - (-35{,}800) = 64{,}941$ ft
49. $(-245) - (-280) = +35$ ft **51.** True **53.** False;
$(+7) - (-3) = +10, (-3) - (+7) = -10$ **55.** True **57.** False;
$|(+9) + (-3)| = +6, |+9| + |-3| = +12$

EXERCISE 2.5

1. +32 **3.** −32 **5.** 0 **7.** +2 **9.** −3 **11.** Not
defined **13.** −14 **15.** −14 **17.** 0 **19.** +3
21. −3 **23.** 0 **25.** −5 **27.** −7 **29.** +2
31. +4 **33.** −4 **35.** +30 **37.** −10 **39.** −8
41. −51 **43.** +17 **45.** −6 **47.** +8 **49.** 0
51. 0 **53.** Not defined **55.** +4 **57.** +6 **59.** −20
61. 0 **63.** (A) +12; (B) +12; (C) +12 **65.** (A) +3;
(B) +3 **67.** (A) −35; (B) −35 **69.** +8 **71.** +8
73. $x = 0$ **75.** No solution

EXERCISE 2.6

1. 5 **3.** −13 **5.** −3 **7.** −2 **9.** −3 **11.** 1
13. −2 **15.** $4x$ **17.** $-4x$ **19.** $-12y$ **21.** $-3x - 3y$
23. $-5x + 3y$ **25.** $6m - 2n$ **27.** $-x + 4y$ **29.** $3x - 2y$
31. $-x + y$ **33.** $-x + 8y$ **35.** $4x - 8y$ **37.** $6xy$
39. $-3x^2y$ **41.** $2x^2 + 2x - 3$ **43.** $2x^2y + 5xy^2 - 6xy$
45. $-2x - y$ **47.** $2t - 20$ **49.** $-3y + 4$ **51.** $-10x$
53. $x - 14$ **55.** $6t^2 - 16t$ **57.** $3x - y$ **59.** $-3x + y$
61. $y + 2z$ **63.** $x - y + z$ **65.** $P = 2x + 2(x - 5) = 4x - 10$
67. 1 **69.** $-8x^2 - 16x$ **71.** $13x^2 - 26x + 10$
73. Value in cents $= 25x + 10(x + 4) = 35x + 40$

EXERCISE 2.7

1. 3 **3.** −3 **5.** −12 **7.** 5 **9.** −3 **11.** −13
13. 8 **15.** −4 **17.** −4 **19.** 3 **21.** 0 **23.** 3
25. 2 **27.** −3 **29.** 7 **31.** 4 **33.** 2 **35.** −4

37. 8   **39.** 7   **41.** 5   **43.** No solution   **45.** 16
47. 16   **49.** −15   **51.** 4   **53.** 6   **55.** No solution
57. $3x = 12$, $x = 4$ **59.** T   **61.** F   **63.** T   **65.** T
67. $a − c = a − c$ identity property of equality
$\quad\quad a = b$ given
$\quad a − c = b − c$ substitution principle

EXERCISE 2.8

1. 25, 26, 27   **3.** 16, 18, 20   **5.** 8 hr   **7.** 13 ft above and 104 ft below   **9.** 239,000 miles   **11.** 7, 9, 11   **13.** 10 ft by 23 ft   **15.** 7 quarters and 10 dimes   **17.** 5 sec   **19.** 22,000 ft   **21.** 2,331, 1,526, 1,918, 2,196, 2,279   **23.** 8 miles   **25.** 70 hours (or 2 days and 22 hr); 1,750 miles   **27.** 130 min (or 2 hr, 10 min)

EXERCISE 2.9
REVIEW EXERCISE

1. −4 (2.2)   **2.** +3 (2.2)   **3.** −5 (2.3)   **4.** −13 (2.3)
5. +6 (2.4)   **6.** −3 (2.4)   **7.** +28 (2.5)   **8.** −18 (2.5)
9. −4 (2.5)   **10.** +6 (2.5)   **11.** 0 (2.5)   **12.** Not defined (2.5)   **13.** −2 (2.3)   **14.** −16 (2.5)   **15.** −8 (2.5)
16. +8 (2.5)   **17.** 0 (2.5)   **18.** −6 (2.5)   **19.** $2x − 8$ (2.6)
20. $5x − 2$ (2.6)   **21.** $2m + 9n$ (2.6)   **22.** $−6x − 18y$ (2.6)
23. $x = −2$ (2.7)   **24.** $x = −7$ (2.7)   **25.** (A) −245; (B) +14,495 (2.1)   **26.** 52, 53, 54 (2.8)   **27.** +12 (2.2)
28. +3 (2.2)   **29.** −2 (2.2)   **30.** −10 (2.4)   **31.** +9 (2.3)
32. +17 (2.4)   **33.** −6 (2.5)   **34.** +12 (2.5)   **35.** 0 (2.5)
36. +34 (2.5)   **37.** 0 (2.5)   **38.** $−2x^2y^2 − 5xy$ (2.6)
39. $4y^3 − 7y^2 + 18y$ (2.6)   **40.** $10x − 24y$ (2.6)
41. $−6x^3y^2 − 10x^2y + 3xy^2$ (2.6)   **42.** $2y − 3$ (2.6)
43. $x − 2y$ (2.6)   **44.** $m = 9$ (2.7)   **45.** $x = 2$ (2.7)
46. 44, 46, 48, 50 (2.8)   **47.** 4 nickels and 5 quarters (2.8)
48. −45 (2.1)   **49.** (A) +15; (B) +5 (2.4)   **50.** (A) +1; (B) +4 (2.5)   **51.** $−11x − 8$ (2.6)   **52.** $\{−9\}$ (2.7)
53. 17 hr (2.8)

PRACTICE TEST

1. (A) −4; (B) +5   **2.** (A) −5; (B) −18   **3.** (A) 0; (B) Not defined   **4.** +1   **5.** $4x − 2y + 5$
6. $−5y^2 − 9y$ **7.** $x = 6$ **8.** $x = −8$
9. (A) $x − 3y$; (B) $x + 2y$ **10.** 35, 37, 39   **11.** 8 nickels and 12 dimes   **12.** 12 km

EXERCISE 3.2

1. a: $−\frac{9}{4}$, b: $−\frac{3}{4}$, c: $\frac{7}{4}$ **3.** $\frac{6}{35}$ **5.** $\frac{28x}{15y}$ **7.** $\frac{3x^2}{2y^3}$ **9.** $\frac{-6}{77}$
or $−\frac{6}{77}$ **11.** $\frac{10}{21}$ **13.** $\frac{21}{25}$ **15.** $\frac{14xy}{15}$ **17.** $\frac{-9}{14}$ **19.** 2
21. 15   **23.** 3   **25.** $9x^2$ **27.** $3x^2$ **29.** $\frac{3}{2}$ **31.** $\frac{-1}{4}$

33. $\dfrac{1}{4y}$ **35.** $\dfrac{4a}{b}$ **37.** $\dfrac{-y^2}{4x}$ **39.** $\frac{2}{3}$ **41.** $\frac{9}{375}$ **43.** 5

45. $\frac{3}{2}$ **47.** $\frac{-3}{4}$ **49.** $\dfrac{1}{z}$ **51.** 4 **53.** $2y^2$ **55.** $\dfrac{3x}{2y}$

57. y **59.** $\dfrac{3ad}{2c}$ **61.** $\dfrac{3v}{2u}$ **63.** $\dfrac{-2x^2}{3y}$ **65.** $\frac{81}{100}$

67. 1 **69.** -2 **71.** $\dfrac{adf}{bce}$

EXERCISE 3.3

1. 2 **3.** $\frac{4}{5}$ **5.** $\frac{7}{8}$ **7.** $\frac{19}{15}$ **9.** $\frac{4}{11}$ **11.** $\frac{10}{11}$ **13.** $\frac{1}{8}$

15. $-\frac{1}{15}$ **17.** $\dfrac{-3}{5xy}$ **19.** $\dfrac{5y}{x}$ **21.** $\dfrac{6}{7y}$ **23.** $\dfrac{7}{6x}$

25. $\dfrac{13x}{6}$ **27.** $\dfrac{9-10x}{15x}$ **29.** $\dfrac{x^2-y^2}{xy}$ **31.** $\dfrac{x-2y}{y}$

33. $\dfrac{5x+3}{x}$ **35.** $\dfrac{1-3x}{xy}$ **37.** $\dfrac{9+8x}{6x^2}$ **39.** $\dfrac{15-2m^2}{24m^3}$

41. $\frac{5}{3}$ **43.** $\dfrac{3x^2-4x-6}{12}$ **45.** $\dfrac{18y-16x+3}{24xy}$

47. $\dfrac{18+4y+3y^2-18y^3}{6y^3}$ **49.** $\dfrac{22y+9}{252}$ **51.** $\dfrac{15x^2+10x-6}{180}$

EXERCISE 3.4

1. 8 **3.** 12 **5.** 6 **7.** -6 **9.** 36 **11.** $-\frac{4}{3}$
13. 20 **15.** 30 **17.** 15 **19.** $-\frac{5}{6}$ **21.** $\frac{27}{5}$ **23.** 9
25. 150 **27.** $\frac{11}{5}$

EXERCISE 3.5

1. $\frac{1}{2}x$ or $\dfrac{x}{2}$ **3.** $\frac{2}{3}x$ or $\dfrac{2x}{3}$ **5.** $\dfrac{x}{3}+2$ **7.** $\dfrac{2x}{3}-8$

9. $\frac{1}{2}(2x-3)$ or $\dfrac{2x-3}{2}$ **11.** (A) $\dfrac{x}{4}+2=\frac{1}{2}$; (B) -6

13. (A) $\dfrac{x}{2}-2=\dfrac{x}{3}$; (B) 12 **15.** (A) $\dfrac{x}{2}-5=\dfrac{x}{3}+3$;

(B) 48 **17.** (A) $\dfrac{2x}{3}+5=\dfrac{x}{4}-10$; (B) -36 **19.** 7.2 meters

21. 75 meters **23.** 9 cm by 27 cm **25.** 84 meters by 24
meters **27.** 45 cm by 11 cm

EXERCISE 3.6

1. 1/4 **3.** 5/1 **5.** 1/3 **7.** 8 **9.** 18 **11.** 4
13. 600 men **15.** 36 cm **17.** 210 **19.** 125 km
21. 2.4 grams **23.** 4 in **25.** $90 per share **27.** 40 kg
29. 5.45 kg **31.** 24.84 miles **33.** 35 ounces **35.** 54.35 yd
37. Approx. 2,390 **39.** 24,000 miles

EXERCISE 3.7

1. $48t = 156$, $t = 3.25\,\text{hr}$ **3.** $12t = 30$, $t = 2.5\,\text{min}$
5. $r \cdot 40 = 220$, $r = \$5.50$ per hr **7.** $r(5.5) = 550$,
$r = 100\,\text{km/hr}$ **9.** $55t + 50t = 630$, $t = 6\,\text{hr}$
11. $20t + 30t = 30{,}000$, $t = 600\,\text{min}$ (10 hr) **13.** $100t + 40t = 630$,
$t = 4.5\,\text{hr}$ **15.** If $t = $ time to catch up, then $50t = 45(t + 1)$ and
$t = 9\,\text{hr}$ **17.** If $t = $ time to complete the job, then
$20t + 30(t - 60) = 30{,}000$; $t = 636\,\text{min}$ (10.6 hr) **19.** If $t = $ time that
assistant worked, then $20(t + 10) + 8t = 1{,}040$; $t = 30\,\text{hr}$ (assistant),
$t + 10 = 40\,\text{hr}$ (chemist) **21.** $\dfrac{d}{3} - \dfrac{d}{5} = 12$, $d = 90$ miles

EXERCISE 3.8

1. 30 nickels, 20 dimes **3.** 3,000 \$10 tickets and 5,000 \$6
tickets **5.** 2 dl **7.** 100 ml **9.** 75 ml of 30% solution and
25 ml of 70% solution **11.** 60 lb of \$3.50-per-pound coffee and 40 lb
of \$4.75-per-pound coffee **13.** \$7,500 at 8 percent and \$2,500 at 12
percent **15.** 5 liters **17.** 70 $\frac{1}{2}$-lb packages and 30 $\frac{1}{3}$-lb
packages

EXERCISE 3.9

1. 45 min **3.** \$210 **5.** 16 **7.** (A) 216 miles; (B) 225
miles **9.** 350 cc **11.** 180 bears **13.** 330 ft
15. (A) 15 in; (B) 20 in; (C) 22.5 in; (D) 24 in; (E) 25 in;
(F) 18 in; (G) 18.75 in **17.** 700 lb **19.** 200,000
miles per sec **21.** 150 cm **23.** $\frac{1}{4}$ hr or 15 min

EXERCISE 3.10
REVIEW EXERCISE

1.

(3.1) **2.** $\dfrac{15x}{8y}$

(3.2) **3.** $\dfrac{6}{5xy}$ (3.2) **4.** $\dfrac{5y}{6}$ (3.3) **5.** $\dfrac{6 - 5xy}{4y}$ (3.3) **6.** $\frac{5}{6}$

(3.4) **7.** $\frac{3}{2}$ (3.4) **8.** 6 (3.4) **9.** 6 (3.4) **10.** 1 (3.4)
11. (A) $\frac{3}{10}x = \frac{2}{5}$; (B) $x = \frac{4}{3}$ (3.5) **12.** 15 cm by 25 cm (3.5)
13. $0.6(30) + x = 0.7(x + 30)$; $x = 10\,\text{ml}$ (3.8) **14.** $\frac{2}{5}x^2$ (3.2)
15. $\dfrac{9y}{10z}$ (3.2) **16.** $\dfrac{9x - 4y}{12x^2y^2}$ (3.3) **17.** $\dfrac{3 - 2x + x^2}{x^2}$ (3.3)
18. $\dfrac{3xz + 18xy - 4yz - 24xyz}{12xyz}$ (3.3) **19.** $\frac{17}{18}$ (3.3) **20.** $\frac{-10}{9}$
(3.4) **21.** -12 (3.4) **22.** 41 (3.4) **23.** $\dfrac{x}{450} = \dfrac{20}{3}$,
$x = 3{,}000$ (3.6) **24.** $\dfrac{x}{40} = \dfrac{1}{2.54}$; $x = 15.75\,\text{in}$ (3.6)
25. $22t + 13t = 2{,}800$; $t = 80\,\text{hr}$ (3.7) **26.** $54t = 48(t + 1)$; $t = 8\,\text{hr}$
(3.7) **27.** $0.4x + 0.7(100 - x) = 0.49(100)$; $x = 70\,\text{dl}$ of 40% solution
and $100 - x = 30\,\text{dl}$ of the 70% solution (3.8) **28.** $\frac{-4}{9}$ (3.2)
29. $\dfrac{27y^2 - 12xy + 25x^2}{90x^2y^2}$ (3.3) **30.** $\frac{-14}{3}$ (3.4)

31. $90t + 110(t - 20) = 6,000$; $t = 41$ min (3.7)
32. $0.4(12 - x) + x = 0.5(12)$; $x = 2$ liters (3.8)

PRACTICE TEST

1. (A) $\dfrac{x}{2y}$; (B) $\dfrac{8x^3}{9y^3}$ **2.** -1 **3.** $\dfrac{8x - 15y^2}{36x^2y^2}$ **4.** 2

5. $x = \frac{10}{3}$ **6.** $x = 35$ **7.** $x = 9$ **8.** $x = -110$

9. $\dfrac{x}{2} = \dfrac{2x}{3} - 5$; $x = 30$ **10.** $0.6x + 0.8(40 - x) = 0.75(40)$; 10 ml of

the 60% solution and 30 ml of the 80% solution **11.** $\dfrac{x}{30} = \dfrac{1}{1.61}$;

$x = 18.63$ miles **12.** $8t + 12(t - 50) = 1,000$; $t = 80$ min

EXERCISE 4.1

1. T **3.** T **5.** T **7.** T **9.** T **11.** T
13. $7 > 5$ **15.** $5 < 7$ **17.** $-7 < -5$ **19.** $-5 > -7$
21. $0 < 8$ **23.** $0 > -8$ **25.** $-7 < 5$ **27.** $-842 < 0$
29. $900 > -1,000$ **31.** $a < d$ **33.** $b > a$ **35.** $e < f$
37. Right
39. (A)

(B)

41. (A)

(B)

43. (A)

(B)

45. (A)

(B)

47. (A)

(B)

49. $0.250\overline{0}$ **51.** $2.5\overline{5}$ **53.** $0.538461\overline{538461}$
55.

57.

59.

EXERCISE 4.2

1. $x > 7$ **3.** $x < -7$ **5.** $x > 4$ **7.** $x \leq -4$
9. $x < -21$ **11.** $x \geq 21$ **13.** $x < 2$ **15.** $x > 2$
17. $x \leq 5$ **19.** $y \leq -2$
21. $x < 10$ ────────○──────→
                              10   x
23. $x \geq 3$ ─────────●───────→
                                    3   x
25. $u \leq -11$ ──────●─────→
                                -11 u
27. $m < \frac{-7}{5}$ ────○──────→
                                   $\frac{-7}{5}$ m
29. $x > -4$ ──────○──────→
                            -4 x
31. $-1 < x < 2$ ────○───○──→
                                  -1   2 x
33. $-2 \leq x \leq 3$ ──●───●──→
                                   -2   3 x
35. $-20 \leq C \leq 20$ ──●──●──→
                                       -20   20 C
37. $14 \leq F \leq 77$ ──●──●──→
                                    14   77 F
39. $-2 < x \leq 3$ ──○───●──→
                                  -2   3 x
41. $2x - 3 \geq -6; \; x \geq \frac{-3}{2}$ **43.** $2 \cdot 10 + 2w < 30; \; w < 5 \text{ cm}$
45. $68 \leq \frac{9}{5}C + 32 \leq 77; \; 20° \leq C \leq 25°$ **47.** $70 \leq \dfrac{MA \cdot 100}{12} \leq 120;$
$8.4 \leq MA \leq 14.4$

EXERCISE 4.3

1. $A(5, 2), B(-2, 3), C(-4, -3), D(3, -2), E(2, 0), F(0, -4)$
3. $A(5, 5), B(8, 2), C(-5, 5), D(-3, 8), E(-5, -6), F(-7, -8), G(5, -5),$
$H(2, -2), I(7, 0), J(-2, 0), K(0, -9), L(0, 4)$
5.

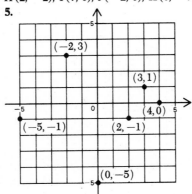

7.

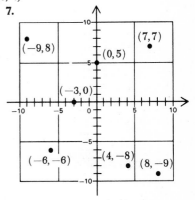

9. $A(2\frac{1}{2}, 1), B(-2\frac{1}{2}, 3\frac{1}{2}), C(-2, -4\frac{1}{2}), D(3\frac{1}{4}, -3), E(1\frac{1}{4}, 2\frac{1}{4}), F(-3\frac{1}{4}, 0),$
$G(1\frac{1}{2}, -4\frac{1}{2})$

11.

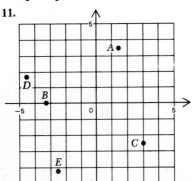

13. (A) III, (B) II, (C) I, (D) IV
15. 5 **17.** 5 **19.** 3
21. 2

EXERCISE 4.4

1.

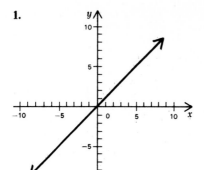

3.

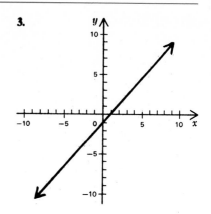

5.

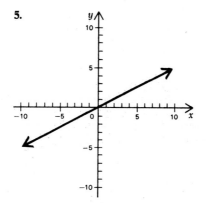

7.

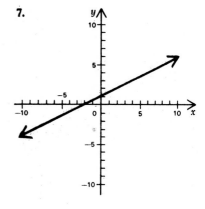

9.

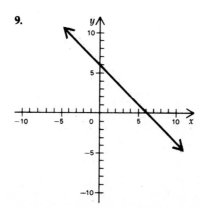

11.

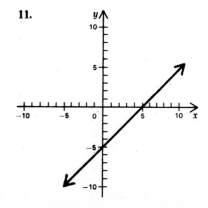

13.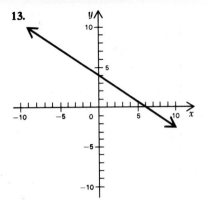

15.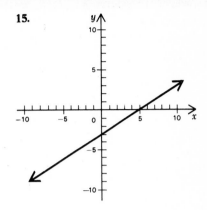

17.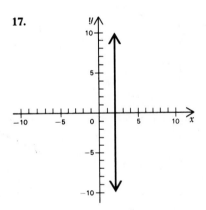

19.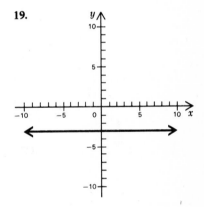

21.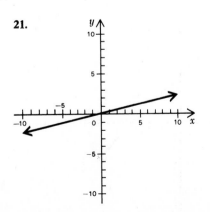

23.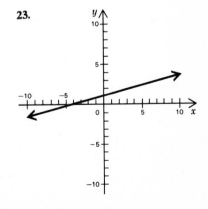

25.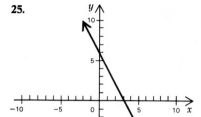

27.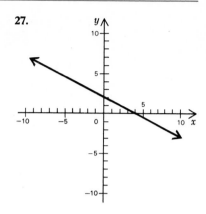

29.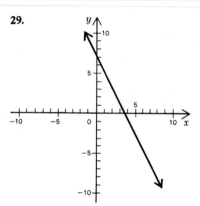

31.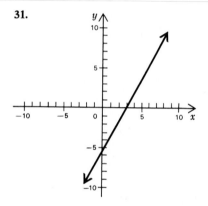

33.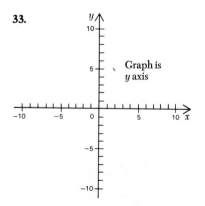

Graph is
y axis

35.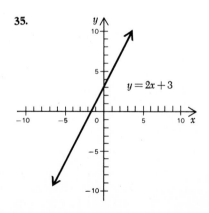

$y = 2x + 3$

37.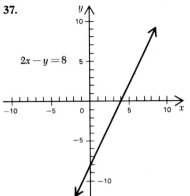

$2x - y = 8$

39.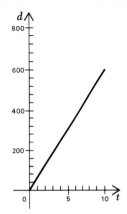

41.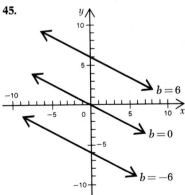

43.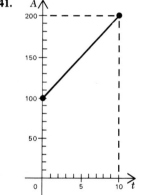

$(2,1)$

$x = 2$
$y = 1$

45.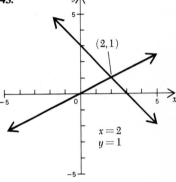

$b = 6$

$b = 0$

$b = -6$

47.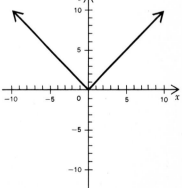

49.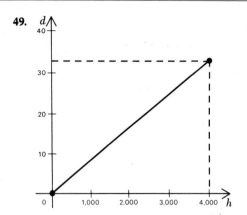

51.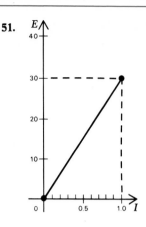

EXERCISE 4.5

1. $x = 3, y = 2$ **3.** $x = 3, y = 2$ **5.** $x = 9, y = 2$
7. $x = 2, y = 4$ **9.** $x = 6, y = 8$ **11.** $x = -4, y = -3$
13. No solution **15.** An infinite number of solutions. Any solution
of one is a solution of the other. **17.** $x + 2y = 4$ becomes
$y = -\frac{1}{2}x + 2$ and $2x + 4y = -8$ becomes $y = -\frac{1}{2}x - 2$. Since the
coefficients of x are the same, the lines are parallel. **19.** If two
lines are parallel and have a point in common, then they coincide.

EXERCISE 4.6

1. $x = 3, y = 2$ **3.** $x = 1, y = 4$ **5.** $x = 2, y = -4$
7. $x = 2, y = -1$ **9.** $x = 3, y = 2$ **11.** $x = -1, y = 2$
13. $x = -2, y = 2$ **15.** $x = -1, y = 2$ **17.** $x = 1,$
$y = -5$ **19.** $p = -\frac{4}{3}, q = 1$ **21.** $m = \frac{3}{2}, n = -\frac{2}{3}$ **23.** No
solution **25.** Infinitely many solutions **27.** $x = \frac{1}{3},$
$y = -2$ **29.** $m = -2, n = 2$ **31.** Limes: 11 cents each;
lemons: 4 cents each **33.** $x = 1, y = 0.2$ **35.** $x = 6, y = 4$
37. $x = -6, y = 12$

EXERCISE 4.7

1. 11 ft and 7 ft **3.** 20 cm by 16 cm **5.** 3,000 $10 tickets and
5,000 $6 tickets **7.** $6,000 at 6 percent and 2,000 at 8 percent
9. 60 lb of $3.50-per-pound coffee and 40 lb of $4.75-per-pound coffee
11. 70 cl water, 50 cl alcohol **13.** 75 ml of the 30% solution and
25 ml of the 70% solution **15.** 500 grams and 300 grams
17. 80 grams of mix 1 and 60 grams of mix 2 **19.** 3 ft from the
42-lb end (9 ft from the 14-lb end) **21.** Primary wave: 25 sec;
secondary wave: 40 sec; distance: 200 km **23.** 16 in and 20 in
25. 30 nickels, 20 dimes **27.** 84 $\frac{1}{4}$-lb packages and 60 $\frac{1}{2}$-lb
packages **29.** 10 test tubes for 1 flask; 3 test tubes for 1 mixing
dish

EXERCISE 4.8
REVIEW EXERCISE

1. **(A)**

2. $x < 6$ *(4.2)*   **3.** $x \geq -3$ *(4.2)*
4. $x \leq -3$ *(4.2)*
5. $-4 \leq x \leq 3$ *(4.2)*

(B) *(4.1)*

6.
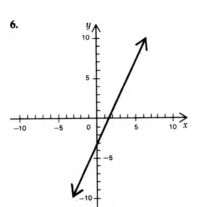

7.
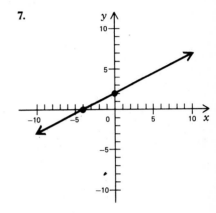

8.
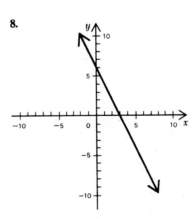

9.
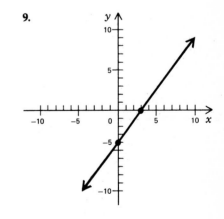

10. $x = 6, y = 1$ *(4.5)*   **11.** $x = 2, y = 1$ *(4.6)*   **12.** $x = -1,$
$y = 2$ *(4.6)*   **13.** $5x - 5 \leq 10; x \leq 3$ *(4.2)*   **14.** $x + y = 30,$
$5x + 10y = 230$; 16 dimes, 14 nickels *(4.7)*   **15.** All are true *(4.1)*
16. All are true *(4.1)*   **17.** $x > -1$ ——○——→ *(4.2)*
                                                    -1 x

18. $x < -\frac{1}{5}$ ——————○——→ *(4.2)*
                          $-\frac{1}{5}$ x

19. $x \geq -8$ ——————●——→ *(4.2)*
                      -8 x

20. $-6 \leq x < 18$ 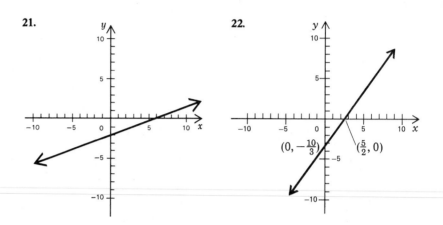 *(4.2)*

21.

22.

$\left(0, -\frac{10}{3}\right)$ $\left(\frac{5}{2}, 0\right)$

23. $x = 3, y = 3$ *(4.5)*

24. $u = 1, v = -2$ *(4.6)*

25. $m = -1, n = -3$ *(4.6)*

26. $59 \leq \frac{9}{5}C + 32 \leq 86$;
$15 \leq C \leq 30$ *(4.2)*

27. $x + y = 6{,}000, \ 0.1x + 0.06y = 440$;
\$2,000 at 10 percent and \$4,000 at 6
percent *(4.7)*

28. $x + y = 100, \ 0.5x + 0.7y = 66$; 20 ml of 50% solution and 80 ml of the
70% solution *(4.7)*   **29.** Infinitely many solutions–same line *(4.5)*

30. No solution *(4.6)*   **31.** $20 \leq \frac{5}{9}(F - 32) \leq 25$; $68 \leq F \leq 77$

(4.2)   **32.** (A) 1.25 hr, 0.75 hr;   (B) 112.5 miles *(4.7)*

PRACTICE TEST

1. (A) T;   (B) F;   (C) T;   (D) T;   (E) T

2. (A)

(B)

3. (A) $x > 6$

(B) $-1 < x \leq 6$

4. $x \leq -3$ **5.** $x > \frac{9}{8}$ **6.**

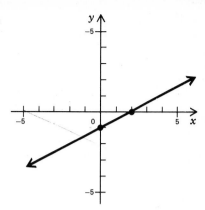

7. $x = 3, y = 3$

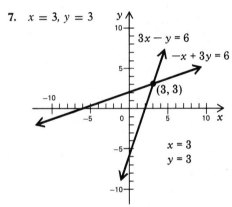

8. $u = -1, v = 3$ **9.** No solution   **10.** $59 \leq F \leq 86$
11. 12 dimes, 13 nickels   **12.** 3 liters of 40% solution, 1 liter of 80% solution

EXERCISE 5.2

1. $5x - 2$ **3.** $5x - 3$ **5.** $6x^2 - x + 3$ **7.** $3x + 4$
9. $-2x + 12$ **11.** $x^2 - 3x$ **13.** $2x^2 + x - 6$
15. $2x^3 - 7x^2 + 13x - 5$ **17.** $x^3 - 6x^2y + 10xy^2 - 3y^3$
19. 1   **21.** 3   **23.** 2   **25.** 6   **27.** $3x^3 + x^2 + x + 6$
29. $-x^3 + 2x^2 - 3x + 7$ **31.** $2x^2 - 2xy + 3y^2$
33. $a^3 + b^3$ **35.** $x^3 + 6x^2y + 12xy^2 + 8y^3$
37. $2x^4 - 5x^3 + 5x^2 + 11x - 10$ **39.** $2x^4 + x^3y - 7x^2y^2 + 5xy^3 - y^4$
41. $-x^2 + 17x - 11$ **43.** $2x^3 - 13x^2 + 25x - 18$

EXERCISE 5.3

1. $x^2 + 3x + 2$ **3.** $y^2 + 7y + 12$ **5.** $x^2 - 9x + 20$
7. $n^2 - 7n + 12$ **9.** $s^2 + 5s - 14$ **11.** $m^2 - 7m - 60$
13. $u^2 - 9$ **15.** $x^2 - 64$ **17.** $y^2 + 16y + 63$
19. $c^2 - 15c + 54$ **21.** $x^2 - 8x - 48$ **23.** $a^2 - b^2$
25. $x^2 + 4xy + 3y^2$ **27.** $3x^2 + 7x + 2$ **29.** $4t^2 - 11t + 6$

312

31. $3y^2 - 2y - 21$ **33.** $2x^2 + xy - 6y^2$ **35.** $6x^2 + x - 2$
37. $9y^2 - 4$ **39.** $5s^2 + 34s - 7$ **41.** $6m^2 - mn - 35n^2$
43. $12n^2 - 13n - 14$ **45.** $6x^2 - 13xy + 6y^2$ **47.** $x^2 + 6x + 9$
49. $4x^2 - 12x + 9$ **51.** $4x^2 - 20xy + 25y^2$
53. $16a^2 + 24ab + 9b^2$ **55.** $x^3 + x^2 - 3x + 1$
57. $6x^3 - 13x^2 + 14x - 12$ **59.** $x^4 - 2x^3 - 4x^2 + 11x - 6$

EXERCISE 5.4

1. $A(2x + 3)$ **3.** $5x(2x + 3)$ **5.** $2u(7u - 3)$
7. $2u(3u - 5v)$ **9.** $5mn(2m - 3n)$ **11.** $2x^2y(x - 3y)$
13. $(x + 2)(3x + 5)$ **15.** $(m - 4)(3m - 2)$ **17.** $(x + y)(x - y)$
19. $3x^2(2x^2 - 3x + 1)$ **21.** $2xy(4x^2 - 3xy + 2y^2)$
23. $4x^2(2x^2 - 3xy + y^2)$ **25.** $(2x + 3)(3x - 5)$ **27.** $(x + 1)(x - 1)$
29. $(2x - 3)(4x - 1)$ **31.** $2x - 2$ **33.** $2x - 8$ **35.** $2u + 1$
37. $3x(x - 1) + 2(x - 1) = (x - 1)(3x + 2)$
39. $3x(x - 4) - 2(x - 4) = (x - 4)(3x - 2)$
41. $4u(2u + 1) - (2u + 1) = (2u + 1)(4u - 1)$
43. $(x - 1)(3x + 2)$ **45.** $(x - 4)(3x - 2)$ **47.** $(2u + 1)(4u - 1)$
49. $2m(m - 4) + 5(m - 4) = (m - 4)(2m + 5)$
51. $3x(2x - 3) - 2(2x - 3) = (2x - 3)(3x - 2)$
53. $3u(u - 4) - (u - 4) = (u - 4)(3u - 1)$
55. $3u(2u + v) - 2v(2u + v) = (2u + v)(3u - 2v)$
57. $3x(2x + y) - 5y(2x + y) = (2x + y)(3x - 5y)$

EXERCISE 5.5

1. $(x + 1)(x + 4)$ **3.** $(x + 2)(x + 3)$ **5.** $(x - 1)(x - 3)$
7. $(x - 2)(x - 5)$ **9.** Not factorable   **11.** Not factorable
13. $(x + 3y)(x + 5y)$ **15.** $(x - 3y)(x - 7y)$ **17.** Not factorable   **19.** $(3x + 1)(x + 2)$ **21.** $(3x - 4)(x - 1)$
23. $(x - 4)(3x - 2)$ **25.** $(3x - 2y)(x - 3y)$ **27.** $(n - 4)(n + 2)$
29. Not factorable   **31.** $(x - 1)(3x + 2)$ **33.** $(x + 6y)(x - 2y)$
35. $(u - 4)(3u + 1)$ **37.** $(3x + 5)(2x - 1)$ **39.** $(3s + 1)(s - 2)$
41. Not factorable   **43.** $(x - 2)(5x + 2)$ **45.** $(2u + v)(3u - 2v)$
47. $(4x - 3)(2x + 3)$ **49.** $(3u - 2v)(u + 3v)$
51. $(u - 4v)(4u - 3v)$ **53.** $(6x + y)(2x - 7y)$
55. $(12x - 5y)(x + 2y)$

EXERCISE 5.6

1. $(3x - 4)(x - 1)$ **3.** Not factorable   **5.** $(2x - 1)(x + 3)$
7. Not factorable   **9.** $(x - 4)(3x - 2)$ **11.** $(3x + 5)(2x - 1)$
13. Not factorable   **15.** $(m - 4)(2m + 5)$ **17.** $(u - 4)(3u + 1)$
19. $(2u + v)(3u - 2v)$ **21.** Not factorable
23. $(4x - 3)(2x + 3)$ **25.** $(4m - 2n)(m + 3n)$ **27.** Not factorable   **29.** $(u - 4v)(4u - 3v)$ **31.** $(6x + y)(2x - 7y)$
33. $(6x + 5y)(3x - 4y)$ **35.** $-13, 13, -8, 8, -7, 7$

313

EXERCISE 5.7

1. $3x^2(2x + 3)$ 3. $u^2(u + 2)(u + 4)$ 5. $x(x - 2)(x - 3)$
7. $(x - 2)(x + 2)$ 9. $(2x - 1)(2x + 1)$ 11. Not factorable
13. $2(x - 2)(x + 2)$ 15. $(3x - 4y)(3x + 4y)$ 17. $3uv^2(2u - v)$
19. $xy(2x - y)(2x + y)$ 21. $3x^2(x^2 + 9)$ 23. $6(x + 2)(x + 4)$
25. $3x(x^2 - 2x + 5)$ 27. Not factorable
29. $4x(3x - 2y)(x + 2y)$ 31. $(x + 3)(x + y)$ 33. $(x - 3)(x - y)$
35. $(2a + b)(c - 3d)$ 37. $(m + n)(2u - v)$
39. $2xy(2x + y)(x + 3y)$ 41. $5y^2(6x + y)(2x - 7y)$
43. $(x - 2)(x^2 + 2x + 4)$ 45. $(x + 3)(x^2 - 3x + 9)$
47. $(x + y)(x^2 - xy + y^2)(x - y)(x^2 + xy + y^2)$

EXERCISE 5.8

1. $x + 2$ 3. $2x - 3$ 5. $2x + 3, R = 5$ 7. $m - 2$
9. $2x + 3$ 11. $2x + 5, R = -2$ 13. $x + 5, R = -2$
15. $x + 2$ 17. $m + 3, R = 2$ 19. $5c - 2, R = 8$
21. $3x + 2, R = -4$ 23. $2y^2 + y - 3$ 25. $x^2 + x + 1$
27. $x^3 - 2x^2 + 4x - 8$ 29. $2y^2 - 5y + 13, R = -27$
31. $2x^3 - 3x^2 - 5, R = 5$ 33. $2x^2 - 3x + 2, R = 4$

**EXERCISE 5.9
REVIEW EXERCISE**

1. $5x^2 + 3x - 4$ (5.2) 2. $2x^2 - x + 7$ (5.2) 3. $3x + 4, R = 2$
(5.8) 4. $6x^2 + 11x - 10$ (5.3) 5. $6u^2 - uv - 12v^2$ (5.3)
6. $3x^3 + x^2 - 3x + 6$ (5.2) 7. $-3x^3 + 3x^2 - 3x - 4$ (5.2)
8. $(x - 7)(x - 2)$ (5.5, 5.6) 9. $(3x - 4)(x - 2)$ (5.5, 5.6) 10. Not
factorable (5.5, 5.6) 11. $2xy(2x - 3y)$ (5.4) 12. $x(x - 2)(x - 3)$
(5.7) 13. $(2u - 3)(2u + 3)$ (5.7) 14. $(x - 1)(x + 3)$ (5.4)
15. Not factorable (5.7) 16. 5, 1, 0 (5.1) 17. $2x^2 + 5x + 5$
(5.2) 18. $x - 4, R = 3$ (5.8) 19. $3x^2 + 2x - 2, R = -2$
(5.8) 20. $4x^3 - 12x^2 + 13x - 6$ (5.2)
21. $27x^4 + 63x^3 - 66x^2 - 28x + 24$ (5.2) 22. $a^3 + b^3$ (5.2)
23. $-2x + 20$ (5.2) 24. $3(u - 2)(u + 2)$ (5.7)
25. $(2x - 3y)(x + y)$ (5.5, 5.6) 26. Not factorable (5.5, 5.6)
27. $3y(2y - 5)(y + 3)$ (5.7) 28. $2x(x^2 - 2xy - 5y^2)$ (5.7)
29. $3xy(4x^2 + 9y^2)$ (5.7) 30. $(x - 1)(x - 3)(x + 3)$ (5.7)
31. $(x - y)(x + 4)$ (5.7) 32. $(x + y)(x - 3)$ (5.7)
33. $(2u - 3)(u + 3)$ (5.4) 34. $(3x + 2)(2x - 1)$ (5.4) 35. 0
(5.2) 36. $x^3 - 3x^2 + 2x + 4, R = -20$ (5.8)
37. $3xy(6x - 5y)(2x + 3y)$ (5.7) 38. $4u^3(3u - 3v - 5v^2)$ (5.7)
39. $2(3a + 2b)(c - 2d)$ (5.7) 40. $(4u - 5v)(3x - 1)$ (5.7)
41. $(2x + 1)(4x^2 - 2x + 1)$ (5.7) 42. $-2, 2, -7, 7$

PRACTICE TEST

1. (A) 1; (B) 4 2. $3x^2 + 2x - 2$ 3. $2x^2 - x + 4$
4. $-7x^2 + x + 11$ 5. $4x^2 + 2x - 3, R = -2$
6. (A) $(y - 2)(y + 2)$; (B) $3m^2(m^2 + 4)$ 7. $(2x - y)(3x - 2y)$
8. $(3x - 2y)(x + 4y)$ 9. Not factorable
10. $3x^2y(6x^2 - 3xy - y^2)$ 11. $(u - 5)(u - v)$ 12. $-5, 5, -7, 7$

EXERCISE 6.2

1. $\frac{x}{3}$ **3.** $\frac{1}{A}$ **5.** $\frac{1}{x+3}$ **7.** $4(y-5)$ **9.** $\frac{x}{3(x+7)^2}$

11. $\frac{x}{2}$ **13.** $3-y$ **15.** $\frac{1}{n}$ **17.** $\frac{2x-1}{3x}$ **19.** $\frac{2x+1}{3x-7}$

21. $\frac{x+2}{2x}$ **23.** $\frac{x-3}{x+3}$ **25.** $\frac{x-2}{x-3}$ **27.** $\frac{x+3}{2x+1}$

29. $3x^2-x+2$ **31.** $\frac{2+m-3m^2}{m}$ **33.** $2m^2-mn+3n^2$

35. $x+2$ **37.** $\frac{2x-3y}{2xy}$ **39.** $\frac{x+2}{x+y}$ **41.** $x-5$

43. $\frac{x+5}{2x}$ **45.** $\frac{x^2+2x+4}{x+2}$

EXERCISE 6.3

1. $\frac{5}{6}$ **3.** 2 **5.** $\frac{2y^3}{9u^2}$ **7.** $\frac{u^2w^2}{25y^2}$ **9.** $\frac{2}{x}$ **11.** $a+1$

13. $\frac{x-2}{2x}$ **15.** $8d^6$ **17.** $\frac{1}{y(x+4)}$ **19.** $\frac{1}{2y-1}$

21. $\frac{x}{x+5}$ **23.** $-3(x-2)$ or $6-3x$ **25.** $\frac{m+2}{m(m-2)}$

27. $\frac{x^2(x+y)}{(x-y)^2}$ **29.** All but one (namely, $x=1$)   **31.** Obtain

$\frac{x^2-(x+2)^2}{(x+1)(x+5)-(x+3)^2}$, which simplifies to $x+1$. Since $x=108{,}641$,
$x+1=108{,}642$, the answer.

EXERCISE 6.4

1. $\frac{3m-1}{2m^2}$ **3.** $\frac{x+1}{x}$ **5.** 2 **7.** $\frac{3}{x}$ **9.** $\frac{1}{2x+3}$

11. $\frac{y-6}{3y(y+3)}$ **13.** $\frac{x-8}{(x+1)(x-2)}$ **15.** $\frac{y^2+8}{8y^3}$

17. $\frac{2}{t-1}$ **19.** $\frac{2x-8}{(2x-3)(x+2)}$ **21.** $\frac{a+2}{a+1}$

23. $\frac{-2}{(x-1)(x+1)^2}$ **25.** $\frac{3y^2-y-18}{(y+2)(y-2)}$ **27.** $\frac{5t-12}{3(t-4)(t+4)}$

29. $\frac{3}{x+3}$ **31.** $\frac{(3x+1)(x+3)}{12x^2(x+1)^2}$ **33.** $\frac{1}{x-1}$

35. $\frac{-17}{15(x-1)}$

EXERCISE 6.5

1. $x=-9$ **3.** $y=2$ **5.** $x=-4$ **7.** $t=4$
9. $L=-4$ **11.** No solution   **13.** $-\frac{6}{5}$ **15.** No solution
17. $x=8$ **19.** $n=\frac{53}{11}$ **21.** $x=1$ **23.** $x=-4$

EXERCISE 6.6

1. $I = A - P$ **3.** $r = \dfrac{d}{t}$ **5.** $t = \dfrac{I}{Pr}$ **7.** $\pi = \dfrac{C}{D}$

9. $x = -\dfrac{b}{a}, \; a \neq 0$ **11.** $t = \dfrac{s + 5}{2}$ **13.** $y = \dfrac{3x - 12}{4}$ or

$y = \tfrac{3}{4}x - 3$ **15.** $E = IR$ **17.** $L = \dfrac{100B}{C}$ **19.** $m_1 = \dfrac{Fd^2}{Gm_2}$

21. $h = \dfrac{2A}{b_1 + b_2}$ **23.** $F = \tfrac{9}{5}C + 32$ **25.** $f = \dfrac{ab}{a + b}$ or

$f = \dfrac{1}{\dfrac{1}{a} + \dfrac{1}{b}}$ **27.** $x = \dfrac{y + 1}{2y - 3}$

EXERCISE 6.7

1. $\tfrac{3}{4}$ **3.** $\tfrac{9}{10}$ **5.** $\tfrac{8}{13}$ **7.** $\tfrac{22}{51}$ **9.** xy **11.** $\dfrac{3xy}{2}$

13. $\dfrac{1}{x - 3}$ **15.** $\dfrac{x + y}{x}$ **17.** $\dfrac{1}{y - x}$ **19.** $\dfrac{x - y}{x + y}$

21. 1 **23.** $-\tfrac{1}{2}$ **25.** $\dfrac{1}{1 - x}$ **27.** $r = \dfrac{2r_R r_G}{r_R + r_G}$

EXERCISE 6.8
REVIEW EXERCISE

1. $\dfrac{3x + 2}{3x}$ *(6.4)*   **2.** $\dfrac{2x + 11}{6x}$ *(6.4)*   **3.** $\dfrac{4x^2 y^2}{9(x - 3)}$ *(6.3)*

4. $\dfrac{(d - 2)^2}{d + 2}$ *(6.3)*   **5.** $\dfrac{x + 1}{2x(3x - 1)}$ *(6.4)*   **6.** $\dfrac{4}{x - 4}$ *(6.4)*   **7.** $\tfrac{3}{10}$

(6.7)   **8.** $\tfrac{28}{9}$ *(6.7)*   **9.** $m = 5$ *(6.5)*   **10.** No solution *(6.5)*

11. $b = \dfrac{2A}{h}$ *(6.6)*   **12.** $\dfrac{y + 2}{y(y - 2)}$ *(6.3)*   **13.** 2 *(6.3)*

14. $\dfrac{-1}{(x + 2)(x + 3)}$ *(6.4)*   **15.** $\dfrac{5x - 3}{6x^2(x - 1)}$ *(6.4)*

16. $\dfrac{4m}{(m - 2)(m + 2)}$ *(6.4)*   **17.** $\dfrac{-2y}{(x - y)^2(x + y)}$ *(6.4)*   **18.** x *(6.7)*

19. $\dfrac{x - y}{x}$ *(6.7)*   **20.** $x = -2$ *(6.5)*   **21.** $x = -5$ *(6.5)*

22. $L = \dfrac{2s - an}{n}$ or $L = \dfrac{2s}{n} - a$ *(6.6)*   **23.** $A = \dfrac{M}{1 + x}$ *(6.6)*

24. $\tfrac{2}{3}$ *(6.5)*   **25.** -1 *(6.3)*   **26.** $\dfrac{y + 4}{2x - y}$ *(6.4)*   **27.** $\dfrac{6 - 3x}{2x}$ *(6.7)*

28. $\dfrac{x - 2}{x - 1}$ *(6.7)*   **29.** No solution *(6.5)*   **30.** $x = \dfrac{3y + 1}{2y - 3}$ *(6.6)*

PRACTICE TEST

1. $\dfrac{10x}{y(x - 1)}$ **2.** $4x^2$ **3.** $\dfrac{x + 1}{3(2x - 1)}$ **4.** $\dfrac{2(x + 6)}{(x - 3)^2}$

5. $\dfrac{x + 6}{6x^2(x - 3)}$ **6.** $y + x$ **7.** $\dfrac{2x - 7}{3x - 8}$ **8.** No solution

9. $x = -5$ **10.** $t = \dfrac{s - 1}{3}$ **11.** $F = \frac{9}{5}C + 32$

12. $x = \dfrac{y + 3}{3y - 4}$

EXERCISE 7.1

1. 12   **3.** 4   **5.** v^6 **7.** 2   **9.** 5   **11.** 3

13. 4   **15.** 7   **17.** 4   **19.** 7   **21.** 4   **23.** 7

25. $6x^{10}$ **27.** $2x^2$ **29.** $\dfrac{2}{u^4}$ **31.** $c^{12}d^{12}$ **33.** $\dfrac{x^6}{y^6}$

35. $6x^9$ **37.** 6×10^{15} **39.** 10^{20} **41.** y^{20} **43.** x^8y^{12}

45. $\dfrac{a^{12}}{b^8}$ **47.** $\dfrac{y^6}{3x^4}$ **49.** $3^3a^9b^6$ **51.** $2x^8y^4$ **53.** $\dfrac{x^6y^3}{8w^6}$

55. $\dfrac{y^3}{16x^4}$ **57.** $\dfrac{-x^2}{32}$ **59.** -1 **61.** $\dfrac{-1}{a^8}$

EXERCISE 7.2

1. 1   **3.** 1   **5.** $\dfrac{1}{2^2}$ **7.** $\dfrac{1}{x^4}$ **9.** 3^2 **11.** x^3

13. 10^2 **15.** x^4 **17.** 1   **19.** 10^{11} **21.** a^{12}

23. $\dfrac{1}{b^8}$ **25.** $\dfrac{1}{10^6}$ **27.** 2^6 **29.** x^{10} **31.** x^3y^2

33. $\dfrac{y^6}{x^4}$ **35.** $\dfrac{y^3}{x^2}$ **37.** 1   **39.** 10^2 **41.** $\dfrac{1}{x}$

43. 10^{17} **45.** 4×10^2 **47.** x^6 **49.** $\dfrac{1}{2^3c^3d^6}$

51. $\dfrac{3^2x^6}{y^4}$ **53.** $\dfrac{x^6}{y^4}$ **55.** $\dfrac{2^6}{3^4}$ **57.** $\dfrac{1}{10^4}$ **59.** $\dfrac{3n^4}{4m^3}$

61. $\dfrac{2x}{y^2}$ **63.** n^2 **65.** $\dfrac{x^{12}}{y^8}$ **76.** $\dfrac{4x^8}{y^6}$ **69.** $\dfrac{1}{(x + y)^2}$

71. $\frac{1}{30}$ **73.** $\frac{144}{7}$ **75.** $\frac{36}{13}$ **77.** $\dfrac{1,000}{11}$

EXERCISE 7.3

1. 6×10 **3.** 6×10^2 **5.** 6×10^5 **7.** 6×10^{-2}

9. 6×10^{-5} **11.** 3.5×10 **13.** 7.2×10^{-1} **15.** 2.7×10^2

17. 3.2×10^{-2} **19.** 5.2×10^3 **21.** 7.2×10^{-4} **23.** 500

25. 0.08   **27.** 6,000,000   **29.** 0.00002   **31.** 7,100

33. 0.00086   **35.** 8,800,000   **37.** 0.0000061

39. 4.27×10^7 **41.** 7.23×10^{-5} **43.** 5.87×10^{12}

45. 3×10^{-23} **47.** 3,460,000,000   **49.** 0.000000623

51. 93,000,000   **53.** 0.000075   **55.** 8×10^2 **57.** 8×10^{-3}

59. 3×10^3 **61.** 3×10^7 **63.** 2×10^4 or 20,000

65. 2×10^{-4} or 0.0002   **67.** 3.3×10^{18} **69.** 10^7; 6×10^8

EXERCISE 7.4

1. 4   **3.** -9 **5.** x **7.** $3m$ **9.** $2\sqrt{2}$ **11.** $x\sqrt{x}$

13. $3y\sqrt{2y}$ **15.** $\frac{1}{2}$ **17.** $-\frac{2}{3}$ **19.** $\frac{1}{x}$ **21.** $\frac{\sqrt{3}}{3}$

23. $\frac{\sqrt{3}}{3}$ **25.** $\frac{\sqrt{x}}{x}$ **27.** $\frac{\sqrt{x}}{x}$ **29.** $5xy^2$ **31.** $2x^2y\sqrt{xy}$

33. $2x^3y^3\sqrt{2x}$ **35.** $\frac{\sqrt{3y}}{3y}$ **37.** $2x\sqrt{2y}$ **39.** $\frac{2}{3}x\sqrt{3xy}$

41. $\frac{\sqrt{6}}{3}$ **43.** $\frac{\sqrt{6mn}}{2n}$ **45.** $\frac{2a\sqrt{3ab}}{3b}$ **47.** 4.242

49. 0.447   **51.** 4.062   **53.** $\frac{\sqrt{2}}{2}$ **55.** In simplest radical

form   **57.** $x\sqrt{x^2-2}$ **59.** Because the square of any real
number cannot be negative   **61.** Yes, no   **63.** $5\frac{1}{2} = \sqrt{5}$
65. If $a^2 = b^2$, then a does not necessarily equal b. For example, let
$a = 2$ and $b = -2$.

EXERCISE 7.5

1. $8\sqrt{2}$ **3.** $3\sqrt{x}$ **5.** $4\sqrt{7}-3\sqrt{5}$ **7.** $-3\sqrt{y}$
9. $4\sqrt{5}$ **11.** $4\sqrt{x}$ **13.** $2\sqrt{2}-2\sqrt{3}$ **15.** $2\sqrt{x}+2\sqrt{y}$
17. $\sqrt{2}$ **19.** $-3\sqrt{3}$ **21.** $2\sqrt{2}+6\sqrt{3}$ **23.** $-\sqrt{x}$
25. $2\sqrt{6}+\sqrt{3}$ **27.** $-\sqrt{6}/6$ **29.** $5\sqrt{2xy}/2$
31. $2\sqrt{3}-\frac{\sqrt{2}}{2}$ **33.** $3\sqrt{2}$

EXERCISE 7.6

1. $4\sqrt{5}+8$ **3.** $10-2\sqrt{2}$ **5.** $2+3\sqrt{2}$ **7.** $5-4\sqrt{5}$
9. $2\sqrt{3}-3$ **11.** $x-3\sqrt{x}$ **13.** $3\sqrt{m}-m$
15. $2\sqrt{3}-\sqrt{6}$ **17.** $5\sqrt{2}+5$ **19.** $2\sqrt{2}-1$
21. $x-\sqrt{x}-6$ **23.** $9+4\sqrt{5}$ **25.** $2-11\sqrt{2}$
27. $6x-13\sqrt{x}+6$ **31.** $\frac{2+\sqrt{2}}{3}$ **33.** $\frac{-1-2\sqrt{5}}{3}$
35. $2-\sqrt{2}$ **37.** $\frac{\sqrt{11}-3}{2}$ **39.** $\frac{\sqrt{5}-1}{2}$ **41.** $\frac{y-3\sqrt{y}}{y-9}$
43. $\frac{7+4\sqrt{3}}{-1}$ or $-7-4\sqrt{3}$ **45.** $\frac{x+5\sqrt{x}+6}{x-9}$

EXERCISE 7.7
REVIEW EXERCISE

1. (A) 16;   (B) $\frac{1}{9}$ (7.1, 7.2)   **2.** (A) 1;   (B) 9 (7.2)   **3.** $\frac{4x^4}{9y^6}$

(7.1)   **4.** $\frac{y^3}{x^2}$ (7.2)   **5.** (A) 6×10^{-2};   (B) 0.06 (7.3)   **6.** -5

(7.4)   **7.** $2xy^2$ (7.4)   **8.** $\frac{5}{y}$ (7.4)   **9.** $-3\sqrt{x}$ (7.5)

10. $5+2\sqrt{5}$ (7.6)   **11.** (A) 9;   (B) $\frac{1}{100}$ (7.2)   **12.** (A) $\frac{1}{27}$;

(B) 1 (7.2)   **13.** $\frac{4x^4}{y^6}$ (7.2)   **14.** $\frac{m^2}{2n^5}$ (7.2)   **15.** (A) 2×10^{-3};

(B) 0.002 *(7.3)*   16. $6x^2y^3\sqrt{y}$ *(7.4)*   17. $\dfrac{\sqrt{2y}}{2y}$ *(7.4)*

18. $\dfrac{\sqrt{6xy}}{2y}$ *(7.4)*   19. $\dfrac{5\sqrt{6}}{6}$ *(7.5)*   20. $1+\sqrt{3}$ *(7.6)*

21. $\dfrac{n^{10}}{9m^{10}}$ *(7.2)*   22. $\dfrac{xy}{x+y}$ *(7.2)*   23. 1.036 cc *(7.3)*

24. $\dfrac{n^2\sqrt{6m}}{3}$ *(7.4)*   25. $2x\sqrt{x^2+4}$ *(7.4)*   26. $\dfrac{x-4\sqrt{x}+4}{x-4}$ *(7.5)*

27. $a^2=b$ *(7.4)*   28. All real numbers *(7.4)*

PRACTICE TEST

1. 2^2 or 4   2. $\dfrac{4x}{y^2}$ 3. $\dfrac{4a^6}{b^4}$ 4. $\dfrac{4a^8}{b^8}$ 5. $\dfrac{uv}{v-u}$

6. 6×10^{17} 7. $2x^2y^2\sqrt{3y}$ 8. $2\sqrt{3x}$ 9. $2\sqrt{2}-5\sqrt{x}$

10. $\dfrac{2\sqrt{15}}{15}$ 11. $2\sqrt{2}-13$ 12. $\dfrac{y+6\sqrt{y}+9}{y-9}$

EXERCISE 8.2

1. 3, 4   3. $-6, 5$ 5. $-4, \frac{2}{3}$ 7. $-\frac{3}{4}, \frac{2}{5}$ 9. $0, \frac{1}{4}$
11. 1, 5   13. 1, 3   15. $-2, 6$ 17. 0, 3   19. 0, 2
21. $-5, 5$ 23. $\frac{1}{2}, -3$ 25. $\frac{2}{3}, 2$ 27. $-4, 8$ 29. $-\frac{2}{3}$,
4   31. Not factorable in the integers   33. 3, -4 35. -3,
3   37. 11 by 3 in   39. 5, -3 41. $\frac{1}{2}, 2$ 43. $\frac{2}{3}$ or $\frac{3}{2}$
45. 1 ft

EXERCISE 8.3

1. ± 4 3. ± 8 5. $\pm\sqrt{3}$ 7. $\pm\sqrt{5}$ 9. $\pm\sqrt{18}$ or
$\pm 3\sqrt{2}$ 11. $\pm\sqrt{12}$ or $\pm 2\sqrt{3}$ 13. $\pm\frac{2}{3}$ 15. $\pm\frac{2}{3}$

17. $\pm\frac{2}{3}$ 19. $\pm\frac{2}{5}$ 21. $\pm\sqrt{\dfrac{3}{4}}$ or $\pm\dfrac{\sqrt{3}}{2}$ 23. $\pm\sqrt{\dfrac{5}{2}}$ or

$\pm\dfrac{\sqrt{10}}{2}$ 25. $\pm\sqrt{\dfrac{1}{3}}$ or $\pm\dfrac{\sqrt{3}}{3}$ 27. $-1, 5$ 29. 3, -7

31. $2\pm\sqrt{3}$ 33. $-1, 2$ 35. No real solution
37. $\dfrac{3\pm\sqrt{6}}{2}$ 39. $b=\pm\sqrt{c^2-a^2}$ 41. 70 mph

EXERCISE 8.4

1. $x^2+4x+4=(x+2)^2$ 3. $x^2-6x+9=(x-3)^2$
5. $x^2+12x+36=(x+6)^2$ 7. $-2\pm\sqrt{2}$ 9. $3\pm 2\sqrt{3}$
11. $x^2+3x+\frac{9}{4}=(x+\frac{3}{2})^2$ 13. $u^2-5u+\frac{25}{4}=(u-\frac{5}{2})^2$

15. $\dfrac{-1\pm\sqrt{5}}{2}$ 17. $\dfrac{5\pm\sqrt{17}}{2}$ 19. No real solution

21. $\dfrac{2\pm\sqrt{2}}{2}$ 23. $\dfrac{-3\pm\sqrt{17}}{4}$ 25. $\dfrac{-m\pm\sqrt{m^2-4n}}{2}$

EXERCISE 8.5

1. $a = 1, b = 4, c = 2$ **3.** $a = 1, b = -3, c = -2$ **5.** $a = 3, b = -2, c = 1$ **7.** $a = 2, b = 3, c = -1$ **9.** $a = 2, b = -5, c = 0$ **11.** $-2 \pm \sqrt{2}$ **13.** $3 \pm 2\sqrt{3}$ **15.** $\dfrac{-3 \pm \sqrt{13}}{2}$

17. $\dfrac{3 \pm \sqrt{3}}{2}$ **19.** $\dfrac{-1 \pm \sqrt{13}}{6}$ **21.** No real solution

23. $\dfrac{3 \pm \sqrt{33}}{4}$ **25.** $\dfrac{4 \pm \sqrt{11}}{5}$ **27.** $\dfrac{2\sqrt{3}}{3}, -\dfrac{\sqrt{3}}{3}$ **29.** T

31. T

EXERCISE 8.6

1. $-2, 3$ **3.** $0, -7$ **5.** ± 4 **7.** $-1 \pm \sqrt{3}$ **9.** $0, 2$

11. $1 \pm \sqrt{2}$ **13.** $\dfrac{3 \pm \sqrt{3}}{2}$ **15.** $0, 1$ **17.** $\frac{1}{3}, -\frac{1}{2}$

19. $\pm 5\sqrt{2}$ **21.** $2 \pm \sqrt{3}$ **23.** No real solution   **25.** $-\frac{2}{3}, 3$

27. $-50, 2$ **29.** $t = \sqrt{\dfrac{2d}{g}}$ **31.** $r = -1 + \sqrt{\dfrac{A}{P}}$

EXERCISE 8.7

1. 8   **3.** 0, 2   **5.** $\frac{1}{3}, 3$ **7.** 20 percent   **9.** 2,000 and 8,000 units   **11.** Approx. 9,854 or 3,146 units   **13.** 20   **15.** 3 meters, 4 meters, 5 meters   **17.** $(1 + \sqrt{2})$ in ≈ 2.414 in   **19.** 50 mph   **21.** 5 mph and 12 mph   **23.** 2 mph

**EXERCISE 8.8
REVIEW EXERCISE**

1. ± 5 (8.3)   **2.** 0, 3 (8.2)   **3.** $-3, \frac{1}{2}$ (8.2)   **4.** 2, 3 (8.2)   **5.** $-3, 5$ (8.2)   **6.** $3x^2 + 4x - 2 = 0; a = 3, b = 4, c = -2$ (8.1)

7. $x = \dfrac{-b \pm \sqrt{b^2 - 4ac}}{2a}$ (8.5)   **8.** $\dfrac{-3 \pm \sqrt{5}}{2}$ (8.5)   **9.** $-5, 2$ (8.2)   **10.** 3, 9 (8.2)   **11.** $\pm 2\sqrt{3}$ (8.3)   **12.** 0, 2 (8.2)

13. $-2, 6$ (8.3)   **14.** $-\frac{1}{3}, 3$ (8.2)   **15.** $\frac{1}{2}, -3$ (8.2)

16. $3 \pm 2\sqrt{3}$ (8.4)   **17.** $\dfrac{1 \pm \sqrt{7}}{3}$ (8.5)   **18.** $5, -4$ (8.2)

19. 6, 12 (8.2)   **20.** 6 in by 5 in (8.7)   **21.** No real solutions (8.3)   **22.** $\dfrac{3 \pm \sqrt{6}}{2}$ (8.3)   **23.** $\frac{3}{4}, \frac{5}{2}$ (8.2)

24. $\dfrac{1 \pm \sqrt{7}}{2}$ (8.4)   **25.** $\dfrac{-3 \pm \sqrt{57}}{6}$ (8.5)   **26.** T (8.5)

PRACTICE TEST

1. $\pm \sqrt{8}$ or $\pm 2\sqrt{2}$ **2.** 0, 2   **3.** $\frac{1}{2}, -2$ **4.** $x = \dfrac{-b \pm \sqrt{b^2 - 4ac}}{2a}$ **5.** $\dfrac{-3 \pm \sqrt{21}}{2}$

6. $\dfrac{2 \pm \sqrt{20}}{8} = \dfrac{1 \pm \sqrt{5}}{4}$ **7.** $x = 2 \pm 2\sqrt{2}$ **8.** $-\frac{2}{3}, 1$

9. $-2, 7$ **10.** $\dfrac{-1 \pm \sqrt{5}}{2}$ **11.** 8 cm by 6 cm

12. $\dfrac{-1 + \sqrt{5}}{2}$

EXERCISE A.1

1. $\frac{8}{15}$ **2.** $\frac{3}{14}$ **3.** $\frac{3}{4}$ **4.** $\frac{9}{16}$ **5.** $\frac{1}{9}$ **6.** $\frac{3}{8}$ **7.** $\frac{5}{2}$
8. $\frac{9}{16}$ **9.** $\frac{2}{3}$ **10.** $\frac{5}{4}$ **11.** $\frac{2}{3}$ **12.** $\frac{4}{7}$ **13.** $\frac{11}{12}$
14. $\frac{11}{10}$ **15.** $\frac{1}{12}$ **16.** $\frac{5}{12}$ **17.** $\frac{19}{24}$ **18.** $\frac{37}{36}$ **19.** $\frac{13}{36}$
20. $\frac{1}{20}$ **21.** $\frac{2}{3}$ **22.** $\frac{3}{4}$ **23.** $\frac{2}{27}$ **24.** $\frac{12}{5}$

EXERCISE A.2

1. 5.8 **2.** 12.3 **3.** 31.662 **4.** 431.335 **5.** 17.73
6. 7.03 **7.** 79.64 **8.** 35.51 **9.** 32.5 **10.** 38.4
11. 0.00852 **12.** 0.01824 **13.** 0.83 **14.** 0.78
15. 0.74 **16.** 0.67 **17.** 38.2 **18.** 15.1 **19.** 7.9
20. 0.6

EXERCISE A.3

1. 0.67 **2.** 0.14 **3.** 0.09 **4.** 0.01 **5.** 2.16
6. 3.08 **7.** 0.006 **8.** 0.001 **9.** 0.074 **10.** 0.028
11. 0.231 **12.** 0.645 **13.** 12 percent **14.** 21 percent
15. 8 percent **16.** 2 percent **17.** 325 percent **18.** 604
percent **19.** 0.7 percent **20.** 0.4 percent **21.** 7.2
percent **22.** 6.9 percent **23.** 40.5 percent **24.** 23.6
percent **25.** 48.36 **26.** 7.2 **27.** 240 **28.** 160
29. 250 **30.** 66 **31.** 1.56 **32.** 1.728 **33.** 0.08
34. 0.56